Arbuscular Mycorrhizas and Stress Tolerance of Plants

Qiang-Sheng Wu
Editor

Arbuscular Mycorrhizas and Stress Tolerance of Plants

Springer

Editor
Qiang-Sheng Wu
College of Horticulture and Gardening
Yangtze University
Jingzhou, Hubei, China

Institute of Root Biology
Yangtze University
Jingzhou, Hubei, China

Department of Chemistry, Faculty of Science
University of Hradec Kralove
Hradec Kralove, Czech Republic

ISBN 978-981-10-4114-3　　ISBN 978-981-10-4115-0　(eBook)
DOI 10.1007/978-981-10-4115-0

Library of Congress Control Number: 2017936364

Printed on acid-free paper

This Springer imprint is published by Springer Nature
The registered company is Springer Nature Singapore Pte Ltd.
The registered company address is: 152 Beach Road, #21-01/04 Gateway East, Singapore 189721, Singapore

Preface

Arbuscular mycorrhizal fungi (AMF), a kind of ancient soil microorganisms, have existed in various ecological soils for almost 400 million years old. These fungi can form mutualistic relationships with over 80% of land's plants, namely, arbuscular mycorrhizas (AMs). This symbiosis is the most common mycorrhizal association in natural ecosystems and partly undertakes the absorption and delivery of mineral nutrients and water from the soil to the host plant by mycorrhizal hyphae. Additionally, AMs represent many benefits for the fungal partner, including enhancement of tolerance to abiotic and biotic stresses, growth improvement, modulation of ecosystem resilience, and maintenance and improvement of soil structure. Therefore, the AMs are critical for plant health, survival, and restoration in native ecosystems and good soil structure.

AMF, as well as plants, are often exposed to all or many of abiotic and biotic stresses, including extreme temperature, pH, drought stress, waterlogging stress, toxic metals, soil pathogens, etc. Studies in the past indicated a quick response of AMF to these stresses by several mechanisms, such as root morphological modification, reactive oxygen species change, osmotic adjustment, direct absorption of water by extraradical hyphae, upregulated expression of relevant stressed genes, release of glomalin-related soil protein, etc. The underlying strategy under mycorrhization to cope with stresses is involved in morphological, physiological, biochemical, and molecular processes, which is a complex network in multiple dimensions. The AMF responses are often associated with homeostatic regulation of internal and external environments.

From 2002, I and my team worked at the mycorrhizal functioning in drought tolerance of plants in China. I realized that mycorrhizal researches on stresses are a hard work. Until now, I have evaluated mycorrhizal roles in alleviating drought stress, temperature stress, waterlogging, and salt stress. I realize that the relevant books are important for mycorrhizal researchers. Hence, I have begun launching the edition of this book in Feb. 2016. The book was approved by Springer in May 2016. Here, I sincerely thank our contributing authors for their outstanding cooperation to write selflessly these valuable chapters. I also thank the editors of Springer, Prasad Gurunadham, Yu Zhu, and Xiao-Jin Huang, who spent much time publishing this book. I really appreciate my PhD tutor, Prof. Ren-Xue Xia, from Huazhong Agricultural University, for introducing me to the mycorrhizal research field.

Finally, I sincerely thank the National Natural Science Foundation of China (31372017; 30800747), the Plan in Scientific and Technological Innovation Team of Outstanding Young, Hubei Provincial Department of Education (T201604), the Key Project from Ministry of Education (211107), the Key Project of Natural Science Foundation of Hubei Province (2012FFA001), and the Science-Technology Project from Hubei Provincial Department of Education (Q20111301) to provide the fund for mycorrhizal researches. I hope that this book can help use AMF more efficiently as a biostimulant to enhance stress tolerance in the host plants.

Jingzhou, Hubei, China Qiang-Sheng Wu

Contents

About the Editor

Qiang-Sheng Wu received his PhD in pomology from Huazhong Agricultural University in 2006. Now, he is working as a professor at the College of Horticulture and Gardening, Yangtze University, and as an administrative director at the Institute of Root Biology, Yangtze University. In addition, Dr. Wu is an invited professor at the Univerzita Hradec Králové. Professor Wu focused on the arbuscular mycorrhizal researches in fruit trees. He concerns the functioning and physiology of arbuscular mycorrhizas in plant and soil, especially in citrus plants. At present, he works at the research in the signaling communication by common mycorrhizal network, the functioning and evaluation of glomalin-related soil protein, the physiology-molecular mechanisms about mycorrhiza-enhanced tolerance of drought stress and salt stress, and the relationship between mycorrhizas and root morphology or root hairs.

Until now, he has published one book titled *Arbuscular Mycorrhizal Research and Application of Horticultural Plants* and five chapters of relevant books. He published more than 60 papers in the international popular journals, such as *Soil Biology and Biochemistry*, *Scientific Reports*, *Mycorrhiza*, *Applied Soil Ecology*, *Soil and Tillage Research*, *Frontiers in Microbiology*, *PLoS One*, *Fungal Ecology*, *Journal of Plant Physiology*, *Journal of Plant Growth Regulation*, *European Journal of Soil Biology*, *Scientia Horticulturae*, etc. He also published more than 30 papers in Chinese journals. He is invited as reviewers to review lots of manuscripts from popular journals, including *Applied Soil Ecology*, *Environmental and Experimental Botany*, *Microbial Ecology*, *Plant Physiology and Biochemistry*, *European Journal of Soil Biology*, *Scientia Horticulturae*, *Plant Biology*, *Journal of Plant Growth Regulation*, *Functional Plant Biology*, etc.

Dr. Wu conducts ten projects, guided two doctors and 18 master's students, served as the editor of eight international journals and the member of Citrus Association in Hubei Province, and obtained the Fourth Scopus Young Researcher Award.

E-mail: wuqiangsh@163.com; wuqiangsheng@yangtzeu.edu.cn

Chapter 1
Arbuscular Mycorrhizas: An Overview

Sajid Mahmood Nadeem, Muhammad Yahya Khan, Muhammad Rashid Waqas, Rana Binyamin, Sohail Akhtar, and Zahir Ahmad Zahir

Abstract Almost every plant in natural ecosystem forms association with fungi either intracellularly as in arbuscular mycorrhizal fungi (AMF), or extracellularly as in ectomycorrhizal fungi. Arbuscular mycorrhizas (AMs) represent the most widespread symbiosis with land plants. The associated fungi colonize the plant roots and reside in the internal tissues of their host plant. This mutualistic association not only plays key role for enhancing plant growth by facilitating the uptake of water and essential nutrients but also protects the plant from adverse soil conditions. The application of mycorrhizal fungi is a promising alternative strategy for sustainable crop production under normal as well as biotic and abiotic stress conditions. The mycorrhizal plants have an improved ability for nutrient uptake and have ability to tolerate stress environments. There are increasing interest for the application of AM fungus for improving plant growth and enhancing crop production. The AM fungus also has positive impact on crop growth by improving soil quality by increasing water infiltration and retention and therefore reducing soil erosion. This review chapter epitomizes the current knowledge on the significance of AM fungus for improving crop production and maintaining agriculture sustainability.

Keywords Agriculture sustainability • Crop production • Mineral • Mycorrhiza

1.1 Introduction

In soil environments, plants are colonized by a number of organisms living in the soil. Some microorganisms, particularly beneficial bacteria and fungi, can improve plant performance by developing symbiotic association with the plant roots. Mycorrhiza is a nonpathogenic symbiotic association between roots and fungi. In

S.M. Nadeem • M.Y. Khan • M.R. Waqas • R. Binyamin • S. Akhtar
University of Agriculture Faisalabad, Sub-Campus Burewala, Vehari, Pakistan

Z.A. Zahir (✉)
Institute of Soil & Environmental Sciences, University of Agriculture, Faisalabad, Pakistan
e-mail: zazahir@yahoo.com

Q.-S. Wu (ed.), *Arbuscular Mycorrhizas and Stress Tolerance of Plants*,
DOI 10.1007/978-981-10-4115-0_1

this association, fungus receives the photosynthetic products prepared by the plant and in turn facilitates the plant by enhancing the availability of major essential nutrients particularly phosphorus and nitrogen. Many plants form symbiotic association with fungus that enhances their nutrient uptake ability, improves photosynthesis, and protects them from diseases and toxicities of heavy metals (HM) (Al-Karaki 2000; Andrade et al. 2009; Mehraban et al. 2009; Zhu et al. 2012, Birhane et al. 2012; Liu et al. 2015).

An arbuscular mycorrhiza is a symbiotic association between plant and fungus, in which the fungus penetrates the root cortical cells of a vascular plant. This association is commonly known as vesicular-arbuscular mycorrhiza (VAM) owing to the formation of particular vesicles (bladder-like structures) and arbuscules (branched finger-like hyphae) after colonizing the root cortical cells. Arbuscular mycorrhiza is the most common type of symbiotic association. It has been suggested that as many as 90% world plant species formed mycorrhizal association.

Arbuscular mycorrhizal fungi can adopt a wide range of soil conditions. Although stress environment having excess salts and oxygen deficiency is harmful for survival and efficiency of mycorrhizal fungus, some fungus is able to persist and colonize the plant root under such adverse conditions (Landwehr et al. 2002; Macek et al. 2012; Talaat and Shawky 2012; Sinclair et al. 2014). Similarly, in semiarid environment, where aggregate stability is one of the most important properties for better plant growth and development, the symbiotic association between plant and fungus enhances the aggregate stability even under stress environment (Caravaca et al. 2006; Wu et al. 2008). The production of glycoprotein glomalin by the fungus is a key factor for improving aggregate stability (Gadkar and Rillig 2006; Kohler et al. 2009). Although the mechanisms by which fungus enhances the water relations of host plant is still not clear yet, it is well documented that arbuscular mycorrhizal fungus improves water uptake by regulating stomatal conductance (Auge et al. 2015). Owing to this water uptake, in addition to providing macro- and micronutrients (Smith and Read 2008; Javaid 2009), mycorrhizal association also helps the plant to survive under water-limited environment (Farahani et al. 2008). In general, arbuscular mycorrhiza is an essential component of sustainable agriculture owing to its vital role in nutrient availability, enzyme production, increasing photosynthesis rate, and enhancing plant tolerance against biotic and abiotic stresses (Adriano-Anaya et al. 2006; Evelin et al. 2009; Wu and Xia 2009; Hu et al. 2010; Kohler et al. 2010).

1.2 Fungal Association with Plants

Fungi are eukaryotic spore-producing organisms, which are usually filamentous and branched and lack chlorophyll. Fungi have cell wall, composed of chitin and glucans but no cellulose. Oomycetes are fungal-like organisms and the majority have cell wall composed of glucans and cellulose, but no chitin. Fungi and oomycetes have both types of relationships with plants, i.e., pathogenic and symbiotic. In fungi, more than 10000 species are pathogenic to plants. Plant pathogenic fungi are parasitic but all parasitic fungi are not pathogenic. Plant parasitic fungi obtain their nutrition from the host plant, but the host in response does not necessarily show disease symptoms, whereas plant pathogenic fungi are parasites, which cause different diseases characterized by specific symptoms. Fungi-like biotroph or obligate parasites can only multiply and grow in association with their specific host plant. Others, which require their host plant for part of their life cycle but can complete their life cycle on dead organic matter also, or in other words they can grow and multiply on living organism as well as on dead organic matter, are called nonobligate parasites. Further, nonobligate parasites are categorized as facultative parasites and facultative saprophytes considering whether these are primarily saprophytes or primarily parasites. Biotrophic fungi obtain nutrients from living tissues of their host plant through specialized cell called haustoria which develop inside the host cells. Necrotrophic fungi first kill the host tissues via different enzymes and toxins and then obtain their nutrients from dead host tissues. Necrotrophic fungi mostly have a wide range of host species, whereas in case of biotrophic fungi, they have a narrow host range as they have developed host-specific and specialized mechanism of parasitism.

Diversity of relationship between fungi and plants varies from pathogenic to symbiotic association. Symbiosis means living together for their mutual benefits. Dialogues at molecular level take place during fungi and plant interactions, which result in host's metabolic modifications, protection against environmental stresses, and providing friendly conditions to symbiont (fungi). Mycorrhizal fungi mostly exist in forest trees and live either on the plant roots or in the plant roots. More than 90% of plant species have symbiotic association with mycorrhizal fungi (Bonfante and Genre 2010). Fruits, vegetables, cereals, and ornamental plants have also shown mycorrhizal growth. This type of fungal association increases absorption surface area of root and benefits the plant in terms of water and nutrient (nitrogen and phosphorus) uptake (Sylvia et al. 2005; Smith and Read 2008) and provides protection from biotic and abiotic stresses (Zhang et al. 2010; Yang et al. 2014). Fungal hyphae are quite thin as compared to plant roots so can easily reach even tiny pores in soil. Plants provide 10–20% sugar (carbon) as a cost of above benefits which was synthesized by plant via photosynthesis (Allen et al. 2011). On the basis of fungi hyphal arrangement within cortical tissues, mycorrhizae are simply classified as ectomycorrhizae and endomycorrhizae. Ectomycorrhizae most of the time dominate in temperate forests and belong to basidiomycetes and some ascomycetes. Ectomycorrhizal relationship exists between 10% of plant species and near 5000 fungal species. Ectomycorrhizae modify plant roots into swollen and profusely branched as compared to non-mycorrhizal roots. The fungal hyphae form a "fungus mantle" around the feeder roots. Fungi penetrate the roots but remain around the cortical cells and form a hartig net. Endomycorrhizae are mostly formed by fungi belonging to the genera *Glomus*, *Gigaspora*, *Funneliformis*, *Entrophospora*, *Acaulospora*, *Archaeospora*, *Scutellospora*, and *Sclerocystis*; however, some basidiomycetes also form endomycorrhizae. Endomycorrhizae dominate in cultivated crops while some are present in forest trees as well. Arbuscular mycorrhizae (AM) symbiotic interaction occurs excessively. Studies showed that AM fungi exists 460 million years before first plants originated (Kistner and Parniske 2002; Bonfante and Genre 2010). Arbuscules are formed by endomycorrhizal hyphae within the plant cortical cells and are highly branched, and mature arbuscules have short life and survive for 4–5 days. Arbuscules are considered as functional site for nutrient exchange (Balestrini et al. 2015). However, not very common but a special type of mycorrhizae also exists which possesses characters of both ectomycorrhizae and endomycorrhizae and is named as ectendomycorrhizae. Ectendomycorrhizae form external mantle with hyphae (ectomycorrhizal character) as well as penetrate host cells for intracellular growth (endomycorrhizal character).

The interaction between AM fungi and plant not only improves tolerance against biotic stresses but also against abiotic problems like drought and salinity (see Chaps. 2 and 4) (Porcel et al. 2011; Auge et al. 2015). Mycorrhizal association with plants not only improves plant growth but also improves soil texture by changing soil particles into stable aggregates that ultimately resists against wind and water erosion (Leifheit et al. 2014, 2015; Rillig et al. 2015). AM fungi also act as filter by binding heavy metals into roots and prevent their translocation toward shoot tissues (Gaur and Adholeya 2004; Cornejo et al. 2013; Tamayo et al. 2014; Meier et al. 2015). Evidence of AM fungi association with metal-tolerant plants (metallophytes) has been reported in rhizosphere of *Viola calaminaria* and *Berkheya coddii* (Tonin et al. 2001; Turnau et al. 2001; Turnau and Mesjasz-Przybylowicz 2003). Mostly reported heavy metal-tolerant AM fungi are *Glomus geosporum* (Sambandan et al. 1992), *Glomus mosseae* (Weissenhorn et al. 1993; Turnau et al. 2001), *Scutellospora dipurpurascens* (Griffioen et al. 1994), and *Glomus claroideum* (Leyval et al. 1995; Del Val et al. 1999). Owing to their ability to tolerate heavy metals and maintain their growth in contaminated soils, AM fungi play a critical role in the phytoremediation of heavy metals. Mycorrhizal association is also helpful to introduce new plant species in an area. This association enables the plant to survive in a new environment. It has been observed that the survival of plants in the absence of this association was not significant in a new environment. For example, attempt of introducing pine in Australia failed due to lack of mycorrhizal fungi (Allen 1991). *Cyperus rotundus* native to Asia (Holm et al. 1977) develops a symbiotic relationship with fungus *Balansia cyperi* native to America (Diehl 1950) when introduced in the American region (Kowalski et al. 2015).

In general, mycorrhizal fungi form symbiotic association with almost all types of plants. This association is common in both normal and stress environments. This symbiotic association provides a number of benefits to the plant including availability of essential nutrients, enhancing water uptake, and promoting growth. This growth promotion takes place not only in normal but also in adverse soil environment.

1.3 Mechanism Action of Arbuscular Mycorrhizal Fungi

There are a number of mechanisms used by mycorrhizal fungus to promote plant growth and development. These mechanisms vary with respect to species and also depend on soil conditions. Some of these mechanisms include production of phytohormones and metabolites including vitamins, amino acids, etc., solubilization of minerals, root colonization, enhancing water uptake, production of osmolytes, and improving soil structure.

Arbuscular mycorrhiza (AM) is a type of endomycorrhizal fungal (fungi of the phylum Glomeromycota) association with the roots of vascular plants in which fungal hyphae penetrate the cortical cells and make some special structures known as arbuscules and sometimes form vesicles (large food-storing hyphal swelling). No significant changes occur in the morphology of roots. Arbuscular mycorrhizal (AM) fungi first establish contact and penetrate young root epidermis. After that AM fungal hyphae spread into root tissues and form mycorrhizal network for nutrient exchange between plant and fungi. Plants secrete sugars, organic acids, and amino acids in the rhizosphere that triggers soil fungi to colonize plant roots and interact with the plant at molecular level. In most cases, endomycorrhizae have both arbuscules and vesicles; that's why it is called VAM (vesicular-arbuscular mycorrhizae) (Parniske 2008).

A cascade of recognition events is involved in establishing a symbiotic relationship of AM fungi with plant roots leading to complete morphological and physiological interactions (Singh 2007). Formation of appressorium on the surface of an epidermal cell is one of the first signs to recognize the interaction between plant and AM fungus (Singh 2007). After being in contact with the plant surface, the surface-bound recognition molecules of fungi and the host plant play their role. Different groups of compounds regulate different phases of colonization (Koske and Gemma 1992). Flavonoids and phenolic compounds are considered as signaling molecules in mycorrhizal associations (Balaji et al. 1995; Nair et al. 1991).

On molecular level, it has been observed that some specific symbiosis genes are responsible for successful symbiotic association which cannot confer resistance to pathogenic infections, thus showing their specificity for symbiosis. When nodulation mutant pea plants (*nod*, which cannot develop nodules) were tried to infect with AM fungi, no colonization occurred. This was because there was no hyphal elongation on some plants, while arbuscules could not be formed on some others, attributable to the deposition of β-1,3-glucan and thickening of cell wall of root cells. It reflects the eliciting of defense system of the host contrary to the case of a compatible host. Interestingly, these mutants were not resistant to pathogenic fungi (Gianinazzi-Pearson 1995).

AM fungi have an extraordinary importance regarding facilitating the plant in nutrient uptake and imparting resistance to biotic and abiotic stresses (Barea et al. 2002). Owing to its ability to improve plant nutrition, this association is very helpful for the plant as plant maintains its growth in nutrient-limited environment. Some selected examples have been reported in Table 1.1 that indicate the potential of

Table 1.1 Improvement in plant nutrition through mycorrhizae

Nutrient	Crop	Mycorrhizae	References
Nitrogen (N)	Pigeon pea (*Cajanus cajan*)	*Glomus mosseae*	Garg and Manchanda (2008)
	River hemp (*Sesbania aegyptiaca*)	*Glomus macrocarpum*	Giri and Mukerji (2004)
	White clover (*Trifolium repens*)	*Glomus mosseae*	Medina et al. (2006)
Phosphorus (P)	Soybean (*Glycine max*)	*Glomus etunicatum*	Sharifi et al. (2007)
	Cotton (*Gossypium arboretum*)	*Glomus mosseae*	Tian et al. (2004)
	Maize (*Zea mays*)	*Glomus mosseae*	Feng et al. (2002)
	Tomato (*Lycopersicon esculentum*)	*Glomus mosseae*	Al-Karaki (2000)
	White clover (*Trifolium repens*)	*Glomus mosseae*	Medina et al. (2006)
	White Clover (*Trifolium repens*)	*Glomus mosseae*	Medina et al. (2006)
Potassium (K)	Soybean (*Glycine max*)	*Glomus etunicatum*	Sharifi et al. (2007)
	White Clover (*Trifolium repens*)	*Glomus mosseae*	Medina et al. (2006)
Calcium (Ca)	Soybean (*Glycine max*)	Glomus etunicatum	Sharifi et al. (2007)
Magnesium (Mg)	River hemp (*Sesbania aegyptiaca*)	Glomus macrocarpum	Giri and Mukerji (2004)
Sulfur (S)	Black locust (*Robinia pseudoacacia*)	*Rhizophagus intraradices*	Yang et al. (2016)
Chloride (Cl)	Cotton (*Gossypium arboretum*)	*Glomus mosseae*	Tian et al. (2004)
Zinc (Zn)	Glycine max	*Glomus etunicatum*	Sharifi et al. (2007)
	Basket willow (*Salix viminalis*)	*Glomus intraradices*	Bissonnette et al. (2010)
	Tomato (*Solanum lycopersicum*)		Watts-Williams et al. (2013)
Copper (Cu)	Elsholtzia splendens	*Gigaspora margarita*, *G. decipiens*, *Scutellospora gilmori*, *Acaulospora* spp.	Wang et al. (2005)

mycorrhizal fungi for improving plant nutrition. AM fungi are involved in facilitating the plants in phosphorus uptake. Mycorrhizal fungi have phosphate transporter for Pi (inorganic phosphate) transportation from soil to plant (Harrison and Van Buuren 1995). AM fungi increase the concentration of P in plants by facilitating its enhanced uptake from the soil by making extensive hyphal growth which allows plants to explore more soil volume than the non-mycorrhizal plants (Ruiz-Lozano and Azcon 2000). Phosphorus uptake was calculated in the roots of *Capsicum annuum* with and without inoculation of AM fungi. The increased P uptake of inoculated plants was mainly because of an up to five times higher P influx of the AM-infected root (Sharif and Claassen 2011). It has been observed in various studies that an inverse correlation exists between the levels of salinity and P concentration in plants. Furthermore, higher P content has been reported in mycorrhizal than non-mycorrhizal *Acacia nilotica* and *Trifolium alexandrinum* plants grown in saline soils with varied levels of salinity (Giri et al. 2007; Shokri and Maadi 2009). It is established that AM symbiosis specifically induces the expression of plant Pi transporters (Xie et al. 2013; Walder et al. 2015). Three Pi transporter (*PT*) genes in tomato (*LePT3*, *LePT4*, and *LePT5*) are upregulated in AM-colonized roots (Nagy et al. 2005). In *Medicago truncatula*, *MtPT4* genes are regulated in root cells to absorb phosphate supplied by mycorrhizae; these genes are crucial for AM symbiosis (Javot et al. 2007).

In addition to Pi transporters, ammonium transporters (*AMT*) are also induced upon AM fungal association with the plant roots. Saline soil conditions interfere with nitrogen (N) uptake from soil and its metabolism (Frechilla et al. 2001). Application of AM fungi can greatly reduce this interference and help in better assimilation of N in the host plant. Nitrogen accumulated in higher amounts in *Sesbania grandiflora* and *S. aegyptiaca* plants having AM fungal association than in the non-mycorrhizal control counterparts (Giri and Mukerji 2004). Genes regarding nitrogen uptake both as organic and inorganic form are identified in ectomycorrhizae and endomycorrhizae (Lucic et al. 2008; Cappellazzo et al. 2008), which activate plant's nitrogen transporters during mycorrhization (Guether et al. 2009). AMTs are located in periarbuscular membrane as demonstrated in soybean and *Medicago* having a role in ammonium transport to cortical cells (Harrison et al. 2002; Kobae et al. 2010; Breuillin-Sessoms et al. 2015). Several Pi and AMT mutant studies reveal that these transporters not only deliver nutrients to the root cells but also trigger some signaling for the better maintenance of arbuscules (Javot et al. 2007; Breuillin-Sessoms et al. 2015).

In addition to Pi and N, other macro- and microelements are also transferred to plants through AM symbiosis. Sulfate transporters involved in the uptake of sulfur from arbuscules have also been identified using the application of laser microdissection technology (Giovannetti et al. 2014). Potassium (K) is one of the macroelements required by plant cell machinery; however, the contribution of AM symbiosis to plant K^+ nutrition has rarely been studied (Garcia and Zimmermann 2014). K is present in soil in abundant quantities but unavailable or available to plants in very low amounts due to its adsorption with other minerals. The element has been reported in AM fungal spores, hyphae, and arbuscules in different studies (Pallon et al. 2007; Olsson et al. 2008; Olsson et al. 2011). Furthermore, the upregulation of a plant K^+ transporter has been reported in mycorrhizal *Lotus japonicus* roots (Guether et al. 2009).

Genetic and agronomic biofortification is the tool to improve the concentration of micronutrients in plants and plant products. The use of AM symbiosis can serve the purpose as an alternative, or in addition. A Zn transporter has been identified (*GintZnT1*) in *Glomus intraradices* (Gonzalez-Guerrero et al. 2005) that is putatively involved in the transport of Zn through hyphae or through the apoplastic space between fungi and plant plasma membrane (Cavagnaro 2008). Lehmann et al. (2014) performed a meta-analysis to analyze the potential role of AM fungi in improving the concentration of Zn in various plant tissues and soil types. It was concluded that AM fungi positively altered the Zn concentration in various plant tissues under distinct environmental conditions. In another meta-analysis performed by the same group, they noted the impact of AMF on crop Cu, Fe, and Mn nutrition. A significant positive effect of AMF was recorded on crop Cu nutrition (average 29%); a mediate effect on Fe was noted only for intermediate experimental duration, and again positive effect on Mn nutrition was noted, but it was only for herbs (Lehmann and Rillig 2015).

Evelin et al. (2009) critically reviewed the mechanisms used by mycorrhizae fungi to induce salt tolerance in plants. They reported that enhanced nutrient acquisition; maintenance of the K/Na ratio; accumulation of proline, polyamines, and carbohydrates; enhancing the activity of antioxidant enzymes; and imparting physiological changes such as photosynthetic efficiency, relative permeability, water status, and abscisic acid accumulation are some of the important mechanisms used by AM fungus in stress conditions. Similarly, the mycorrhizae protect the plant from drought stress by maintaining water content in the plant tissue. It is evident from the work of (Porcel and Ruiz-Lozano 2004) that AM symbiosis enhanced osmotic adjustment in roots which maintained a water potential gradient favorable to the water entrance from soil into the roots.

Similarly, mycorrhizae enable the plant to maintain plant growth in contaminated environment by improving nutritional status and reducing metal uptake ((Toler et al. 2005; Andrade et al. 2008). Some general mechanisms used by fungus to alleviate the negative impact of heavy metals include metal immobilization, metal chelation, better root colonization, and compartmentalization inside fungal cells (Kaldorf et al. 1999; Vodnik et al. 2008; Redon et al. 2009; Yang et al. 2016) (see Chap. 7).

1.4 Interaction with Other Soil Microbes

In soil environment, plant-fungal interactions occur in the mycorrhizosphere that is a zone surrounding the root and fungal hyphae (Johansson et al. 2004). In this zone, interaction between fungus and other microbial population also takes place. The synergistic and antagonistic nature of these interactions depends upon the microbial strains/species involved. For example, in a study conducted by Jäderlund et al. (2008), it was observed that bacterial strain affected the colonization ability of AM fungus. They reported that the AM fungus *G. intraradices* enhanced the dry weight of wheat plant when applied singly or in co-inoculation with *P. fluorescens*; however, the effect was nonsignificant when the fungus was applied with *P. brasiliensis*. The synergistic interaction not only enhanced the growth of plant but also caused positive effect on population of each other (Yusran et al. 2009). Keeping in view the synergistic association between bacteria and AM fungus, a number of studies have been conducted to evaluate the potential of co-inoculation for improving plant growth and development. Some selected examples have been reported in Table 1.2.

Table 1.2 Synergism between AMF and PGPB for mitigating various stresses on different plants

AMF	PGPB	Plant	Stress	Effect	References
Rhizophagus irregularis	*Variovorax paradoxus* 5C – 2	*Solanum lycopersicum*	Drought	Improved CO_2-fixation capacity and root hydraulic conductivity	Calvo-Polanco et al. (2016)
Rhizophagus intraradices	*Paenibacillus mucilaginosus*	*Citrus trifoliata*	P-deficient soil	Significantly increased the root colonization and plant growth. Nitrogen and phosphorus uptake significantly increased by inoculation	Wang et al. (2016)
Glomus fasciculatum	*Exiguobacterium oxidotolerans*	*Mentha arvensis* L.	Salt-stressed soils	Improved plant growth and better AMF colonization and mitigation of salinity stress	Bharti et al. (2016)
Glomus intraradices KA and *Gigaspora margarita* AA	*Burkholderia cepacia* 4684 and *Bacillus subtilis* 7612	*Solanum lycopersicum*	Root-knot nematode (*Meloidogyne* spp.)-infested soil	Reduced galling and nematode multiplication, increased dry biomass of plant	Siddiqui and Akhtar (2009)
Rhizophagus irregularis	*Chryseobacterium humi* ECP37T and *Pseudomonas* reactants EDP28	*Zea mays*	Heavy metals (mine-contaminated soil)	Increased root and shoot biomass as well as shoot elongation. Improved substantially P accumulation in roots	Moreira et al. (2016)

(continued)

Table 1.2 (continued)

AMF	PGPB	Plant	Stress	Effect	References
Claroideoglomus etunicatum, *Rhizophagus intraradices*, and *Funneliformis mosseae*	*Bacillus subtilis*	*Acacia gerrardii*	Salt stress	Enhanced N, P, K, Mg, and Ca contents and phosphatase activities, reduced Na and Cl concentration and protection of the plant from ion toxicity	Hashem et al. (2016)
Septoglomus constrictum, *Diversispora aunantia*, *Archaespora trappei*, *Glomus versiforme*, and *Paraglomus occultum*	*Bacillus thuringiensis*	*Zea mays*	Drought	Decreased oxidative damage to lipids and accumulation of proline induced by drought	Armada et al. (2015)
Glomus intraradices	*Acinetobacter* sp.	*Avena sativa*	Saline-alkali soil contaminated with petroleum	Increased dry weight and stem height, enhanced activities of antioxidant enzymes, decreased MDA and free proline contents, improved soil quality by increasing the activities of soil enzyme such as urease, sucrase, and dehydrogenase	Xun et al. (2015)
Acaulospora laevis, *Glomus geosporum*, *Glomus mosseae*, and *Scutellospora armeniaca*	*Rhizobium leguminosarum* bv. *viciae*	*Vicia faba* L.	Alkaline soil	Increased number and mass of nodules, nitrogenase activity, leghemoglobin content of nodule, mycorrhizal colonization, dry mass of root and shoot	Abd-Alla et al. (2014)

Rhizophagus irregularis	*Bacillus subtilis*	*Trigonella foenum-graecum*	Drought	Improved plant weight and decreased ACC concentration, enhanced nodulation and arbuscular mycorrhizal fungi colonization in the plants	Barnawal et al. (2013)
Glomus intraradices JJ291	*Pseudomonas fluorescens* F113 and *Azospirillum lipoferum* CRT1	*Zea mays*	Reduced fertilization	Qualitative and quantitative modifications of root secondary metabolites, particularly benzoxazinoids and diethylphthalate	Walker et al. (2012)
Glomus spp.	*Rhizobium meliloti*	*Medicago arborea*	Semiarid conditions	Improved shoot and root biomass production, increased beneficial effect of dual symbiotic colonization of AMF and *Rhizobium*	Valdenegro et al. (2001)

The results showed that dual inoculation was not only helpful for improving plant growth under normal but also under stress environments.

It has been suggested that bacteria associated with mycorrhizosphere are involved in the establishment of arbuscular fungi and ultimately promote plant growth (Larsen et al. 2009). Enhanced root colonization of clover by AM fungus *G. mosseae* has been observed in the presence of bacterial strain *Paenibacillus brasiliensis* (Artursson 2005). Similarly, the co-inoculation of *G. mosseae* with *Trichoderma* spp. increased the yield and seed quality of soybean (Egberongbe et al. 2010). Also, significant increase in tomato growth has been observed by the application of *P. fluorescens* with *G. mosseae* BEG12 (Gamalero et al. 2004).

Bacteria associated with AM fungus stimulate the hyphal growth and facilitate the root penetration by the fungus. It is generally occurring by the production of certain compounds from the bacteria that enhances the rate of root exudates. Increased root exudation activates the fungal hyphae and improves root colonization (Barea et al. 2005). The production of exopolysaccharides by certain bacterial strains contributes to soil aggregation and increased water retention in the rhizosphere (Kaci et al. 2005). Bacteria stimulate the hyphal growth by enhancing cell permeability and facilitate root penetration by the fungus (Jeffries et al. 2003). In bacterial-fungal interaction, bacteria enhance the activity of AM fungus, and mycorrhizal fungus helps the bacteria to fix nitrogen and to solubilize phosphorus from insoluble source. The co-inoculation of mycorrhizae with more than one strain has also shown significant impact on plant growth. Matias et al. (2009) evaluated the effectiveness of inoculation with mycorrhizae, rhizobia, and PGPR in terms of solubilization of nutrients and growth of *Centrosema coriaceum* (Leguminosae) and *Tibouchina multiflora* (Melastomataceae). They demonstrated that plants inoculated with mycorrhizal fungi and/or rhizobium showed a significant improvement in the plant growth, survival index, physical and chemical soil properties, leaf nitrogen (N) and phosphorous (P), and soil P.

1.5 Enhancing Crop Production and Inducing Stress Tolerance

The abiotic stresses such as drought, salinity, and extreme temperature regimes are linked by the fact that they all decrease the availability of water and nutrients to plant cells. The production of crops and their sustainability is mostly influenced by abiotic stresses, such as drought (Saleem et al. 2007), salinity (Nadeem et al. 2007; Al-Khaliel 2010), extreme temperature (Canci and Toker 2009), nutrient deficiency (Zhang et al. 2000; Ho et al. 2004; Arevalo et al. 2005), organic contaminant (Arshad et al. 2007), heavy metals (Kumar et al. 2008), etc. The mycorrhizal inoculation proved useful for promoting plant growth and development. Some selected examples have been presented in Table 1.3.

Table 1.3 Impact of mycorrhizae on plant growth under stress conditions

Type of stress	Plant	Mycorrhizae	Mechanism used	References
Drought	Tomato (*Solanum lycopersicum*)	*Glomus spp.*	Improving water and nutrient absorption	Kuswandi and Sugiyarto (2015)
	Australian pine tree (*Casuarina equisetifolia*)	*Glomus caledonium*, *Glomus versiforme*	Enhancing the concentrations of P nutrition, soluble sugar, soluble protein and activities of Peroxidase	Zhang et al. (2010)
	White clover (*Trifolium repens*)	*Glomus mosseae*	Better root colonization	Medina et al. (2006)
	Orchard grass (*Dactylis glomerata*)	*Glomus intraradices*, *Glomus mosseae*	Improved water content	Kyriazopoulos et al. (2014)
	Seaberry (*Hippophae rhamnoides*)	*Glomus mosseae*, *Glomus geosporum*	Alkaline phosphatase activity	Ming and Hui (1999)
	Black locust (*Robinia pseudoacacia*)		Regulating plant growth, leaf water status, photosynthesis, and nutrient concentration	Yang et al. (2014)
Salt	Strawberry (*Fragaria* × *ananassa*)	*Funneliformis caledonius*, *F. mosseae and Rhizophagus irregularis*	Better root colonization	Sinclair et al. (2014)
	Peanut (*Arachis hypogaea*)	*Glomus mosseae*	Improved plant nutrition	Al-Khaliel (2010)
	Wheat (*Triticum aestivum*)	*Glomus mosseae*	Improved root colonization	Ibrahim et al. (2011)
	Pigeonpea (*Cajanus cajan*)	*Glomus mosseae*	Enhanced antioxidant enzyme activities and improved nitrogen-fixing efficiency, reduced membrane permeability and lipid peroxidation	Garg and Manchanda (2008)
	Pepper (*Capsicum annum* cv. 11B14)	*Glomus clarum*	Improved nutrition and reduced membrane leakage	Kaya et al. (2009)
	Sesbania aegyptiaca and *Sesbania grandiflora*	*Glomus macrocarpum*	Reduced Na uptake and increased P, N and Mg absorption	Giri and Mukerji (2004)

(continued)

Table 1.3 (continued)

Type of stress	Plant	Mycorrhizae	Mechanism used	References
	Maize (*Zea mays*)	*Glomus mosseae*	Enhanced soluble sugars content and electrolyte concentrations	Feng et al. (2002)
	Cotton (*Gossypium arboretum* L)	*Glomus mosseae*	Enhanced phosphorus concentrations, and reduced sodium and chloride uptake	Tian et al. (2004)
	Orange (*Poncirus trifoliata*)	*Glomus mosseae*	Reduced malondialdehyde and hydrogen peroxide contents	Wu et al. (2010)
Organic pollutants	Clover (*Trifolium repens*), ryegrass (*Lolium multiflorum*)	*Glomus mosseae*	Enhanced P uptake	Joner and Leyval, (2001)
	Ryegrass (*Lolium multiflorum*)	*Glomus intraradices*	Better root colonization and P uptake	Alarcón et al. (2008)
	Barley (*H. vulgare)*	*Glomus hoi*	Improving activities of antioxidant enzymes	Khalvati et al. (2010)
Heavy metals	White clover (*Trifolium repens*)	*Glomus mosseae*	Enhanced soil enzymatic activities in sugar beet waste and rock phosphate treated soil	Medina et al. (2006)
	Carrot (*Daucus carota*)	*Glomus intraradices* and *Gigaspora margarita*	High sorption capacity for heavy metals	Janouskova and Vosatka (2005)
	Basket willow (*Salix viminalis*)	*G. intraradices*		Bissonnette et al. (2010)
	Black locust (*Robinia pseudoacacia*)	*Rhizophagus intraradices*	Improved plant biomass causing positive impact on photosynthesis and macronutrient acquisition	Yang et al. (2016)
	Back locust (*Robinia pseudoacacia*)	*Funneliformis mosseae and Rhizophagus intraradices*	Better root colonization	Yang et al. 2015

(continued)

Table 1.3 (continued)

Type of stress	Plant	Mycorrhizae	Mechanism used	References
	Elsholtzia splendens	*Gigaspora margarita*, *G. decipiens*, *Scutellospora gilmori*, *Acaulospora* spp.	Improved P nutrition	Wang et al. (2005)
	Zenia insignis	*Glomus mosseae*, *Glomus intraradices*	Improved P nutrition and changed heavy metal uptake	Li et al. (2014)
Soil compaction	Wheat (*Triticum aestivum* L.)	*Glomus mosseae*	Improving root shoot growth	Miransari et al. (2008)
	Corn (*Zea mays*)	*Glomus etunicatum*, *G. mosseae*, *G. intraradices*	Enhanced root growth and absorption area	Miransari (2013)

If a single abiotic stress is to be recognized as the most regular in limiting the growth of crop roots worldwide, it is most probably low water supply (Boyer 1982; Araus et al. 2002). Water deficiency affects root growth (Arshad et al. 2007), root length density (Siopongco et al. 2005), and lateral root development (Banoc et al. 2000). Among other abiotic stresses, particularly salinity notably affects root along with plant growth and crop yields (Zahran 1999; Duzan et al. 2004), rhizobial colonization of roots (Tu 1981; McKay and Djordjevic 1993), and nitrogen-fixing ability of roots in case of legumes (Bolanos et al. 2003). Extreme temperature and nutrient deficiency are becoming increasingly considerable in limiting growth of root system. Globally, low temperature also is a major limitation of root and plant growth, and this has a major impact on grasses (Tester and Bacic 2005). Recently, stresses like organic contaminants and heavy metals achieved an attraction with detrimental impacts in denting the root and plant growth. Studies revealed that Cd affected a change in root hair density of crops (Kuriakose and Prasad 2008) and Cd-induced water and oxidative stress that is responsible for the root growth inhibition, probably through an accelerated differentiation of root tissues (Tamás et al. 2008).

Under such circumstances the most crucial attribute of plants is the ability of them to alter their root architecture, so that plant can develop root system which helps them to sustain growth and development under proceeding stress (Richards and Passioura 1981; Lynch 1995; El-Hafid et al. 1998; Kashiwagi et al. 2005). For example, one of the adaptive responses of plants to drought conditions is the development of deep and extensive root systems (Fukai and Cooper 1995; Serraj et al. 2004), which includes thick roots (Price et al. 2000), increased root length density (Siopongco et al. 2005), and ability of the roots to proliferate quickly (Suralta and

Yamauchi 2008) as a result of the plasticity in lateral root development (Azhiri-Sigari et al. 2000; Banoc et al. 2000; Kamoshita et al. 2000; Kondo et al. 2003). Mycorrhiza mitigates the water-deficit stress conditions by enhancing the active absorptive surface area which ultimately stimulates the uptake of water. This elevated water uptake tends to improve the drought tolerance by increasing the soil water movement into the roots of plants along with osmotic regulations (Ruiz-Lozano 2003). During the onset of drought stress, mycorrhizae help the plant in postponing the decline in water potential (Porcel and Ruiz-Lozano 2004). Furthermore, mycorrhizae extend the root zone environment by strengthening the soil aggregation and ultimately enhance the overall soil water condition to mitigate drought stress (Rillig et al. 2002).

Several researchers have reported the role of mycorrhizae in plant protection under salt stress. The positive symbiosis existing between fungus and plant exhibits substantial response in enhanced nutrient uptake, osmoregulator accumulation, higher photosynthetic rates, and augmented water use efficiency which undoubtedly suggests salinity alleviation by mycorrhizae (Marulanda et al. 2003, 2007; Wu et al. 2007). It has also been observed that this growth enchantment depends upon the mycorrhizal species and plant species. Daei et al. (2009) found that mycorrhizae alleviated the negative impact of salinity on wheat growth by reducing the uptake of Na and Cl as well as improving the nutrient uptake. They found that among three mycorrhizae species (*Glomus etunicatum*, *G. mosseae*, *G. intraradices*), *G. etunicatum* performed better. Similarly, the impact of mycorrhizae on wheat genotypes was also variable. The mycorrhizae significantly increased the growth and nutrient uptake of mutated Tabasi line compared to Roshan and Kavir genotypes.

The enhanced use of xenobiotics or synthetically produced organic chemicals in recent past created several environmental issues regarding air, water, and soil. The accelerated use of pesticides and fertilizers in agriculture has drawn serious attention worldwide which have deteriorated the farm soil and ground- and surface water and subsequently have entered our food chain to open severe health hazards (Chaudhry et al. 2005).

Worldwide, the influence of heavy metals (HM) has increased at a rapid speed mainly due to industrialization. This situation alarms about the serious land degradation occurring over the world and particularly in the environment of arid and semiarid regions (Giri et al. 2003a, b; Al-Karaki 2006). The entry of HM in ecosystem is obvious because of mining processes, deposition of agrochemicals (pesticides/fertilizers), and anthropogenic activities (Liu et al. 1997). Higher concentration of HM in the soil such as Zn, Cu, Pb, As, Cd, and Cr has detrimental impact on soil and plant. Mycorrhizae also help in alleviating the negative impacts of heavy metals and maintaining the plant growth and yield in contaminated environment (Chen et al. 2005; Andrade et al. 2008; Turnau et al. 2008; Nadeem et al. 2014).

In general, mycorrhizal association protects the plant from adverse conditions. Improvement in plant growth under stress environment has been reported by various workers. It indicates their potential for promoting plant growth and development under stress conditions.

1.6 Conclusion and Future Prospects

It is evident from above discussion that mycorrhizal association is very effective for promoting plant growth and development. There are various mechanisms used by fungus to enable the plant to maintain its growth in a better way in normal as well as in stress environment. These mechanisms differ among mycorrhizal strains and also depend upon the plant species. The mycorrhizal fungus is equally effective when used singly or in combination with other microbes.

Although dual inoculation of PGPR and mycorrhizae proved useful for improving plant growth, however, there are still certain aspects which need critical consideration, and one of them is its effectiveness under natural field conditions. Also, the specificity of strain in this regard is also important so that maximum benefits can be obtained from dual inoculation. It is also observed that most of the studies are conducted under abiotic stresses; therefore, the impact of mycorrhizae for improving plant growth under biotic stress is also needed to evaluate. Variable response obtained by mycorrhizae under different environmental stresses also needs further investigations. Another aspect is to generate transgenic plants encoding the genes of particular traits of mycorrhizae. Preliminary studies mostly conducted under controlled conditions showed that transgenic plants have the ability to withstand the stress conditions. However, further work is needed on this aspect particularly under natural soil environments.

References

Abd-Alla MH, El-Enany AW, Nafady NA et al (2014) Synergistic interaction of *Rhizobium leguminosarum* bv. viciae and arbuscular mycorrhizal fungi as a plant growth promoting biofertilizers for faba bean (*Vicia faba* L.) in alkaline soil. Microbiol Res 169:49–58

Alarcón A, Davies FT Jr, Autenrieth RL, Zuberer DA (2008) Arbuscular mycorrhiza and petroleum-degrading microorganisms enhance phytoremediation of petroleum-contaminated soil. Int J Phytorem 10:251–263

Al-Karaki GN (2000) Growth of mycorrhizal tomato and mineral acquisition under salt stress. Mycorrhiza 10:51–54

Al-Karaki GN (2006) Nursery inoculation of tomato with arbuscular mycorrhizal fungi and subsequent performance under irrigation with saline water. Sci Hortic 109:1–7

Al-Khaliel AS (2010) Effect of salinity stress on mycorrhizal association and growth response of peanut infected by *Glomus mosseae*. Plant Soil Environ 56:318–324

Allen MF (1991) The ecology of mycorrhizae. Cambridge University Press, New York, p 184

Allen MF, Swenson W, Querejeta JI et al (2011) Ecology of mycorrhizae: a conceptual framework for complex interactions among plants and fungi. Phys Ther 41:271–300

Araus JL, Slafer GA, Reynolds MP et al (2002) Plant breeding and drought in C3 cereals: what should we breed for? Ann Bot 89:925–940

Armada E, Azcon R, Lopez-Castillo OM et al (2015) Autochthonous arbuscular mycorrhizal fungi and *Bacillus thuringiensis* from a degraded Mediterranean area can be used to improve physiological traits and performance of a plant of agronomic interest under drought conditions. Plant Physiol Biochem 90:64–74

Arshad M, Saleem M, Hussain S (2007) Perspectives of bacterial ACC deaminase in phytoremediation. Trends Biotechnol 25:356–362

Auge RM, Toler HD, Saxton AM (2015) Arbuscular mycorrhizal symbiosis alters stomatal conductance of host plants more under drought than under amply watered conditions: a meta-analysis. Mycorrhiza 25:13–24

Azhiri-Sigari T, Yamauchi A, Kamoshita A et al (2000) Genotypic variation in response of rainfed lowland rice to drought and rewatering. Plant Prot Sci 3:180–188

Balaji B, Paulin MJ, Vierheilig H et al (1995) Responses of an arbuscular mycorrhizal fungus, *Gigaspora margarita*, to exudates and volatiles from Ri T-DNA-transformed roots of nonmycorrhizal and mycorrhizal mutants of *Pisum sativum* L. sparkle. Exp Mycol 19:275–283

Balestrini R, Lumini E, Borriello R et al (2015) Plant-soil biota interactions. In: Paul EA (ed) Soil microbiology. Ecology and Biochemistry. Academic/Elsevier, London, pp 311–338

Banoc DM, Yamauchi A, Kamoshita A et al (2000) Genotypic variations in response of lateral root development to fluctuating soil moisture in rice. Plant Prot Sci 3:335–343

Barea J, Azcon-Aguilar C, Azcon R (2002) Interactions between mycorrhizal fungi and rhizosphere micro-organisms within the context of sustainable soil-plant systems. In: Gange AC, Brown VK (eds) Multitrophic interactions in terrestrial systems. Cambridge University Press, Cambridge, pp 65–68

Barea J, Pozo M, Azcon R et al (2005) Microbial co-operation in the rhizosphere. J Exp Bot 56:1761–1767

Barnawal D, Maji D, Bharti N et al (2013) ACC deaminase-containing Bacillus Subtilis reduces stress ethylene-induced damage and improves mycorrhizal colonization and rhizobial nodulation in *Trigonella foenum-graecum* under drought stress. J Plant Growth Regul 32:809–822

Bharti N, Barnawal D, Shukla S et al (2016) Integrated application of *Exiguobacterium oxidotolerans*, *Glomus fasciculatum*, and vermicompost improves growth, yield and quality of *Mentha arvensis* in salt-stressed soils. Ind Crop Prod 83:717–728

Birhane E, Sterck FJ, Fetene M et al (2012) Arbuscular mycorrhizal fungi enhance photosynthesis, water use efficiency, and growth of frankincense seedlings under pulsed water availability conditions. Oecologia 169:895–904

Bissonnette L, St-Arnaud M, Labrecque M (2010) Phytoextraction of heavy metals by two Salicaceae clones in symbiosis with arbuscular mycorrhizal fungi during the second year of a field trial. Plant Soil 332:55–67

Bolanos L, El-Hamdaoui A, Bonilla I (2003) Recovery of development and functionality of nodules and plant growth in salt-stressed *Pisum sativum-Rhizobium leguminosarum* symbiosis by boron and calcium. J Plant Physiol 160:1493–1497

Bonfante P, Genre A (2010) Mechanisms underlying beneficial plant-fungus interactions in mycorrhizal symbiosis. Nat Commun 1:1–11

Boyer JS (1982) Plant productivity and environment. Science 218:443–448

Breuillin-Sessoms F, Floss DS, Gomez SK et al (2015) Suppression of arbuscule degeneration in *Medicago truncatula* phosphate transporter4 mutants is dependent on the ammonium transporter 2 family protein AMT2; 3. Plant Cell 27:1352–1366

Calvo-Polanco M, Sanchez-Romera B, Aroca R et al (2016) Exploring the use of recombinant inbred lines in combination with beneficial microbial inoculants (AM fungus and PGPR) to improve drought stress tolerance in tomato. Environ Exp Bot 131:47–57

Canci H, Toker C (2009) Evaluation of yield criteria for drought and heat resistance in chickpea (*Cicer arietinum* L.) J Agron Crop Sci 195:47–54

Cappellazzo G, Lanfranco L, Fitz M et al (2008) Characterization of an amino acid permease from the endomycorrhizal fungus *Glomus mosseae*. Plant Physiol 147:429–437

Cavagnaro TR (2008) The role of arbuscular mycorrhizas in improving plant zinc nutrition under low soil zinc concentrations: a review. Plant Soil 304:315–325

Chaudhry Q, Blom-Zandstra M, Gupta SK et al (2005) Utilising the synergy between plants and rhizosphere microorganisms to enhance breakdown of organic pollutants in the environment (15 pp). Environ Sci Pollut Res 12:34–48

Chen X, Wu C, Tang J et al (2005) Arbuscular mycorrhizae enhance metal lead uptake and growth of host plants under a sand culture experiment. Chemosphere 60:665–671

Cornejo P, Perez-Tienda J, Meier S et al (2013) Copper compartmentalization in spores as a survival strategy of arbuscular mycorrhizal fungi in Cu-polluted environments. Soil Biol Biochem 57:925–928

Daei G, Ardekani MR, Rejali F et al (2009) Alleviation of salinity stress on wheat yield, yield components, and nutrient uptake using arbuscular mycorrhizal fungi under field conditions. J Plant Physiol 166:617–625

Del Val C, Barea JM, Azcon-Aguilar C (1999) Diversity of arbuscular mycorrhizal fungus populations in heavy metal contaminated soils. Appl Environ Microbiol 99:718–723

Diehl WW (1950) Balansia and the Balansiae in America, Agriculture monograph No. 4. USDA, Washington, DC

Duzan HM, Zhou X, Souleimanov A et al (2004) Perception of *Bradyrhizobium japonicum* nod factor by soybean [*Glycine max* (L.) Merr.] root hairs under abiotic stress conditions. J Exp Bot 55:2641–1646

El Hafid R, Smith DH, Karrou M et al (1998) Physiological responses of spring durum wheat cultivars to early-season drought in a Mediterranean environment. Ann Bot 81:363–370

Evelin H, Kapoor R, Giri B (2009) Arbuscular mycorrhizal fungi in alleviation of salt stress: a review. Ann Bot 104(7):1263–1280

Feng G, Zhang F, Li X et al (2002) Improved tolerance of maize plants to salt stress by arbuscular mycorrhiza is related to higher accumulation of soluble sugars in roots. Mycorrhiza 12:185–190

Frechilla S, Lasa B, Ibarretxe L et al (2001) Pea responses to saline stress is affected by the source of nitrogen nutrition (ammonium or nitrate). Plant Growth Regul 35:171–179

Fukai S, Cooper M (1995) Development of drought-resistant cultivars using physiomorphological traits in rice. Field Crop Res 40:67–86

Garcia K, Zimmermann SD (2014) The role of mycorrhizal associations in plant potassium nutrition. Front Plant Sci 5:337

Garg N, Manchanda G (2008) Effect of arbuscular mycorrhizal inoculation on salt-induced nodule senescence in *Cajanus cajan* (pigeonpea). J Plant Growth Regul 27:115–124

Gaur A, Adholeya A (2004) Prospects of arbuscular mycorrhizal fungi in phytoremediation of heavy metal contaminated soils. Curr Sci 86:528–534

Gianinazzi-Pearson V (1995) Morphofunctional compatibility in interactions between roots and arbuscular mycorrhizal fungi: molecular mechanisms, genes and genes expression. In: Kohmoto K, Singh RP, Singh US (eds) Diseases. Peragmon Press, Oxford, pp 251–263

Giovannetti M, Tolosano M, Volpe V et al (2014) Identification and functional characterization of a sulfate transporter induced by both sulfur starvation and mycorrhiza formation in *Lotus japonicus*. New Phytol 204:609–619

Giri B, Mukerji K (2004) Mycorrhizal inoculant alleviates salt stress in *Sesbania aegyptiaca* and *Sesbania grandiflora* under field conditions: evidence for reduced sodium and improved magnesium uptake. Mycorrhiza 14:307–312

Giri B, Kapoor R, Mukerji KG (2003) Influence of arbuscular mycorrhizal fungi and salinity on growth, biomass, and mineral nutrition of *Acacia auriculiformis*. Biol Fertil Soils 38:170–175

Giri B, Kapoor R, Mukerji KG (2007) Improved tolerance of *Acacia nilotica* to salt stress by arbuscular mycorrhiza, *Glomus fasciculatum*, may be partly related to elevated Kþ/Naþ ratios in root and shoot tissues. Microb Ecol 54:753–760

Gonzalez-Guerrero M, Azcón-Aguilar C, Mooney M et al (2005) Characterization of a *Glomus intraradices* gene encoding a putative Zn transporter of the cation diffusion facilitator family. Fungal Genet Biol 42:130–140

Griffioen WAJ, Iestwaart JH, Ernst WHO (1994) Mycorrhizal infection of *Agrostis capillaris* population on a copper contaminated soil. Plant Soil 158:83–89

Guether M, Neuhäuser B, Balestrini R et al (2009) A mycorrhizal-specific ammonium transporter from *Lotus japonicus* acquires nitrogen released by arbuscular mycorrhizal fungi. Plant Physiol 150:73–83

Harrison MJ, van Buuren ML (1995) From the mycorrhizal fungus *Glomus versiforme*. Nature 378

Harrison MJ, Dewbre GR, Liu J (2002) A phosphate transporter from *Medicago truncatula* involved in the acquisition of phosphate released by arbuscular mycorrhizal fungi. Plant Cell 14:2413–2429

Hashem A, Abd_Allah EF, Alqarawi AA et al (2016) The interaction between arbuscular mycorrhizal fungi and endophytic bacteria enhances plant growth of *Acacia gerrardii* under salt stress. Front Microbiol 7:1089

Ibrahim AH, Abdel-Fattah GM, Eman FM et al (2011) Arbuscular mycorrhizal fungi and spermine alleviate the adverse effects of salinity stress on electrolyte leakage and productivity of wheat plants. Phyton 51:261

Jäderlund L, Arthurson V, Granhall U et al (2008) Specific interactions between arbuscular mycorrhizal fungi and plant growth-promoting bacteria: as revealed by different combinations. FEMS Microbiol Lett 287:174–180

Janouskova M, Vosatka M (2005) Response to cadmium of *Daucus carota* hairy roots dual cultures with *Glomus intraradices* or *Gigaspora margarita*. Mycorrhiza 15:217–224

Javot H, Penmetsa RV, Terzaghi N et al (2007) A *Medicago truncatula* phosphate transporter indispensable for the arbuscular mycorrhizal symbiosis. Proc Natl Acad Sci 104:1720–1725

Joner EJ, Leyval C (2001) Influence of arbuscular mycorrhiza on clover and ryegrass grown together in a soil spiked with polycyclic aromatic hydrocarbons. Mycorrhiza 10:155–159

Kaldorf M, Kuhn AJ, Schroder WH et al (1999) Selective element deposits in maize colonized by a heavy metal tolerance conferring arbuscular mycorrhizal fungus. J Plant Physiol 154:718–728

Kamoshita A, Wade LJ, Yamauchi A (2000) Genotypic variation in response of rainfed lowland rice to drought and rewatering. III. Water extraction during the drought period. Plant Prot Sci 3:189–196

Kashiwagi J, Krishnamurthy L, Upadhyaya HD et al (2005) Genetic variability of drought-avoidance root traits in the mini-core germplasm collection of chickpea (*Cicer arietinum* L.) Euphytica 146:213–222

Kaya C, Ashraf M, Sonmez O et al (2009) The influence of arbuscular mycorrhizal colonisation on key growth parameters and fruit yield of pepper plants grown at high salinity. Sci Hortic 121:1–6

Khalvati M, Bartha B, Dupigny A et al (2010) Arbuscular mycorrhizal association is beneficial for growth and detoxification of xenobiotics of barley under drought stress. J Soils Sediments 10:54–64

Kistner C, Parniske M (2002) Evolution of signal transduction in intracellular symbiosis. Trends Plant Sci 7:511–518

Kobae Y, Tamura Y, Takai S et al (2010) Localized expression of arbuscular mycorrhiza-inducible ammonium transporters in soybean. Plant Cell Physiol 51:1411–1415

Koske R, Gemma J (1992) Fungal reactions to plants prior to mycorrhizal formation. In: Allen MF (ed) Mycorrhizal functioning: an integrative plant-fungal process. Chapman and Hall, NewYork, pp 3–27

Kumar KV, Singh N, Behl HM et al (2008) Influence of plant growth promoting bacteria and its mutant on heavy metal toxicity in *Brassica juncea* grown in fly ash amended soil. Chemosphere 72:678–683

Kuriakose SV, Prasad MN (2008) Cadmium stress affects seed germination and seedling growth in *Sorghum bicolor* (L.) Moench by changing the activities of hydrolyzing enzymes. Plant Growth Regul 54:143–156

Kuswandi PC, Sugiyarto L (2015) Application of mycorrhiza on planting media of two tomato varieties to increase vegetable productivity in drought condition. J Sains Dasar 4:17–22

Kyriazopoulos AP, Orfanoudakis M, Abraham EM et al (2014) Effects of arbuscular mycorrhiza fungi on growth characteristics of *Dactylis glomerata* L. under drought stress conditions. Not Bot Horti Agrobo 42:132–137

Lehmann A, Rillig MC (2015) Arbuscular mycorrhizal contribution to copper, manganese and iron nutrient concentrations in crops-a meta-analysis. Soil Biol Biochem 81:147–158

Lehmann A, Veresoglou SD, Leifheit EF et al (2014) Arbuscular mycorrhizal influence on zinc nutrition in crop plants–a meta-analysis. Soil Biol Biochem 69:123–131

Leifheit EF, Veresoglou SD, Lehmann A et al (2014) Multiple factors influence the role of arbuscular mycorrhizal fungi in soil aggregation-a meta-analysis. Plant Soil 374:523–537

Leifheit EF, Verbruggen E, Rillig MC (2015) Arbuscular mycorrhizal fungi reduce decomposition of woody plant litter while increasing soil aggregation. Soil Biol Biochem 81:323–328

Leyval C, Singh BR, Joner EJ (1995) Occurrence and infectivity of arbuscular mycorrhizal fungi in some Norwegian soils influenced by heavy metals and soil properties. Water Air Soil Pollut 84:203–216

Li X, Peng XW, Wu SL et al (2014) Effect of arbuscular mycorrhizae on growth, heavy metal uptake and accumulation of *Zenia insignis* Chun seedlings. Huan Jing Ke Xue 35:3142–3148

Liu XF, Supek F, Nelson N et al (1997) Negative control of heavy metal uptake by the Saccharomyces Cerevisiae BSD2 gene. J Biol Chem 272:11763–11769

Liu T, Sheng M, Wang CY et al (2015) Impact of arbuscular mycorrhizal fungi on the growth, water status, and photosynthesis of hybrid poplar under drought stress and recovery. Photosynthetica 53:250–258

Lucic E, Fourrey C, Kohler A et al (2008) A gene repertoire for nitrogen transporters in *Laccaria bicolor*. New Phytol 180:343–364

Lynch J (1995) Root architecture and plant productivity. Plant Physiol 109:7–13

Marulanda A, Azcon R, Ruiz-Lozano JM (2003) Contribution of six arbuscular mycorrhizal fungal isolates to water uptake by *Lactuca sativa* plants under drought stress. Physiol Plant 119:526–533

Marulanda A, Porcel R, Barea JM et al (2007) Drought tolerance and antioxidant activities in lavender plants colonized by native drought-tolerant or drought-sensitive *Glomus* species. Microb Ecol 54:543–552

Matias SR, Pagano MC, Muzzi FC et al (2009) Effect of rhizobia, mycorrhizal fungi and phosphate-solubilizing microorganisms in the rhizosphere of native plants used to recover an iron ore area in Brazil. Eur J Soil Biol 45:259–266

McKay IA, Djordjevic MA (1993) Production and excretion of nod metabolites by *Rhizobium leguminosarum* bv. Trifolii are disrupted by the same environmental factors that reduce nodulation in the field. Appl Environ Microbiol 59:3385–3392

Medina A, Vassileva M, Barea JM et al (2006) The growth-enhancement of clover by Aspergillus-treated sugar beet waste and *Glomus mosseae* inoculation in Zn contaminated soil. Appl Soil Ecol 33:87–98

Meier S, Cornejo P, Cartes P et al (2015) Interactive effect between Cu-adapted arbuscular mycorrhizal fungi and biotreated agrowaste residue to improve the nutritional status of *Oenothera picensis* growing in Cu-polluted soils. J Plant Nutr Soil Sci 178:126–135

Ming T, Hui C (1999) Effects of arbuscular mycorrhizal fungi alkaline phosphatase activities on *Hippophae rhamnoides* drought-resistance under water stress conditions. Trees 14:113–115

Miransari M (2013) Corn (*Zea mays* L.) growth as affected by soil compaction and arbuscular mycorrhizal fungi. J Plant Nutr 36:1853–1867

Miransari M, Bahrami HA, Rejali F et al (2008) Using arbuscular mycorrhiza to alleviate the stress of soil compaction on wheat (*Triticum aestivum* L.) growth. Soil Biol Biochem 40:1197–1206

Moreira H, Pereira SI, Marques AP et al (2016) Mine land valorization through energy maize production enhanced by the application of plant growth-promoting rhizobacteria and arbuscular mycorrhizal fungi. Environ Sci Pollut Res 23:6940–6950

Nadeem SM, Zahir ZA, Naveed M et al (2007) Preliminary investigations on inducing salt tolerance in maize through inoculation with rhizobacteria containing ACC deaminase activity. Can J Microbiol 53:1141–1149

Nadeem SM, Ahmad M, Zahir ZA et al (2014) The role of mycorrhizae and plant growth promoting rhizobacteria (PGPR) in improving crop productivity under stressful environment. Biotechnol Adv 32:429–448

Nagy R, Karandashov V, Chague V et al (2005) The characterization of novel mycorrhiza specific phosphate transporters from *Lycopersicon esculentum* and *Solanum tuberosum* uncovers functional redundancy in symbiotic phosphate transport in solanaceous species. The Plant J 42:236–250

Nair MG, Safir GR, Siqueira JO (1991) Isolation and identification of vesicular-arbuscular mycorrhiza-stimulatory compounds from clover (*Trifolium repens*) roots. Appl Environ Microbiol 57:434–439

Olsson PA, Hammer EC, Wallander H et al (2008) Phosphorus availability influences elemental uptake in the mycorrhizal fungus *Glomus intraradices*, as revealed by particle-induced X-ray emission analysis. Appl Environ Microbiol 74:4144–4148

Olsson PA, Hammer EC, Pallon J et al (2011) Elemental composition in vesicles of an arbuscular mycorrhizal fungus, as revealed by PIXE analysis. Fungal Biol 115:643–648

Pallon J, Wallander H, Hammer E et al (2007) Symbiotic fungi that are essential for plant nutrient uptake investigated with NMP. Nucl Instrum Methods Phy Res Sec B: Beam Interact Mater At 260:149–152

Parniske M (2008) Arbuscular mycorrhiza: the mother of plant root endosymbioses. Nat Rev Microbiol 6:763–775

Porcel R, Ruiz-Lozano JM (2004) Arbuscular mycorrhizal influence on leaf water potential, solute accumulation, and oxidative stress in soybean plants subjected to drought stress. J Exp Bot 55:1743–1750

Porcel R, Aroca R, Ruiz-Lozano JM (2011) Salinity stress alleviation using arbuscular mycorrhizal fungi. A review. Agron Sustain Dev 32:181–200

Price AH, Steele KA, Moore BJ et al (2000) A combined RFLP and AFLP linkage map of upland rice (*Oryza sativa* L.) used to identify QTLs for root-penetration ability. Theor Appl Genet 100:49–56

Redon P, Beguiristain T, Leyval C (2009) Differential effects of AM fungal isolates on *Medicago truncatula* growth and metal uptake in a multi metallic (Cd, Zn, Pb) contaminated agricultural soil. Mycorrhiza 19:187–195

Richards RA, Passioura JB (1981) Seminal root morphology and water use of wheat II. Genetic variation. Crop Sci 21:253–255

Rillig MC, Wright SF, Eviner VT (2002) The role of arbuscular mycorrhizal fungi and glomalin in soil aggregation: comparing effects of five plant species. Plant Soil 238:325–333

Rillig MC, Aguilar-Trigueros CA, Bergmann J et al (2015) Plant root and mycorrhizal fungal traits for understanding soil aggregation. New Phytol 205:1385–1388

Ruiz-Lozano JM (2003) Arbuscular mycorrhizal symbiosis and alleviation of osmotic stress. New perspectives for molecular studies Mycorrhiza 13:309–317

Ruiz-Lozano JM, Azcon R (2000) Symbiotic efficiency and infectivity of an autochthonous arbuscular mycorrhizal *Glomus* sp. from saline soils and *Glomus deserticola* under salinity. Mycorrhiza 10:137–143

Saleem M, Arshad M, Hussain S et al (2007) Perspective of plant growth promoting rhizobacteria (PGPR) containing ACC deaminase in stress agriculture. J Ind Microbiol Biotechnol 34:635–648

Sambandan K, Kannan K, Raman N (1992) Distribution of vesicular- arbuscular mycorrhizal fungi in heavy metal polluted soils of Tamil Nadu. J Environ Biol 13:159–167

Serraj R, Krishnamurthy L, Kashiwagi J et al (2004) Variation in root traits of chickpea (*Cicer arietinum* L.) grown under terminal drought. Field Crop Res 88:115–127

Sharif M, Claassen N (2011) Action mechanisms of arbuscular mycorrhizal fungi in phosphorus uptake by *Capsicum annuum* L. Pedosphere 21:502–511

Sharifi M, Ghorbanli M, Ebrahimzadeh H (2007) Improved growth of salinity-stressed soybean after inoculation with salt pre-treated mycorrhizal fungi. J Plant Physiol 164:1144–1151

Shokri S, Maadi B (2009) Effects of arbuscular mycorrhizal fungus on the mineral nutrition and yield of *Trifolium alexandrinum* plants under salinity stress. J Agron 8:79–83
Siddiqui ZA, Akhtar MS (2009) Effects of antagonistic fungi, plant growth-promoting rhizobacteria, and arbuscular mycorrhizal fungi alone and in combination on the reproduction of Meloidogyne incognita and growth of tomato. J Gen Plant Pathol 75:144–153
Sinclair G, Charest C, Dalpe Y et al (2014) Influence of colonization by arbuscular mycorrhizal fungi on three strawberry cultivars under salty conditions. Agric Food Sci 23:146–158
Singh A (2007) Molecular basis of plant-symbiotic fungi interaction: an overview. Sci World 5:115–131
Siopongco JD, Yamauchi A, Salekdeh H et al (2005) Root growth and water extraction response of doubled-haploid rice lines to drought and rewatering during the vegetative stage. Plant Prot Sci 8:497–508
Smith SE, Read DJ (2008) Mycorrhizal symbiosis, 3rd edn. Academic Press, London
Suralta RR, Yamauchi A (2008) Root growth, aerenchyma development, and oxygen transport in rice genotypes subjected to drought and waterlogging. Environ Exp Bot 64:75–82
Sylvia D, Fuhrmann J, Hartel P et al (2005) Principles and applications of soil microbiology. Upper Saddle River, NJ, Pearson
Talaat NB, Shawky BT (2012) Influence of arbuscular mycorrhizae on root colonization, growth and productivity of two wheat cultivars under salt stress. Arch Agron Soil Sci 58:85–100
Tamás L, Dudíkova J, Durčekova K et al (2008) Alterations of the gene expression, lipid peroxidation, proline and thiol content along the barley root exposed to cadmium. J Plant Physiol 165:1193–1203
Tamayo E, Gomez-Gallego T, Azcón-Aguilar C et al (2014) Genome-wide analysis of copper, iron and zinc transporters in the arbuscular mycorrhizal fungus *Rhizophagus irregularis*. Plant Traffic Transp 5:1–13
Tester M, Bacic A (2005) Abiotic stress tolerance in grasses; from model plants to crop plants. Plant Physiol 137:791–793
Tian CY, Feng G, Li XL et al (2004) Different effects of arbuscular mycorrhizal fungal isolates from saline or non-saline soil on salinity tolerance of plants. Appl Soil Ecol 26:143–148
Toler HD, Morton JB, Cumming JR (2005) Growth and metal accumulation of mycorrhizal sorghum exposed to elevated copper and zinc. Water Air Soil Pollut 164:155–172
Tonin C, Vandenkoornhuyse P, Joner EJ et al (2001) Assessment of arbuscular mycorrhizal fungi diversity in the rhizosphere of *Viola calaminaria* and effect of these fungi on heavy metal uptake by clover. Mycorrhiza 10:161–168
Tu JC (1981) Effect of salinity on rhizobium-root-hair interaction, nodulation and growth of soybean. Can J Plant Sci 61:231–239
Turnau K, Mesjasz-Przybylowicz J (2003) Arbuscular mycorrhiza of *Berkheya coddii* and other Ni-hyperaccumulating members of Asteraceae from ultramafic soils in South Africa. Mycorrhiza 13:185–190
Turnau K, Ryszka P, Gianinazzi-Pearson V, van Tuinen D (2001) Identification of arbuscular mycorrhizal fungi in soils and roots of plants colonizing zinc wastes in southern Poland. Mycorrhiza 10:169–174
Turnau K, Anielska T, Ryszka P et al (2008) Establishment of arbuscular mycorrhizal plants originating from xerothermic grasslands on heavy metal rich industrial wastes–new solution for waste revegetation. Plant Soil 305:267–280
Valdenegro M, Barea JM, Azcon R (2001) Influence of arbuscular-mycorrhizal fungi, rhizobium meliloti strains and PGPR inoculation on the growth of *Medicago arborea* used as model legume for re-vegetation and biological reactivation in a semi-arid mediterranean area. Plant Growth Regul 34:233–240
Vodnik D, Grcman H, Macek I et al (2008) The contribution of glomalin-related soil protein to Pb and Zn sequestration in polluted soil. Sci Total Environ 392:130–136

Walder F, Brule D, Koegel S et al (2015) Plant phosphorus acquisition in a common mycorrhizal network: regulation of phosphate transporter genes of the Pht1 family in sorghum and flax. New Phytol 205:1632–1645

Walker V, Couillerot O, Von Felten A et al (2012) Variation of secondary metabolite levels in maize seedling roots induced by inoculation with *Azospirillum*, *Pseudomonas* and *Glomus* consortium under field conditions. Plant Soil 356:151–163

Wang F, Lin X, Yin R (2005) Heavy metal uptake by arbuscular mycorrhizas of *Elsholtzia splendens* and the potential for phytoremediation of contaminated soil. Plant Soil 269:225–232

Wang P, Wu SH, Wen MX et al (2016) Effects of combined inoculation with *Rhizophagus intraradices* and *Paenibacillus mucilaginosus* on plant growth, root morphology, and physiological status of trifoliate orange (*Poncirus trifoliata* L. Raf.) seedlings under different levels of phosphorus. Sci Hortic 205:97–105

Watts-Williams SJ, Patti AF, Cavagnaro TR (2013) Arbuscular mycorrhizas are beneficial under both deficient and toxic soil zinc conditions. Plant Soil 371:299–312

Weissenhorn I, Leyval C, Berthelin J (1993) Cd-tolerant arbuscular mycorrhizal (AM) fungi from heavy-metal polluted soils. Plant Soil 157:247–256

Wu QS, Zou YN, Liu W et al (2010) Alleviation of salt stress in citrus seedlings inoculated with mycorrhiza: changes in leaf antioxidant defense systems. Plant Soil Environ 56:470–475

Xie X, Huang W, Liu F et al (2013) Functional analysis of the novel mycorrhiza-specific phosphate transporter AsPT1 and PHT1 family from *Astragalus sinicus* during the arbuscular mycorrhizal symbiosis. New Phytol 198:836–852

Xun F, Xie B, Liu S et al (2015) Effect of plant growth-promoting bacteria (PGPR) and arbuscular mycorrhizal fungi (AMF) inoculation on oats in saline-alkali soil contaminated by petroleum to enhance phytoremediation. Environ Sci Pollut Res 22:598–608

Yang Y, Tang M, Sulpice R et al (2014) Arbuscular mycorrhizal fungi alter fractal dimension characteristics of *Robinia pseudoacacia* L. seedlings through regulating plant growth, leaf water status, photosynthesis, and nutrient concentration under drought stress. J Plant Growth Regul 33:612–625

Yang Y, Song Y, Scheller HV et al (2015) Community structure of arbuscular mycorrhizal fungi associated with *Robinia pseudoacacia* in uncontaminated and heavy metal contaminated soils. Soil Biol Biochem 86:146–158

Yang Y, Liang Y, Han X et al (2016) The roles of arbuscular mycorrhizal fungi (AMF) in phytoremediation and tree-herb interactions in Pb contaminated soil. Sci Report 6:20469

Zahran HH (1999) Rhizobium-legume symbiosis and nitrogen fixation under severe conditions and in an arid climate. Microbiol Mol Biol Rev 63:968–989

Zhang Y, Zhong CL, Chen Y et al (2010) Improving drought tolerance of *Casuarina equisetifolia* seedlings by arbuscular mycorrhizas under glasshouse conditions. New For 40:261–271

Chapter 2
Arbuscular Mycorrhizal Fungi and Tolerance of Drought Stress in Plants

Qiang-Sheng Wu and Ying-Ning Zou

Abstract Drought stress has strong inhibition in plant growth and crop production. Arbuscular mycorrhizal fungi (AMF) can colonize the roots of 80% of land's plants to establish arbuscular mycorrhizal symbiosis. A relative short-term soil drought did not appear to discourage root AMF colonization, whereas a long-term soil drought intensity considerably decreased root colonization and hyphal growth in the soil. Such change in mycorrhizal development still strongly stimulated the improvement of plant growth and increased plant survival under drought stress. AMF had shown to enhance drought tolerance in various plants. Firstly, mycorrhizal plants could adapt the drought stress in morphology, especially leaf epicuticular wax and root morphology. And mycorrhizal plants possessed direct pathway of water uptake by extraradical hyphae. In addition, AMF enhanced drought tolerance of the host plant through physiological mechanisms in nutrient uptake and biochemical mechanisms regarding hormones, osmotic adjustment, and antioxidant systems. AMF also released glomalin into soil, defined as glomalin-related soil protein, to improve soil structure, thereby regulating water relations of plant/soil. Molecule mechanisms about expression of relevant stressed genes were clarified a bit more detail. Future perspectives in this field are provided.

Keywords Extraradical hyphae • Mycorrhizas • Mycorrhizal colonization • Water stress • Water uptake

---

Q.-S. Wu (✉)
College of Horticulture and Gardening, Yangtze University, Jingzhou, Hubei 434025, China

Institute of Root Biology, Yangtze University, Jingzhou, Hubei 434025, China

Department of Chemistry, Faculty of Science, University of Hradec Kralove, Hradec Kralove 50003, Czech Republic
e-mail: wuqiangsh@163.com

Y.-N. Zou
College of Horticulture and Gardening, Yangtze University, Jingzhou, Hubei 434025, China

Institute of Root Biology, Yangtze University, Jingzhou, Hubei 434025, China

Q.-S. Wu (ed.), *Arbuscular Mycorrhizas and Stress Tolerance of Plants*,
DOI 10.1007/978-981-10-4115-0_2

2.1 Introduction

Drought stress is considered as one of the most serious abiotic stresses that limits plant growth and reduces crop production in numerous regions worldwide. It is estimated that near one-third of soils are subjected to drought stress, thus, for supporting normal plant development (Calvo-Polanco et al. 2016). Many factors including a lack of rainfall capacity, irregular rainfall distribution, drought intensity and duration, and progression rate of stress are response for water scarcity (Kolenc et al. 2016). Drought stress often causes lower soil water potential, inducing cell dehydration, ultimately resulting in inhibiting cell expansion and division, leaf size, stem elongation, root proliferation, disturbed stomatal oscillations, plant water, nutrient uptake, and water use efficiency (Kaushai and Wani 2016). Certainly, plants also develop sophisticated and complex mechanisms in morphological, physiological, and biochemical characteristics, dividing into escape, avoiding, and drought tolerance, to cope with drought stress (Khoyerdi et al. 2016).

Arbuscular mycorrhizal fungi, belonging to phylum Glomeromycota, are soil inhabitants, where AMF can colonize the roots of 80% of land's plants. Mycorrhizal characteristics are defined as the mutual benefits: AMF provide the host plant with essential nutrients (especially P) and water, and photosynthates are transported into endosymbiotic AMF for its development. As stated by Requena et al. (2007), compatible roots are colonized by the hyphae, produced by soil propagules of AMF, asexual spores, or mycorrhizal roots. In general, mycorrhizal hyphae colonize roots by means of an appressorium and then penetrate into the root cortex, finally forming distinct morphological structures: intercellular and intracellular hyphae, coils, and arbuscules (Fig. 2.1). Moreover, extraradical mycorrhizal of colonized roots will further explore the soil in mineral nutrients and even also colonize the roots of other neighbor plants, establishing common mycorrhizal network (CMN).

The AM symbiosis has shown many beneficial roles in plant growth, nutrition absorption, root architecture, flowering, and stress tolerance (Pozo et al. 2015). Studies indicated that AM symbiosis is able to enhance drought tolerance of plants (Augé 2001; Augé and Moore 2005), but the exact mechanisms are not fully known. This work was begun firstly by Safir et al. (1971), who found that AMs reduced the resistance to water uptake by soybean, possibly attributable to changes in root resistance. Safir's works initiated the research between AM fungi and water relations of plants. In the next 40 years, new perspectives are achieved recently in terms of advances in molecular, biochemical, and physiological techniques.

The objectives of this chapter are to highlight the effects of drought stress on mycorrhizal development and to discuss these possible mechanisms regarding AMF-induced drought tolerance enhancement in plants.

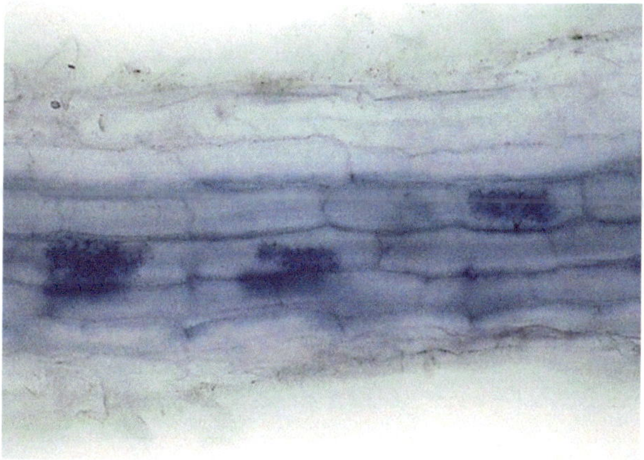

Fig. 2.1 Arbuscules in white clover colonized by *Rhizoglomus intraradices*

2.2 Effects of Drought Stress on Mycorrhizal Development

Drought stress strongly affects mycorrhizal development in roots and soils. Augé (2001) proposed, based on >150 literatures, that relative short-term soil drought did not appear to favor or discourage root AMF colonization and longer-term soil drought decreased AMF colonization. So it is more common in increased levels of root AMF colonization in response to soil drought stress than in decreased levels, which is related to reductions in plant P levels (Finlay et al. 2008).

In fact, spores of AMF live in soils, and soil moisture can affect spore germination and development. Studies had confirmed that soil water status strongly altered spore behavior. Spores are surrounded by soils, where soil water availability is one of the important factors modulating the contact zone between the spore and the soil (Tommerup 1984). Tiny changes in the matric potential of the soil surrounding a spore result in either recession of water from smaller spores or movement into larger spores with consequential changes in the degree of contact between spores and soil water, and in the conductivity of water in the capillaries at the spore-soil interface. So spore germination is favored in soil at or above field capacity but is decreased with decreasing soil water potentials below field capacity (Daniels and Trappe 1980). As reported by Griffin (1972), spore germination increased with the time lapsing before hydration (<1.4 Mpa, soil water potential), then decreased between −1.4 and −5 Mpa, and stopped between −5 and −10 Mpa. As reviewed by Wu et al. (2013), drought stress considerably inhibited the root mycorrhizal development in citrus plants. Root AMF colonization in adult pomelo (*Citrus grandis* Osbeck cv. Shatianyou) trees increased from a drying soil to well-watered soil and decreased with the increase of soil water status, due to low O_2 concentration (Xue 2004). As a result, soil waterlogging generally inhibited mycorrhizal development in roots and soil (Wu et al. 2013).

Moreover, part of AMF species can quickly adapt to soil drought and thus still keep relatively higher root AMF colonization, which is beneficial for survival and growth of the host plant (Nasim 2010). For example, three *Glomus* species, *G. macrocarpum*, *G. clarum*, and *G. etunicatum*, exhibited considerable tolerance to soil drying (Sylvia and Schenck 1983). If the spores of AMF were treated by storage in different soil water potentials, *Glomus mosseae* and *G. deserticola* under −0.04 Mpa showed better infectivity than *G. fasciculatum* under −0.8 Mpa, indicating that surrounding of spores would have strong function on its efficiency in root colonization (Wu et al. 2013). Douds and Schenck (1991) reported that spore germination of *Gigaspora margarita* was independent on soil water content, while germination of *G. intraradices*, *G. mosseae*, and *Acaulospora longula* was inhibited by matric potentials between −0.50 and −2.20 Mpa. In a word, soil wetting and drying cycles are the most factors affecting spore survival and germination and subsequent infectivity of AMF (Giovannetti et al. 2010).

However, other studies also showed an increased AMF colonization on *Sorghum bicolor* with drier soils (Sieverding 1981) or no differences of root AMF colonization for winter wheat between wet and dry treatments (Allen and Boosalis 1983). Besides root mycorrhizal colonization, low soil moisture also strongly inhibits soil extraradical hyphal length (Neumann et al. 2009).

2.3 AM Effects on Plant Growth and Transplanted Survival Under Drought Stress

2.3.1 Plant Growth

In general, soil drought stress strongly inhibited plant growth, while AMF inoculation considerably mitigated the negative effects by drought stress in various plants, including sugarcane, citrus, mung bean, apple, tomato, maize, wheat, wild jujube, trifoliate orange (Fig. 2.2), etc. As reported by He et al. (1999), inoculation with *G. mosseae*, *G.* spp., and *G. caledonium* showed 1.99, 1.95, and 1.80 times higher biomass of mung bean than non-AMF treatment under 12% soil water content conditions. Moreover, the AMF-stimulated plant growth enhancement may be more important in the host plant under drought stress than under well-watered conditions (Sánchez-Díaz and Honrubia 1994; Zou et al. 2015a). As outlined by Pozo et al. (2015), stress-induced abscisic acid (ABA) and strigolactones may participate in the promotion of AM functioning. In addition, other explanations including increase of P nutrition (Bethlenfalvay et al. 1988; Sweatt and Davies 1984), water uptake by hyphae (Zou et al. 2015b), and increase of root length (Bryla and Duniway 1997) should be included.

Fig. 2.2 Plant growth of trifoliate orange inoculated with *Diversispora versiformis* (AMF) under well-watered (WW) and drought stress (DS)

2.3.2 Plant Survival at Transplanting

Plant survival at transplanting is an important factor in crop production. Earlier studies showed that AMF inoculation strongly enhanced plant survival at transplanting under drought stress conditions. In *Casuarina equisetifolia* seedlings grown in the condition of greenhouse, plant survival of the seedlings inoculated with *Glomus caledonium* Gc90068, *G. versiforme* Gv9004, and *G. caledonium* Gc90036 increased by 37%, 23%, and 17%, respectively, compared with non-AMF seedlings exposed to drought stress without watering for 7 days (Zhang et al. 2010). Wu et al. (2006) also found 8% higher plant survival in trifoliate orange colonized by *F. mosseae* than non-AMF colonization after water stress and rewatering. Similar results were reported in Wyoming big sagebrush (*Artemisia tridentata* Nutt. ssp. *wyomingensis* Beetle and Young) seedlings grown in the process of −2.5 to −3.8 Mpa soil environment (Stahl et al. 1998). In a word, the increase of plant survival in transplanting by AMF will help to rebuild ecosystems under drought stress.

At weaning and hardening stages of micropagated *Syngonium podophyllum* plantlets, inoculation with a mixed AMF inocula relatively increased plant survival than non-AMF inoculation, while the AMF effect was high in weaning stage than in hardened stage (Gaur and Adholeya 1999). This may be due to a more effective root architecture for water and nutritional absorption and the developed external hyphae in the soil (Graham et al. 1982; Berta et al. 1990; Wu et al. 2011b).

2.4 Potential Mechanisms Regarding AMF-Enhanced Drought Tolerance in Plants

Since Safir et al. (1971) reported AMF effects on water uptake of soybean, investigations of AM symbiosis in plant species under drought stress conditions have lasted for 45 years, and many mechanisms have been outlined to clarify AMF roles in drought tolerance in the levels of morphology, physiology, biochemistry, and molecule.

2.4.1 *Morphological Adaptation*

Studies showed that AMF colonization can modulate morphological adaptation to enhance drought tolerance of the host plant. Earlier studies revealed that less epicuticular wax and lower cuticle weight were found in leaves of AM rose plants than non-AM plants during drought acclimation (Henderson and Davies 1990). The lack of wax in AM rose during drought acclimation would be ascribed to the tendency to abscise leaves. Another explanation is an increased ability of AM plants to absorb water. Morphological studies further showed more starch storage in palisade mesophyll in AM rose under low P, while less starch quantity was found in palisade layer, near vascular bundle in non-AM rose under high P (Augé et al. 1987). Mycorrhizal plants would recover more quickly from wilting than non-mycorrhizal plants after drought recovery (Gemma et al. 1997). Even so, AMF inoculation did not considerably affect stomatal density and guard cell size than non-AMF treatment (Henderson and Davies 1990).

In addition to leaf morphological adaptation, root morphological adaptation is also a strategy by mycorrhization under drought stress. Recently, Liu et al. (2016) analyzed the changes in root morphology of trifoliate orange infected with or without *F. mosseae*. The results showed significantly higher root total length, projected area, surface area, average diameter, volume, and number of first-, second-, and third-order lateral roots in AMF trifoliate orange seedlings under well-watered and drought stress conditions, as compared with non-AMF seedlings (Fig. 2.3). Such better root morphological adaptation caused by mycorrhization can provide more exploration of soil volume to absorb water and nutrients from the soil (Comas et al. 2013), thereby potentially enhancing drought tolerance of the host plant. The AMF-induced root morphological changes may be due to the regulation of endogenous polyamine metabolism and phytohormone equilibrium, especially root putrescine synthetases by activation of arginine decarboxylase and ornithine decarboxylase, as well as root indole-3-acetic acid (IAA) (Wu et al. 2012; Liu et al. 2016).

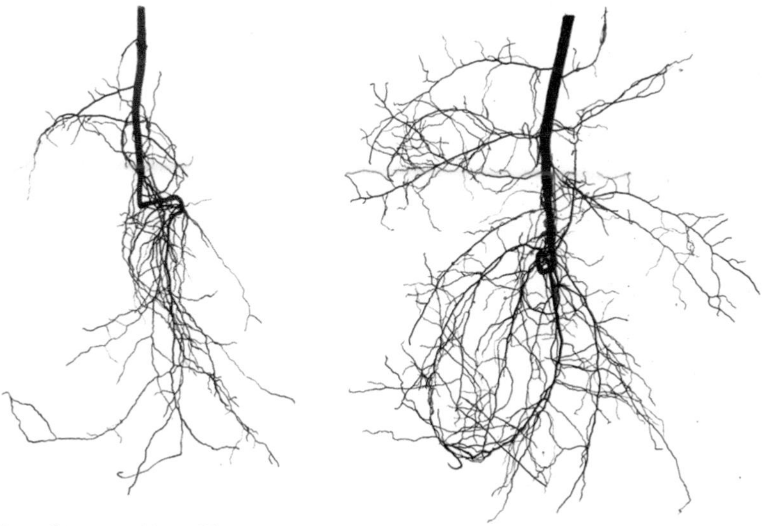

Fig. 2.3 Root morphology of trifoliate orange inoculated with *Funneliformis mosseae* under drought stress

2.4.2 Water Uptake Directly by Mycorrhizal Extraradical Hyphae

Mycorrhizal hyphae have an important function on uptake and transport of water from the bulk soil to the host plant (Egerton-Warburton et al. 2007). George et al. (1992) reported that mycorrhizal hyphae had the ability to transport considerable quantities of phosphate and nitrogen to the plant from soil zones but no evidence for a significant direct water transport by AM hyphae to plants. To examine the long-distance transport of water through mycorrhizal hyphae, Faber et al. (1991) designed a rhizobox system, consisting of plant + hyphae chamber and hyphae chamber, which confirmed an active role in water transport by mycorrhizal hyphae. Subsequently, Ruiz-Lozano and Azcón (1995) designed a simple split-root-hyphae system chamber to evaluate water uptake by extraradical hyphae. Zou et al. (2015b) also reported that disruption of mycorrhizal hyphae in a two-chambered rootbox resulted in a decrease in leaf water potential in trifoliate orange. It is well known that mycorrhizal hyphae have a diameter of 2–5 μm that penetrates soil pores inaccessible to root hairs (Sánchez-Díaz and Honrubia 1994), thereby extending root contacted zones more access to available water (Khan 2003). Due to the few or no septa in mycorrhizal hyphae, mycorrhizal hyphae can be used as a highway to transport water from hyphae to plant cells, in company with phosphate (Wu and Zou 2009c). It is estimated that water transport in 10-mm-diameter hyphae would be 5.4 nl/h under a 0.5 MPa gradient conditions (Allen 2006).

As proposed by Allen (2007), mycorrhizal hyphae have bidirectional flows. In addition to water uptake, mycorrhizal hyphae also contributed to water redistribution with rhizosphere through reverse flow, namely, transfer of water from the root to AMF (Egerton-Warburton et al. 2007).

2.4.3 Physiological Mechanisms in Nutrient Uptake

Earlier studies showed that the increase in water uptake by AM was a consequence of P increase, an indirect response (Safir et al. 1972). Subsequently, many studies also confirmed the correlation of drought tolerance with improved nutrition, especially P (Nelsen and Safir 1982a; Bethlenfalvay et al. 1988; Fitter 1988; Yano-Melo et al. 1999). For instance, addition of P fertilizer to non-AM control plants essentially eliminated the differences in resistance to water transport (Nelsen and Safir 1982a, b). The increases in transpiration rate and stomatal conduction in AMF-inoculated sunflower plants ascribed to P-mediated improvement (Koide 1993). Interestingly, mycorrhizal effect responses to N, P, K, Ca, Fe, Mn, and Zn were higher under drought stress than under well-watered conditions (Wu and Zou 2009b). It seems that mycorrhizal effect on nutrient uptake is more important under drought stress than under well-watered conditions. The study of Wu et al. (2011a) reported that though active, functional and total hyphae were decreased by drought stress in trifoliate orange; these hyphal activities under drought stress still helped host plants to sustain greater nutritional (especially P) uptake and water transport. Hence, AMF-enhanced nutrient absorption is an important physiological mechanism in drought tolerance of the host plant caused by mycorrhization.

Other researches considered that AM plants had different sizes than non-AM plants under drought stress, which did not directly judge the results originated from nutrient differences. They proposed equal in size between AM and non-AM plants (Davies et al. 1992). Graham et al. (1987) confirmed no significant difference in water relations between AM citrus plants and non-AM plants fertilized with soluble P. So AMF-induced enhancement of nutrient absorption may be an indirect mechanism in AMF functioning on drought tolerance of plants.

2.4.4 Biochemical Mechanisms Regarding Hormones, Osmotic Adjustment, and Antioxidant Systems

Studies indicated higher Spd and Spm levels in *Glomus fasciculatum*-infected *Medicago sativa* under drought stress than in non-AMF-infected plants. Meanwhile, Spd level was considerably correlated with proline accumulation (Goicoechea et al. 1998). AMF inoculation with *Gigaspora margarita* significantly decreased ethylene level and 1-aminocyclopropane-1-carboxylic acid (precursor of ethylene) concentration in the root of papaya under drought stress conditions (Cruz et al. 2000). This suggests that AMF would postpone consenescence of the host plants. In trifoliate orange seedlings, AMF inoculation with *F. mosseae* significantly increased the levels of indole-3-acetic acid (IAA), abscisic acid (ABA), methyl jasmonate (MeJA), and zeatin riboside (ZR) levels under well-watered conditions and IAA, ABA, MeJA, ZR, and brassinosteroids (BRs) levels under drought stress conditions, respectively (Liu et al. 2016). Such changes in phytohormones by AMF would provide the clue to enhance drought tolerance in the host plant.

In addition to phytohormones, enhancement of osmotic adjustment is considered to be an important component of drought tolerance in AM plants. Earlier, Augé et al. (1986) reported that AMF decreased leaf osmotic potential at both the full-turgor and turgor-loss points, in company with an increase in pressure potential at full turgor, thereby maintaining greater leaf water status of the host plant. Wu and his collaborators further analyzed the changes regarding osmolytes under drought stress (Wu and Xia 2006; Wu et al. 2007). These results revealed that AM citrus plants under drought stress had higher soluble sugar in leaf, soluble starch and total nonstructural carbohydrate in leaf and root, glucose in root, sucrose in leaf and root, and K^+ and Ca^{2+} in leaf and root, as compared with non-AM plants (Wu et al. 2007). These positive responses of osmolytes to AMF colonization protect and stabilize macromolecules and structures from damage by greater capacity of osmotic adjustment, contributing to maintaining a water potential gradient and water absorption from soil into roots, thereby enhancing drought tolerance of plants (Martinez et al. 2004; Zhang et al. 2015).

Proline, a primary osmolyte in the process of osmotic adjustment, induced two diverse changes by AMF inoculation under drought stress. Higher proline accumulation in *Lactuca sativa* (Azcón et al. 1996), *Macadamia tetraphylla* (Yooyongwech et al. 2013), and *Oryza sativa* (Ruíz-Sánchez et al. 2011) under drought stress and *Prunes persica* (Tuo et al. 2015) under waterlogged stress was found in AMF plants than non-AMF plants. More proline accumulation in AMF plants would provide greater capacity of osmotic adjustment to cope with drought stress. However, lower accumulation of proline in *Erythrina variegata* (Manoharan et al. 2010), *Knautia arvensis* (Doubková et al. 2013), pistachio (Abbaspour et al. 2012), and *Poncirus trifoliata* (Zou et al. 2013) was often found in AMF than in non-AMF seedlings under drought stress. The lower proline level in AMF plants is derived from the integration of an inhibition of glutamate synthetic pathway of proline but not ornithine pathway, with an enhancement of proline degradation (Zou et al. 2013). Such lower proline in AMF plants also reflects the less damage by drought stress, because of greater water status in AMF plants, thereby providing an avoidance of drought stress. It concludes that proline changes by mycorrhization would be a response for tolerance or avoidance of drought (Augé and Moore 2005).

Changes in reactive oxygen species (ROS) and antioxidant-protected systems by mycorrhization have been reviewed by Wu et al. (2014) in details, as well as in Chap. 10. AM plants possessed higher antioxidant enzyme activities and nonenzymatic antioxidant concentrations, which may serve to protect the organism against oxidative damage, thereby enhancing drought tolerance (Wu et al. 2006). Here, our team reported that AMF inoculation with *F. mosseae* induced significantly higher net H_2O_2 effluxes in roots, especially in the root meristem zone of trifoliate orange under drought stress (Zou et al. 2015a). Possibly, mycorrhizal extraradical hyphae participate in the H_2O_2 efflux, because mycorrhizal hyphae possess functional aquaporins, which can transport H_2O and H_2O_2 (Zou et al. 2015a). On the other hand, AMF colonization significantly increased net Ca^{2+} influxes in root elongation zone of trifoliate orange under drought stress, which is a downstream component in the H_2O_2 signaling pathway. The Ca^{2+} influxes were significantly correlated with root H_2O_2 effluxes/concentrations, suggesting that AMF-mediated Ca^{2+} influxes can result in less H_2O_2 production.

In another work conducted by Huang et al. (2014), AMF colonization resulted in an enhancement of calmodulin (CaM) levels, which might participate in regulation of superoxide dismutase (SOD) activity with a CaM-binding protein, as well as CAT activity. The study of Huang et al. (2014) also showed that leaf Cu/Zn-SOD and Mn-SOD activity was stimulated in an increased trend by mycorrhization under drought stress, but not well-watered conditions, suggesting that drought stress profoundly stimulated AM to trigger the overexpression of SOD isozymes, potentially leading to a low accumulation of ROS in the host plant.

Correlation studies revealed that root colonization and arbuscule, but not entry point and vesicle, were significantly correlated with ROS metabolism (Wu and Zou 2009a).

2.4.5 *Improvement of Soil Structure by AMF-Released Glomalin-Related Soil Protein*

Glomalin, an immunoreactive glycoprotein, is released exclusively by mycorrhizal hyphae and spores of AMF (Wright et al. 1996). In nature, glomalin is very stable and high persisted in the soil and also an alkaline-soluble protein, defined as glomalin-related soil protein (GRSP) by the Bradford assay (Rillig et al. 2001; Rillig 2004). The study of Augé et al. (2001) revealed that AM soils possessed better water-stable aggregates, conferring a higher soil moisture. The work conducted by Wu et al. (2008) further indicated relatively higher distribution of water-stable aggregates in citrus rhizosphere, which is highly correlated with total glomalin-related soil protein concentration. As reported by Zou et al. (2014), in total glomalin-related soil protein (T-GRSP) and easily extractable glomalin-related soil protein (EE-GRSP), only T-GRSP was significantly negatively correlated with soil and leaf soil potential, indicating that T-GRSP is more active under drought stress than EE-GRSP (Zou et al. 2016). GRSP generally coats on fungal hyphae and forms a hydrophobic layer in the aggregate surface, thereby reducing water loss within soil aggregates (Nichols 2008). Therefore, mycorrhizal soils possessed well-structured soils through AMF-released GRSP production, thereby keeping comparatively higher available water than poorly structured non-mycorrhizal soils under drought stress (Augé 2001). Moreover, GRSP-induced aggregate stability was more conspicuous under drought stress than under salinity stress (Kohler et al. 2009).

2.4.6 Molecule Mechanisms

Drought stress generally regulated two groups of stressed inducible genes to respond to the stress. Meanwhile, first group functions in stress tolerance, including late embryogenesis abundant (LEA) proteins, osmotin, mRNA-binding proteins, key enzymes for osmolyte biosynthesis, water and ion channels, ROS-scavenging enzymes, etc. (Yokota et al. 2006). The second group contains protein factors involved in the regulation of signal transduction and gene expression that probably function in stress response: protein kinases, transcription factors, and enzymes in phospholipid metabolism (Shinozaki and Yamaguchi-Shinozaki 1999). Studies showed two regulated patterns in AMF-induced stressed genes that responded to drought stress: downregulation or upregulation. Ruiz-Lozano et al. (2001) firstly reported downregulated expression in *Mn-sods* and *Fe-sod* genes under ample water conditions and upregulated expression in *Mn-sod II* gene under drought stress. A downregulated pattern in plasma membrane aquaporins gene and Δ^1-pyrroline-5-carboxylate synthetase (*p5cs*, a key proline synthetase) gene in soybean and lettuce was found in AM plants under drought stress conditions (Porcel et al. 2004, 2006a). Another work also revealed that the combination of exogenous ABA and AM symbiosis strongly inhibited the expression of plasma membrane intrinsic proteins (PIP) aquaporin gene as water conservation in the host plant, allowing the host plant to maintain higher leaf relative water status (Ruiz-Lozano et al. 2009). In *Robinia pseudoacacia* seedlings, root expression of *RpTIP1;1* was induced by AMF under well-watered condition but was downregulated by AMF under drought stress condition (He et al. 2016). This indicates the expression of *RpAQP* genes in mycorrhizal plants dependent on soil water condition and plant tissue. Alternatively, reduction of PIP gene expression could be compensated by changing the abundance or the activity of other aquaporins. However, the PIP gene expression was increased under salt stress, suggesting that effect of AM on PIP gene expression depends on the intrinsic properties of the osmotic stress itself (Ruiz-Lozano and Aroca 2010).

Other studies showed upregulated expression in relevant genes by mycorrhization. A gene from *G. intraradices* encoding a binding protein (GiBiP, a molecular chaperone) (Porcel et al. 2007) or cloned from an aquaporin (water channel) gene (Aroca et al. 2007) was upregulated by drought stress in some mycorrhizal plants. Porcel et al. (2006b) showed the involvement of a group of proteins (*G. intraradices* 14-3-3 gene) under drought stress by mycorrhization to regulate signaling pathways and effector proteins. In trifoliate orange, mRNA abundance of four genes involved in reactive oxygen species homeostasis and oxidative stress battling was higher in the AM plants when compared with the NAM plants (Fan and Liu 2011).

Liu et al. (2007) used a 16,000-feature oligonucleotide array to explore transcriptional changes triggered in *Medicago truncatula* by *Gigaspora gigantea* and *G. intraradices*. In roots, *Gigaspora gigantea* upregulated 107 genes and downregulated 50 genes, and 56 genes were induced in the two AMF species. Such changes in gene expression provide clear clue to screen differential genes and understand molecule mechanisms.

2.5 Future Perspectives

Future research in this field will have to concern the mechanisms as the following:

1. Using the noninvasive micro-test technique (NMT) to monitor dynamic changes in specific ions/molecules (including K^+, Na^+, Ca^{2+}, H^+, Cl^-, Mg^{2+}, H_2O_2, IAA, and glucose) noninvasively after AMF inoculation.
2. Detecting ROS accumulation in roots and AM structure to further confirm the functioning of AM on ROS accumulation, as well as the functioning of ROS on AM development.
3. Combining of AMF with other plant growth-promoting rhizobacteria can be used in further works to clarify the synergy effect on drought tolerance of the host plant.
4. Analyzing the role of GRSP in soil structure and subsequent improving soil/plant water relations.
5. Utilizing RNA-seq technique to understand changes in metabolic pathways and to screen differentiated expressed genes in whole genes, which are confirmed by qRT-PCR for the relative expression.
6. Clarifying the perspectives in the study of aquaporins under drought stress, as well as other stress conditions in both drought tolerance and AM symbiosis.

Acknowledgments This study was supported by the Plan in Scientific and Technological Innovation Team of Outstanding Young, Hubei Provincial Department of Education (T201604) and the National Natural Scientific Foundation of China (31101513).

References

Abbaspour H, Saeidi-Sar S, Afshari H et al (2012) Tolerance of mycorrhiza infected pistachio (*Pistacia vera* L.) seedlings to drought stress under glasshouse conditions. J Plant Physiol 169:704–709

Allen MF (2006) Water dynamics of mycorrhizas in arid soils. In: Gadd GM (ed) Fungi in biogeochemical cycles. Cambridge University Press, Cambridge, pp 74–97

Allen MF (2007) Mycorrhizal fungi: highways for water and nutrients in arid soils. Vadose Zone J 6:291–297

Allen MF, Boosalis MG (1983) Effects of two species of VA mycorrhizal fungi on drought tolerance of winter wheat. New Phytol 93:67–76

Aroca R, Porcel R, Ruiz-Lozano JM (2007) How does arbuscular mycorrhizal symbiosis regulate root hydraulic properties and plasma membrane aquaporins in *Phaseolus vulgaris* under drought, cold, or salinity stresses? New Phytol 173:808–816

Augé RM (2001) Water relations, drought and vesicular-arbuscular mycorrhizal symbiosis. Mycorrhiza 11:3–42

Augé RM, Moore JL (2005) Arbuscular mycorrhizal symbiosis and plant drought resistance. In: Mehrotra VS (ed) Mycorrhiza: role and applications. Allied Publishers, Nagpur, pp 136–162

Augé RM, Schekel KA, Wample RL (1986) Osmotic adjustment in leaves of VA mycorrhizal nonmycorrhizal rose plants in response to drought stress. Plant Physiol 82:765–770

Augé RM, Stodola AJW, Tims JE et al (2001) Moisture retention properties of a mycorrhizal soil. Plant Soil 230:87–97

Azcón R, Gómez M, Tobar R (1996) Physiological and nutritional responses by *Lactuca sativa* L. to nitrogen sources and mycorrhizal fungi under drought conditions. Biol Fertil Soils 22:156–161

Berta G, Fusconi A, Trotti A et al (1990) Morphogenetic modifications induced by the mycorrhizal fungus *Glomus* strain E3 in the root system of *Allium porrum*. New Phytol 114:207–215

Bethlenfalvay GJ, Brown MS, Ames RN et al (1988) Effects of drought on host and endophyte development in mycorrhizal soybeans in relation to water use and phosphate uptake. Physiol Plant 72:565–571

Bryla DR, Duniway JM (1997) Effects of mycorrhizal infection on drought tolerance and recovery in safflower and wheat. Plant Soil 197:95–103

Calvo-Polanco M, Sánchez-Romera B, Aroca R et al (2016) Exploring the use of recombinant inbred lines in combination with beneficial microbial inoculants (AM fungus and PGPR) to improve drought stress tolerance in tomato. Environ Exp Bot 131:47–57

Comas LH, Becker SR, Von Mark VC et al (2013) Root traits contributing to plant productivity under drought. Front Plant Sci 4:442

Cruz AF, Ishii T, Kadoya K (2000) Effects of arbuscular mycorrhizal fungi on tree growth, leaf water potential, and levels of 1-aminocyclopropane-1-carboxylic acid and ethylene in the roots of papaya under water-stress conditions. Mycorrhiza 10:121–123

Daniels BA, Trappe JM (1980) Factors affecting spore germination of the vesicular-arbuscular mycorrhizal fungus, *Glomus epigaeus*. Mycologia 72:457–471

Davies FT, Potter J, Linderman RG (1992) Mycorrhiza and repeated drought exposure affect drought resistance and extraradical hyphae development of pepper plants independent of plant size and nutrient content. J Plant Physiol 139:289–294

Doubková P, Vlasáková E, Sudová R (2013) Arbuscular mycorrhizal symbiosis alleviates drought stress imposed on *Knautia arvensis* plants in serpentine soil. Plant Soil 370:149–161

Douds DD, Schenck NC (1991) Germination and hyphal growth of VAM fungi during and after storage in soil at five matric potentials. Soil Biol Biochem 23:177–183

Egerton-Warburton LM, Querejeta JI, Allen MF (2007) Common mycorrhizal networks provide a potential pathway for the transfer of hydraulically lifted water between plants. J Exp Bot 58:1473–1483

Faber BA, Zasoski RJ, Munns DN et al (1991) A method for measuring hyphal nutrient and water uptake in mycorrhizal plants. Can J Bot 69:87–94

Fan QJ, Liu JH (2011) Colonization with arbuscular mycorrhizal fungus affects growth, drought tolerance and expression of stress-responsive genes in *Poncirus trifoliata*. Acta Physiol Plant 33:1533–1542

Finlay RD, Lindahl BD, Taylor AFS (2008) Responses of mycorrhizal fungi to stress. In Br Mycol Soc Symp Ser, vol 27. Stress in yeast and filamentous fungi. Academic Press, p 201–219

Fitter AH (1988) Water relations of red clover *Trifolium pratense* L. as affected by VA mycorrhizal infection and phosphorus supply before and during drought. J Exp Bot 39:595–603

Gaur A, Adholeya A (1999) Mycorrhizal effects on the acclimatization, survival, growth and chlorophyll of micropropagated Syngonium and Dracaena inoculated at weaning and hardening stages. Mycorrhiza 9:215–219

Gemma JN, Koske RE, Roberts EM et al (1997) Mycorrhizal fungi improve drought resistance in creeping bentgrass. J Turfgrass Sci 73:15–29

George E, Häussler KU, Vetterlein D et al (1992) Water and nutrient translocation by hyphae of *Glomus mosseae*. Can J Bot 70:2130–2137

Giovannetti M, Avio L, Sbrana C (2010) Fungal spore germination and pre-symbiotic mycelial growth – physiological and genetic aspects. In: Koltai H, Kapulnik Y (eds) Arbuscular mycorrhizas: physiology and function. Springer, New York, pp 3–32

Goicoechea N, Szalai G, Antolín MC et al (1998) Influence of arbuscular mycorrhizae and rhizobium on free polyamines and proline levels in water-stressed alfalfa. J Plant Physiol 153:706–711

Graham JH, Linderman RG, Menge JA (1982) Development of external hyphae by different isolates of mycorrhizal *Glomus* spp. in relation to root colonization and growth of Troyer citrange. New Phytol 91:83–189

Graham JH, Syvertsen JP, Smith ML (1987) Water relations of mycorrhizal and phosphorus-fertilized non-mycorrhizal *Citrus* under drought stress. New Phytol 105:411–419

Griffin DM (1972) Ecology of soil fungi. Chapman & Hall, London

He X, Zhao J, Li S (1999) Effects of water stress and VA mycorrhizal fungi on the growth of mung bean. Acta Agric Nudeatae Sin 14:290–294

He F, Zhang HQ, Tang M (2016) Aquaporin gene expression and physiological responses of *Robinia pseudoacacia* L. to the mycorrhizal fungus *Rhizophagus irregularis* and drought stress. Mycorrhiza 26:311–323

Henderson JC, Davies FT (1990) Drought acclimation and the morphology of mycorrhizal *Rosa hybrida* L cv Ferdy is independent of leaf elemental content. New Phytol 115:503–510

Huang YM, Srivastava AK, Zou YN et al (2014) Mycorrhizal-induced calmodulin mediated changes in antioxidant enzymes and growth response of drought-stressed trifoliate orange. Front Microbiol 5:682

Kaushai M, Wani SP (2016) Rhizobacterial-plant interactions: strategies ensuring plant growth promotion under drought and salinity stress. Agric Ecosyst Environ 231:68–78

Khan SA (2003) Interaction of vesicular arbuscular mycorrhizae, hormones and drought in soybeans. Dissertation, The City University of New York, New York

Khoyerdi FF, Shamshiri MH, Estaji A (2016) Changes in some physiological and osmotic parameters of several pistachio genotypes under drought stress. Sci Hortic 198:44–51

Kohler J, Caravaca F, Roldán A (2009) Effect of drought on the stability of rhizosphere soil aggregates of *Lactuca sativa* grown in a degraded soil inoculated with PGPR and AM fungi. Appl Soil Ecol 42:160–165

Koide R (1993) Physiology of the mycorrhizal plant. Adv Plant Path 9:33–54

Kolenc Z, Vodnik D, Mandelc S et al (2016) Hop (*Humulus lupulus* L.) response mechanisms in drought stress: proteomic analysis with physiology. Plant Physiol Biochem 105:67–78

Liu JY, Maldonado-Mendoza I, Lopez-Meyer M et al (2007) Arbuscular mycorrhizal symbiosis is accompanied by local and systemic alterations in gene expression and an increase in disease resistance in the shoots. Plant J 50:529–544

Liu J, Guo C, Chen ZL et al (2016) Mycorrhizal inoculation modulates root morphology and root phytohormone responses in trifoliate orange under drought stress. Emirates J Food Agric 28:251–256

Manoharan PT, Shanmugaiah V, Balasubramanian N et al (2010) Influence of AM fungi on the growth and physiological status of *Erythrina variegata* Linn. Grown under different water stress conditions. Eur J Soil Biol 46:151–156

Martinez J-P, Lutts S, Schanck A et al (2004) Is osmotic adjustment required for drought stress resistance in the Mediterranean shrub *Atriplex halimus* L? J Plant Physiol 161:1041–1051

Nasim G (2010) The role of arbuscular mycorrhizae in inducing resistance to drought and salinity stress in crops. In: Ashraf M, Ozturk M, Ahmad MSA (eds) Plant adaptation and phytoremediation. Springer, Berlin, pp 119–141

Nelsen CE, Safir GR (1982a) Increased drought tolerance of mycorrhizal onion plants caused by improved phosphorous nutrition. Planta 154:407–413

Nelsen CE, Safir GR (1982b) The water relations of well-watered, mycorrhizal, and nonmycorrhizal onion plants. J Am Soc Hortic Sci 107:271–274

Neumann E, Schmid B, Romheld V et al (2009) Extraradical development and contribution to plant performance of an arbuscular mycorrhizal symbiosis exposed to complete or partial rootzone drying. Mycorrhiza 20:13–23

Nichols KA (2008) Indirect contributions of AM fungi and soil aggregation to plant growth and protection. In: Siddiqui ZA, Akhtar MS, Futai K (eds) Mycorrhizae: sustainable agriculture and forestry. Springer Science, Berlin, pp 177–194

Porcel R, Azcón R, Ruiz-Lozano JM (2004) Evaluation of the role of genes encoding for Δ^1-pyrroline-5-carboxylate synthetase (P5CS) during drought stress in arbuscular mycorrhizal *Glycine max* and *Lactuca sativa* plants. Physiol Mol Plant Pathol 65:211–221

Porcel R, Aroca R, Azcón R et al (2006a) PIP aquaporin gene expression in arbuscular mycorrhizal *Glycine max* and *Lactuca sativa* plants in relation to drought stress tolerance. Plant Mol Biol 60:389–404

Porcel R, Aroca R, Cano C et al (2006b) Identification of a gene from the arbuscular mycorrhizal fungus *Glomus intraradices* encoding for a 14-3-3 protein that is up-regulated by drought stress during the AM symbiosis. Microb Ecol 52:575–582

Porcel R, Aroca R, Cano C et al (2007) A gene from the arbuscular mycorrhizal fungus *Glomus intraradices* encoding a binding protein is up-regulated by drought stress in some mycorrhizal plants. Environ Exp Bot 60:251–256

Pozo MJ, López-Ráez JA, Azcón-Aguilar C et al (2015) Phytohormones as integrators of environmental signals in the regulation of mycorrhizal symbioses. New Phytol 205:1431–1436

Requena N, Serrano E, Ocon A et al (2007) Plant signals and fungal perception during arbuscular mycorrhiza establishment. Phytochem 68:33–40

Rillig MC (2004) Arbuscular mycorrhizae, glomalin, and soil aggregation. Can J Soil Sci 84:355–363

Rillig MC, Wright SF, Nichols KA et al (2001) Large contribution of arbuscular mycorrhizal fungi to soil carbon pools in tropical forest soils. Plant Soil 233:167–177

Ruiz-Lozano JM, Aroca R (2010) Modulation of aquaporin genes by the arbuscular mycorrhizal symbiosis in relation to osmotic stress tolerance. In: Seckbach J, Grube M (eds) Symbioses and stress. Springer, Berlin, pp 357–374

Ruiz-Lozano JM, Azcón R (1995) Hyphal contribution to water uptake in mycorrhizal plants as affected by the fungal species and water status. Physiol Plant 95:472–478

Ruiz-Lozano JM, Collados C, Barea JM et al (2001) Cloning of cDNAs encoding SODs from lettuce plants which show differential regulation by arbuscular mycorrhizal symbiosis and by drought stress. J Exp Bot 52:2241–2242

Ruiz-Lozano JM, del Mar AM, Bárzana G et al (2009) Exogenous ABA accentuates the differences in root hydraulic properties between mycorrhizal and non mycorrhizal maize plants through regulation of PIP aquaporins. Plant Mol Biol 70:565–579

Ruíz-Sánchez M, Armada E, Muñoz Y et al (2011) Azospirillum and arbuscular mycorrhizal colonization enhance rice growth and physiological traits under well-watered and drought conditions. J Plant Physiol 168:1031–1037

Safir GR, Boyer JS, Gerdemann JW (1971) Mycorrhizal enhancement of water transport in soybean. Science 172:581–583

Safir GR, Boyer JS, Gerdemann JW (1972) Nutrient status and mycorrhizal enhancement of water transport in soybean. Plant Physiol 49:700–703

Sánchez-Díaz M, Honrubia M (1994) Water relations and alleviation of drought stress in mycorrhizal plants. In: Gianinazzi S, Schüepp H (eds) Impact of arbuscular mycorrhizas on sustainable agriculture and natural ecosystems. Birkhäuser Verlag, Basel, pp 167–178

Shinozaki K, Yamaguchi-Shinozaki K (1999) Molecular responses to drought stress. In: Shinozaki K, Yamaguchi-Shinozaki K (eds) Molecular responses to cold, drought, heat and salt stress in higher plants. Landes Company, Austin, pp 11–28

Sieverding E (1981) Influence of soil water regimes in VA mycorrhiza: I. Effect on plant growth, water utilization and development of mycorrhiza. Z Acker Pflanzenbau 150:400–411

Stahl PD, Schuman GE, Frost SM et al (1998) Arbuscular mycorrhizae and water stress tolerance of Wyoming big sagebrush seedlings. Soil Sci Soc Am J 62:1309–1313

Sweatt MR, Davies FT (1984) Mycorrhizae, water relations, growth and nutrient uptake of geranium grown under moderately high phosphorus regimes. J Am Soc Hortic Sci 109:210–213

Sylvia DM, Schenck NC (1983) Germination of chlamydospores of three *Glomus* species as affected by soil matric potential and fungal contamination. Mycologia 75:30–35

Tommerup IC (1984) Effect of soil water potential on spore germination by vesicular- arbuscular mycorrhizal fungi. Trans Br Mycol Soc 3:193–202

Tuo XQ, Li S, Wu QS et al (2015) Alleviation of waterlogged stress in peach seedlings inoculated with *Funneliformis mosseae*: changes in chlorophyll and proline metabolism. Sci Hortic 197:130–134

Wright SF, Franke-Snyder M, Morton JB et al (1996) Time-course study and partial characterization of a protein on hyphae of arbuscular mycorrhizal fungi during active colonization of roots. Plant Soil 181:193–203

Wu QS, Xia RX (2006) Arbuscular mycorrhizal fungi influence growth, osmotic adjustment and photosynthesis of citrus under well-watered and water stress conditions. J Plant Physiol 163:417–425

Wu QS, Zou YN (2009a) Mycorrhiza has a direct effect on reactive oxygen metabolism of drought-stressed citrus. Plant Soil Environ 55:436–442

Wu QS, Zou YN (2009b) Mycorrhizal influence on nutrient uptake of citrus exposed to drought stress. Philipp Agric Scientist 92:33–38

Wu QS, Zou YN (2009c) Arbuscular mycorrhizas improve water relations of plants exposed to drought. In: Hemantaranjan A (ed) Advances in plant physiology, vol 11. Scientific Publishers, Jodhpur, pp 23–52

Wu QS, Xia RX, Zou YN (2006) Reactive oxygen metabolism in mycorrhizal and non-mycorrhizal citrus (*Poncirus trifoliata*) seedlings subjected to water stress. J Plant Physiol 163:1101–1110

Wu QS, Xia RX, Zou YN et al (2007) Osmotic solute responses of mycorrhizal citrus (*Poncirus trifoliata*) seedlings to drought stress. Acta Physiol Plant 29:543–549

Wu QS, Xia RX, Zou YN (2008) Improved soil structure and citrus growth after inoculation with three arbuscular mycorrhizal fungi under drought stress. Eur J Soil Biol 44:122–128

Wu QS, Zou YN, He XH (2011a) Differences of hyphal and soil phosphatase activities in drought-stressed mycorrhizal trifoliate orange (*Poncirus trifoliata*) seedlings. Sci Hortic 129:294–298

Wu QS, Zou YN, Wang GY (2011b) Arbuscular mycorrhizal fungi and acclimatization of micropropagated citrus. Commun Soil Sci Plant Anal 42:1825–1832

Wu QS, He XH, Zou YN et al (2012) Arbuscular mycorrhizas alter root system architecture of *Citrus tangerine* through regulating metabolism of endogenous polyamines. Plant Growth Regul 68:27–35

Wu QS, Srivastava AK, Zou YN (2013) AMF-induced tolerance to drought stress in citrus: a review. Sci Hortic 164:77–87

Wu QS, Zou YN, Abd-Allah EF (2014) Mycorrhizal association and ROS in plants. In: Ahmad P (ed) Oxidative damage to plants. Elsevier Inc, Amsterdam, pp 453–475

Xue HQ (2004) Water ecophysiology effects of arbuscular mycorrhizal fungi on *Citrus grandis* L. Osbeck cv. Shatianyou in Changshou. Dissertation, Southwest Agriculture University, Changshou

Yano-Melo AM, Saggin OJ, Lima JM et al (1999) Effect of arbuscular mycorrhizal fungi on the acclimatization of micropropagated banana plantlets. Mycorrhiza 9:119–123

Yokota A, Takahara K, Akashi K (2006) Water stress. In: Reddy KJ, Rao KV, Raghavendra AS (eds) Physiology and molecular biology of stress tolerance in plants. Springer, Dordrecht, pp 15–39

Yooyongwech S, Phaukinsang N, Chaum S et al (2013) Arbuscular mycorrhiza improved growth performance in *Macadamia tetraphylla* L. grown under water deficit stress involves soluble sugar and proline accumulation. Plant Growth Regul 69:285–293

Zhang Y, Zhong CL, Chen Y et al (2010) Improving drought tolerance of *Casuarina equisetifolia* seedlings by arbuscular mycorrhizas under glasshouse conditions. New For 40:261–267

Zhang Y, Yao Q, Li J et al (2015) Contribution of an arbuscular mycorrhizal fungus to growth and physiology of loquat (*Eriobotrya japonica*) plants subjected to drought stress. Mycol Prog 14:84

Zou YN, Wu QS, Huang YM et al (2013) Mycorrhizal-mediated lower proline accumulation in *Poncirus trifoliata* under water deficit derives from the integration of inhibition of proline synthesis with increase of proline degradation. PLoS One 8:e80568

Zou YN, Srivastava AK, Wu QS et al (2014) Glomalin-related soil protein and water relations in mycorrhizal citrus (*Citrus tangerina*) during soil water deficit. Arch Agron Soil Sci 60:1103–1114

Zou YN, Huang YM, Wu QS et al (2015a) Mycorrhiza-induced lower oxidative burst is related with higher antioxidant enzyme activities, net H_2O_2 effluxes, and Ca^{2+} influxes in trifoliate orange roots under drought stress. Mycorrhiza 25:143–152

Zou YN, Srivastava AK, Ni QD et al (2015b) Disruption of mycorrhizal extraradical mycelium and changes in leaf water status and soil aggregate stability in rootbox-grown trifoliate orange. Front Microbiol 6:203

Zou YN, Srivastava AK, Wu QS (2016) Glomalin: a potential soil conditioner for perennial fruits. Int J Agric Biol 18:293–297

Chapter 3
Arbuscular Mycorrhizal Fungi and Tolerance of Waterlogging Stress in Plants

Faisal Danu Tuheteru and Qiang-Sheng Wu

Abstract Waterlogging is an environmental factor that negatively affects the survival, growth, and development of plants. The same effect also occurs against the presence of fungi in aquatic ecosystems such as arbuscular mycorrhizal fungi (AMF). AMF has been reported as having a symbiosis with various types of aquatic and wetland plants. The existence of AMF has also been found in various types of both permanent and seasonal wetlands including lake and stream flooding, mangrove, salt marsh, river, riparian and floodplain, peat swamp forest, and other wetlands. *Glomeraceae* is a family with most types, and some of them are found in all types of wetlands, for example, *Funneliformis mosseae*, *Rhizophagus fasciculatus*, and *F. geosporus*. Colonization, spore density, species richness, and diversity of AMF in a puddle condition are influenced by many factors, including the availability of oxygen, seasonal changes, the availability of P, water depth, type of AMF, and other types of vegetation. The AMF presence can promote the growth and biomass of plants through improved nutritional status and potential adjustment and may accelerate the succession in early pioneer vegetation in some types of aquatic habitat.

Keywords Waterlogging • *Funneliformis mosseae* • Lake and stream • Temperate

---

F.D. Tuheteru (✉)
Department of Forestry, Faculty of Forestry and Environmental Science,
Halu Oleo University, Kendari, Southeast Sulawesi 93121, Indonesia
e-mail: faisaldanu_28@yahoo.com

Q.-S. Wu
College of Horticulture and Gardening, Yangtze University, Jingzhou, Hubei 434025, China

Institute of Root Biology, Yangtze University, Jingzhou, Hubei 434025, China

Department of Chemistry, Faculty of Science, University of Hradec Kralove,
Hradec Kralove 50003, Czech Republic
e-mail: wuqiangsh@163.com

Q.-S. Wu (ed.), *Arbuscular Mycorrhizas and Stress Tolerance of Plants*,
DOI 10.1007/978-981-10-4115-0_3

3.1 Introduction

Waterlogging is an abiotic stress that affects the growth, production, and distribution of plants (Kozlowski 1984; Tanaka et al. 2011). Waterlogging often causes hypoxic and anoxic conditions (Elzenga and van Veen 2010) which led to a decrease in root hydraulic conductivity and reduce stomatal aperture with implications for the internal water deficit and lower photosynthetic capacity and nutrient availability (Kozlowski and Pallardy 1997; Ashraf 2012). In the inundation conditions, plants can be adapted through the modification of the morphology, anatomy, and physiology. Plant roots will form adventitious roots and lenticels (Kozlowski and Pallardy 1997; Ashraf 2012). Its main function is to maintain respiration and mediate the cycle of O_2 in roots (Yin et al. 2012). The availability of O_2 allegedly encouraged the activity and development of soil microbes such as arbuscular mycorrhizal fungi (AMF).

Arbuscular mycorrhizal fungi (AMF) are an obligate fungus of the phylum Glomeromycota and have a wide distribution of terrestrial to aquatic ecosystems (Smith and Read 2008) as well as symbiotic with 70–90% of land species (Wang and Qiu 2006). There are 214 types of AMF belonging to 19 genera, 13 families, 1 order, and 1 class (Liu et al. 2009). Many studies have reported that AMF increased the tolerance of crops in various biotic and abiotic stresses such as drought (Augé 2001), root pathogens (Akhtar and Siddiqui 2008), heavy metals (Husna et al. 2016), salinity (Wu et al. 2010), and waterlogging (Fougnies et al. 2007).

AMF has been reported in symbiosis with the roots of aquatic and wetland plants in both the permanent and seasonal flooding conditions of waterlogging. AMF is found in submerged plant root (Clayton and Bagyaraj 1984; Tanner and Clayton 1985a; Beck-Nielsen and Madsen 2001; Sraj-Krzic et al. 2006) and emergent plants (Khan and Belik 1995; Stevens and Peterson 1996; White and Charvat 1999) on various types of aquatic systems such as mangrove (Wang et al. 2010; D'Souza and Rodrigues 2013); salt marsh (Rozema et al. 1986); river, riparian, and floodplain (Lodge 1989; de Marins et al. 2009); lake and stream (Søndergaard and Laegaard 1977); peat swamp forest (Tawaraya et al. 2003); and other wetlands (Keeley 1980; Turner and Friese 1998; Escudero and Mendoza 2005; Wolfe et al. 2007). Ecology research study and the presence of the AMF in aquatic systems have been carried out by Khan and Belik (1995), Khan (1995, 2004), Shah (2014), and D'Souza (2016).

AMF symbiosis with aquatic plants is still little understood. Therefore, this paper was written for the review of related topics, including the existence of AMF (including recaps of the number of types of AMF) on several types of wetland, adaptation mechanism of AMF in puddle condition, the factors that affect colonization, the abundance of AMF spores, and the role of the AMF to the tolerance of plants in puddle conditions.

3.2 Occurrence of AMF with Aquatic and Wetland Plants

Based on the review of about 48 related literatures, we found 101 species/strains of AMF occurred at various wetland types including 19 genera and 9 families. *Glomeraceae*, a family with most types or about 53% of the total species, were found. Some AMF types are reportedly widespread in a variety of wetlands including *Gigaspora margarita*, *Glomus macrocarpum*, *Funneliformis mosseae*, *Rhizophagus fasciculatus*, *R. intraradices*, and *Claroideoglomus claroideum*. In addition, there are likewise certain kinds that are only found in certain wetlands. *Archaeospora trappei*, *Acaulospora nicolsonii*, *A. rugosa*, *Glomus spurcum*, *Rhizophagus tenuis*, *R. melanum*, *Claroideoglomus luteum*, *C. drummondii*, and *Paraglomus majewskii* are only found in river, riparian, and floodplain site (Table 3.1).

3.2.1 Mangrove

Some studies have reported AMF symbiosis with mangroves in India and China. In India, the research was done in Pichavaram forest (Gupta et al. 2002); Ganges River Estuary (Sengupta and Chaudhuri 2002); Great Nicobar of West Bengal (Kothamasi et al. 2006); the Sundarbans of West Bengal (Kumar and Ghose 2008); Goa, West India (D'Souza and Rodrigues 2013); and Bhitarkanika mangrove forests on the east coast of Orissa (Gupta et al. 2016). In China, the research was done in saline-alkaline soil of Yellow River Delta (Wang et al. 2004) and Pearl River Estuary, Southern China (Wang et al. 2010, 2011). In a mangrove habitat, ± 80 kinds of AMF from 15 genera and 5 families have been found. Family *Glomeraceae* were the dominant type of *Funneliformis mosseae*, *Rhizophagus fasciculatus*, *R. intraradices*, and *F. geosporus*. In addition to *Glomeraceae*, *Acaulosporaceae* was also widespread like *Acaulospora foveata* and *A. scrobiculata* (Wang et al. 2004; Kumar and Ghose 2008; Wang et al. 2010, 2011; D'Souza and Rodrigues 2013; Gupta et al. 2016). Gupta et al. (2016) reported some types of AMF which are well adapted to the high-salinity conditions such as *Entrophospora infrequens*, *Gigaspora gigantea*, *F. geosporum*, *R. intraradices*, *G. ambisporum*, *G. candida*, *G. pallidum*, *R. intraradices*, and *F. geosporus* that were found in high-salinity zone, and the extensive distribution indicates possible fungal adaptation to salt tolerance.

Colonization and AMF spore abundance in plant roots tend to vary and low mangrove. The percentage of AMF colonization in roots of mangrove plants ranges from 0 to 68%. AMF structures are commonly found in plant roots including vesicles, hyphal coils, and internal hyphae (Kothamasi et al. 2006; Wang et al. 2010). AMF colonizes some mangrove genera including *Acanthus*, *Acrostichum*, *Aegiceras*, *Aglaia*, *Avicennia*, *Bruguiera*, *Ceriops*, *Derris*, *Kandelia*, *Rhizophora*, *Sonneratia*, and *Xylocarpus*. Nevertheless, there is consistency between the different colonization sites on the same plant species. AMF structure was found in the roots of

Table 3.1 List of AMF species on aquatic and wetland plant

Order/family/genus/species	Mg	SM	RI/FP	LS	OW	Site/region	References
Archaeosporales							
Ambiosporaceae/Ambiospora							
A. gerdemanii	+					IN, CN, US, DE	2,10,20,38
A. leptoticha	+					IN, CN, US	2,10,38
Archaeosporaceae/Archaeospora							
A. schenckii				+		IN	30,31
A. trappei			+			US, NA, BR	24–26,38
A. myriocarpa		+	+	+		US, DE	20,25
Diversisporales							
Acaulosporaceae/Acaulospora							
A. bireticulata	+					IN, CN	2,4,8
A. delicata	+		+			IN, US, NA, BR	4,8,10,24–26
A. denticulata	+	+		+		IN, CN	2,21
A. foveata	+				+	IN, CN, CR	2,4,6,8,10,40
A. koskei					+	US	38
A. lacunosa	+					IN	4
A. laevis	+	+	+	+		IN, CN, US, AU, BR	2,8,18,26,27,38
A. longula	+		+			IN, CN, AU, BR, SE	4,8,21,40
A. mellea	+				+	IN	18
A. morrowiae		+	+	+	+	IN, NA, BR, CR	18,24–26,40
A. nicolsonii			+			BR	26
A. rugosa			+			BR	26
A. scrobiculata	+		+		+	IN, CN, US, AU, NA, BR	2,4,5,8,10,24–26, 30,35,43
A. spinosa	+	+		+	+	IN, CR	4,8,21,40
A. sporocarpia		+		+		IN	18
A. tuberculata	+				+	CN, AU, CR	2,30,35,40

A. undulata					+	AU	42
A. elegans	+					CN	2
Kuklospora							
K. columbiana	+		+			IN, US, AU, BR	10,26,30,31,38
Diversisporaceae/Diversispora							
D. aurantium	+					CN	9
D. eburnea	+		+			IN, CN, US, NA	4,10,24,25
D. versiforme		+		+		IR, DE	13,20
Entrophosporaceae/Entrophospora							
E. infrequens	+		+			IN, US	8,10,25
Gigasporaceae							
Cetraspora							
C. pellucida	+				+	IN, CR	4,10,40
Dentisculata							
D. cerradensis					+	AU	43
D. heterogama					+	US	38
D. reticulata					+	AU	43
Gigaspora							
Gi. albida	+				+	IN, AU	8,37
Gi. decipiens	+					IN	4,18
Gi. gigantea		+		+		IN, CN, AU	18,21,30,35
Gi. margarita	+	+	+	+	+	IN, US, AU, CR, AT	4,11,18,22,23,40,41
Gi. rosea		+		+		IN	18
Racocetra							
R. castanea					+	CT	17
R. gregaria	+					IN	8
R. persica	+					IN	4

(continued)

Table 3.1 (continued)

Order/family/genus/species	Mg	SM	RI/FP	LS	OW	Site/region	References
R. verrucosa		+		+		IN	19
R. weresubiae	+					IN	8
Scutellospora							
S. auriglоba		+		+	+	IN, AU	18,43
S. calospora	+	+	+	+		IN, CN, US, AU	4,18,22,30,35
S. dipapillosa	+					IN	4
S. dipurpurescens	+					IN	8
S. nigra		+		+		IN	18
Pacisporaceae/Pacispora							
P. scintillans		+				DE	20
Glomerales							
Glomeraceae							
Glomus							
G. aggregatum	+	+		+	+	IN, CN, AU	2,4,5,8,18,21,30,31,35,37
G. ambisporum	+	+	+	+		IN, CN, AU	2,10,18,27,30,31
G. arborense		+		+		IN	21
G. australe	+					IN	7
G. candida	+					IN	7
G. citricola	+					IN	7
G. delhiense	+					CN	2
G. formosanum	+					IN,	8,10
G. glomerulatum	+					IN, CN	2,10
G. heterosporum	+					IN	7
G. hoi	+	+		+		CN, DE	2,20
G. hyderabadensis	+					IN	8
G. invermaium		+	+	+	+	IN, DE, AU	20,21,27,43

G. macrocarpum	+	+	+	+	+	IN, NA, BR, PK	1,7,11,24–27,39
G. maculosum	+					IN	8,10
G. magnicaule	+					IN	3,7
G. melanosporum	+					CN	2
G. microaggregatum	+	+	+	+		IN, CN, US, DE, AU, NA	2,20,21,24,25,30,31
G. microcarpum	+				+	IN, AU, PK	1,30,32,39
G. multicaulis	+	+		+		IN	1,4,8,11
G. nanolumen	+					IN	8
G. pansihalos	+	+		+		CN, DE	2,20
G. pellicium	+					IN	7
G. pulvinatum	+					IN	7
G. pustulatum	+					IN, CN, PK	2,10
G. reticulatum	+				+	IN, CN,	2,10,39
G. tenebrosum	+					CN	2
G. trimurates	+					IN	7
G. tortuosum			+		+	US, CR	25,40
G. spurcum			+			US	24,25
Funneliformis							
F. caledonius	+				+	IN, CN, AU	2,10,30,32,44
F. coronatus	+	+		+		IN	4,20
F. dimorphicus	+	+		+		IN, CN	2,6,21
F. geosporus	+	+		+		IN, CN CE, PT, HU, DE, AU	2,4,5,8,10,14–16,18,20,30,35
F. halonatus	+	+		+		IN, DE	10,20
F. monosporus	+					CN	6
F. mosseae	+	+	+	+	+	IN, CN, US, IR, PT, DE, AU, NA, BR, SE, PK, AT	1,2,4–6,8,9,13,15, 18,20,21, 23–26,30, 33,34,35, 39,41
F. verruculosus	+					IN	4

(continued)

Table 3.1 (continued)

Order/family/genus/species	Mg	SM	RI/FP	LS	OW	Site/region	References
Sclerocystis							
S. coremioides	+	+		+		IN	4,18
S. rubiformis	+	+	+	+		IN, CN, DE, AU	5,8,18,20,23,30,35
S. pachycaulis						IN, AU	30,31
S. sinuosa	+					IN	4
G. tawanense	+					IN	8
Rhizophagus							
R. clarus (syn. G. clarum)	+	+		+	+	IN, CN, AU	2,4,8,21,30,35,43
R. diaphanus	+	+		+		IN	10,21
R. fasciculatus	+	+	+	+	+	IN, CN, US, AR, DE AU,PK, AT	1,3–5,8,10–12,18,20–23,25,30,31,35,38,41,42
R. intraradices	+	+	+	+	+	IN, CN, US, IR, AR, DE, AU, NA, SE	2,4–6,8,10,13, 17,18,20,21,24,25,30,31,33,37,42
R. irregulare	+		+			CN, PL	6,28
R. manihotis	+				+	CN, AU	2,37
R. tenuis			+			AU	29
R. vesiculiferus		+		+		DE	20
R. melanum			+			NO	36
Septoglomus							
S. constrictum	+	+	+	+	+	IN, CN, DE, PL	2,8,10,18,20,28
S. deserticola	+	+	+	+		IN, CN, US	2,18,25
Claroideglomeraceae Claroideoglomus							
C. claroideum	+	+	+	+	+	IN, CN, US, DE, AU, BR, PL, SE	2,4,9,10,19–21,25, 26,28,30,31,33,35,37
C. etunicatum	+	+	+	+		IN, CN, IR, DE, AU, NA, SE, AT	2,4,8–10,13,18,20, 21,24,25,30,32,33,38,41

C. lamellosum	+		+			IN, CN, BR	4,9,10,26
C. luteum			+			US, BR	24–26
C. drummondii			+			PL	28
Paraglomerales							
Paraglomeraceae							
Paraglomus							
P. albidum	+					IN, CN, AU	2,4,10,130,32
P. laccatum		+		+		DE	20
P. lacteum	+					IN	4
P. occultum		+	+	+		IN, US, DE, AU, NA	4,10,18,20,24,25,30,31
P. majewskii			+			PL	28

Remarks

(a) Mg (mangrove): [1]Sengupta and Chaudhuri (2002), [2]Wang et al. (2004), [3]Kothamasi et al. (2006), [4]Kumar and Ghose (2008),[5]Wang et al. (2010), [6]Wang et al. (2011), [7]Gupta et al. (2002), [8]D'Sousa and Rodrigues (2013), [9]Wang et al. (2015), and [10]Gupta et al. (2016)

(b) SM (salt marsh): [11]Sengupta and Chaudhuri (1990), [12]Brown and Bledsoe (1996), [13]Aligharzadech et al. (2001), [14]Hildebrandt et al. (2001), [15]Carvalho et al. (2001), [16]Landwehr et al. (2002), [17]Sannazzaro et al. (2004), [18]Mathur et al. (2007), [19]Radhika and Rodrigues (2007), [20]Wilde et al. (2009), and [21]Choudhury et al. (2010)

(c) RI/FP (river, riparian, and floodplain): [22]Lodge (1989), [23]Khan (1993), [24]Kennedy et al. (2002), [25]Bauchamp et al. (2006), [26]De Marins et al. (2009), [27]Seerangan and Thangavelu (2014), [28]Nobis et al. (2015), and [29]Orchard et al. (2016)

(d) LS (lake and stream): [30]Anderson et al. (1984), [31]Ragupathy et al. (1990), [32]Stenlund and Charvat (1994), [33]Nielsen et al. (2004), [34]Wirsel (2004), [35]Kai and Zhiwei (2006), and [36]Sudova et al. (2015)

(e) OW (other wetlands): [37]Wetzel and van der Valk (1996), [38]Miller and Bever (1999), [39]Bajwa et al. (2001), [40]Lovelock et al. (2003), [41]Fuch and Haselwandter (2004), [42]Escudero and mendoza (2005), [43]Brundrett and Ashwath (2013), and [44]Yang et al. (2016)

(f) Site/region: IN (India), CN (China), US (USA), IR (Iran), CE (Central Europe), PT (Portugal), HU (Hungary), AR (Argentina), DE (Germany), AU (Australia), NA (North America), BR (Brazil), PL (Poland), SE (Sweden), NO (Norway), PK (Pakistan), CR (Costa Rica), and AT (Austria)

Avicennia alba of the Ganges River Estuary in India (Sengupta and Chaudhuri 2002) and Goa, West India (D'Souza and Rodrigues 2013), but no colonization of AMF in the Sundarbans of West Bengal (Kumar and Ghose 2008).

3.2.2 Salt Marsh

Besides mangroves, symbiosis of AMF was also noted in plant roots of salt marsh. Studies of the AMF in the salt marsh land had been carried out in Argentina (Sannazzaro et al. 2004), Central Europe (Hildebrandt et al. 2001), China (Zhang et al. 2014), Germany (Wilde et al. 2009), India (Radhika and Rodrigues 2007; Choudhury et al. 2010), Iran (Aliasgharzadeh et al. 2001), the Netherlands (Rozema et al. 1986; Wilde et al. 2009), Pakistan (Khan 1974), Portugal (Carvalho et al. 2001), Slovenia (Likar et al. 2009), and the USA (Mason 1928; Cooke and Lefor 1990; Hoefnagels et al. 1993; Brown and Bledsoe 1996; Bohrer et al. 2004). About 29 types of AMF, belonging to 3 genera and 6 families in salt marsh plants, were found. *R. intraradices*, *F. mosseae*, and *F. geosporus* were dominant and widespread in the salt marsh habitat. *F. geosporus* was reported predominantly in salt marsh in Central Europe (Hildebrandt et al. 2001), Portugal (Carvalho et al. 2001), and the Netherlands (Wilde et al. 2009). *R. intraradices* was dominant and well adapted to saline soil and may play an important role in plant resistance (Sannazzaro et al. 2004).

Hildebrandt et al. (2001) reported mycorrhizal colonization (6–90%) and high spore density in the Central European salt marsh. Brown and Bledsoe (1996) reported approximately 20% AMF colonization in the USA. AMF inoculum potential is low on salt marsh in the Ganges River of India (Sengupta and Chaudhuri 1990). Rozema et al. (1986) also reported that AMF colonization in *Aster tripolium* under waterlogged conditions was depressed in low saline. Several species in salt marsh of North Carolina, namely, *Spartina patens*, *S. cynosuroides*, *Distichlis spicata*, and *Juncus roemerianus*, could be colonized by AMF (Hoefnagels et al. 1993). Choudhury et al. (2010) reported that approximately 92% of plants in marshy and shoreline in Deepar Beel Ramsar Site of Assam, India, were colonized by AMF. *Plantago maritima*, *P. coronopus*, *Artemisia maritima*, *Glaux maritima*, and *Oenanthe lachenalii* were strongly colonized by the mycorrhizae (Hildebrandt et al. 2001).

3.2.3 River, Riparian, and Floodplain

The literature review found 33 types of AMF from 13 genera and 6 families with between 1 and 27 kinds of species richness. Family *Glomeraceae* including the dominant species have been widespread in several locations including *F. mosseae*, *G. macrocarpum*, *C. luteum*, *C. claroideum*, and *R. intraradices*. Kennedy et al. (2002)

observed that *R. intraradices* and *F. mosseae* had a higher frequency of attendance at the San Pedro River in southeastern Arizona. *R. intraradices* was also predominantly found in the Verde River, Arizona (Beauchamp et al. 2006). In contrast to previous researches, de Marins et al. (2009) reported that *Acaulospora laevis* and *C. luteum* were predominantly found in the floodplain of the upper Paraná River, Brazil; *Septoglomus constrictum* in the San River Valley, southeastern Poland (Nobis et al. 2015); and *Gigaspora margarita* in New South Wales, Australia (Khan 1993) and North Carolina, USA (Lodge 1989).

Kennedy et al. (2002) reported that AMF colonization greatly influenced the growth stage, and there was a positive correlation between AMF colonization by soil moisture on the kind of *Sporobolus wrightii* along the San Pedro River in southeastern Arizona. Beauchamp et al. (2006) reported that the forest age, distance, and elevation from above the active channel, annual species cover, perennial species richness, and exchangeable potassium concentration all played a role in structuring the AMF community in the Verde River, Arizona. Orchard et al. (2016) reported a lower AMF colonization for *Lolium rigidum* in lower/riparian landscape zone in the Peel-Harvey catchment, Western Australia, and GlassHouse Research has indicated that there is a decrease in root colonization of AMF for *Lolium rigidum* in waterlogging conditions. Spore abundance increased from depositional to establish island regions and is strongly influenced by sediments (Harner et al. 2011). Harner et al. (2009) reported that there are AMF propagules in the sediment after the inundation. The propagules were allegedly dispersed by flooding (water) and deposited in the sediment. Spore density and infection levels appeared to be lower in severely eroded sites in a river floodplain (Nakatsubo et al. 1994). Nakatsubo et al. (1994) stated that river flooding may affect mycorrhizae in two ways, thus by decreasing propagule density (erosion) or by causing anaerobic conditions (submergence).

Zhao et al. (2016) reported 50 species of riparian plants colonized by AMF in the Jialing River and the Yangtze River (Southwest China) and a high AMF colonization in the Yangtze River than the Jialing River. They also alleged that the plant community of the Three Gorges Reservoir had a strong influence on the alleged existence of mycorrhizae. Becerra et al. (2009) reported the first colonization of the AMF at the root of *Salix humboldtiana* in two riparian in central Argentina. According to the study of Stevens et al. (2010), AMF structure was found in the roots of 35 species of plant that grows in the floodplain of the Fork of the Trinity River, Texas, USA. AMF colonization was low (less than 1%) in the root of *Chosenia arbutifolia* and *Salix sachalinensis* in the Satsunai River floodplain, Hokkaido, Japan (Hashimoto and Higuchi 2003).

3.2.4 *Lake and Stream*

Studies in the past showed that AMF on lake and stream habitats has been conducted in several countries including China (Kai and Zhiwei 2006; Wang et al. 2016), the Czech Republic (Sudova et al. 2011), Denmark (Søndergaard and

Laegaard 1977; Beck-Nielsen and Madsen 2001), Germany (Wirsel 2004), India (Ragupathy et al. 1990), Italy (Borriello et al. 2015), Kenya (Othira et al. 2014), New Zealand (Clayton and Bagyaraj 1984), the Netherlands (Baar et al. 2011), Norway (Baar et al. 2011; Sudova et al. 2015), Portugal (Oliveira et al. 2001), Scotland (Farmer 1985), Slovenia (Sraj-Krzic et al. 2006; Dolinar and Gaberščk 2010), Sweden (Nielsen et al. 2004; Moora et al. 2016), and the USA (Anderson et al. 1984). Twenty-four types of AMF from 11 genera and 4 families of lake and stream were identified by morphological and molecular DNA analysis. *F. mosseae* and *R. fasciculatus* including dominant species were found (Wirsel 2004; Kai and Zhiwei 2006). AMF species richness found in lake and stream habitat varied from 1 to 21 based on the type of site. Søndergaard and Laegaard (1977) were the first researchers to report the AMF symbiosis with aquatic plants *L. uniflora* and *Lobelia dortmanna* of Danish oligotrophic softwater lakes with low P concentration. Sudová et al. (Sudova et al. 2015) found *Rhizoglomus melanum* as a kind of new AMF associated with submerged plant in freshwater lake Avsjoen in Norway.

AMF communities on lake and stream habitat greatly influenced the type of lake and environmental factors (land). Borriello et al. (2015) have reported that factors such as soil availability of N; the levels of N, P, and Mg in soil; and soil compaction contribute to the AMF community. According to Ragupathy et al. (1990), the substrate of wet soil and aquatic sediment affects the distribution of AMF. *Glomus aggregatum*, *R. fasciculatus*, and *P. occultum* were found on the second substrate. According to Wang et al. (2016), the intensity of flooding and adaptation of host plants also affects the AMF community. Intensive inundation inhibits colonization and diversity level of AMF. Moora et al. (2016) also reported that the intensity of colonization and species richness vary along environmental gradients, in which it's low in mesotrophic but high in oligotrophic lake. Studies from Baar et al. (2011) suggest that the AMF communities are more diverse in the lake with dissolved P concentration and lower than the lake with high P (oligotrophic lakes). They are negatively correlated with high organic matter and nutrient content and low redox potential (Wigand et al. 1998; Sudova et al. 2011; Møller et al. 2013). Nonetheless, Clayton and Bagyaraj (1984), Farmer (1985), and Wigand et al. (1998) found no correlation between AMF colonization on plants with lake trophic status.

3.2.5 Other Wetlands

Tawaraya et al. (2003) had conducted the studies of peat swamp forest in Central Kalimantan, Indonesia. They observed that 77% of tree (17 of 22 tree species) plants were colonized by AMF. Four types of symbiotic association with AMF in the Pentland Lake in southern boreal Alberta, Canada (Thormann et al. 1999), were also recorded. Researches in other wetlands have also been done between prairie pothole wetland (Wetzel and van der Valk 1996), fed wetland (Turner et al. 2000), and grassland (Miller and Bever 1999; Escudero and Mendoza 2005).

3.3 Mechanisms About AMF Adaptation to Waterlogging

Based on literatures, several adaptation mechanisms of AMF in puddle conditions are proposed. Such mechanisms include (1) the transport of oxygen from the top (stem, wild roots, and lenticels) to the roots for their aerenchyma, (2) AMF is concentrated near the roots, (3) some types of AMF can particularly grow at low oxygen, (4) AMF which survives becomes effective when the ground began to dry, (5) more structure of vesicles in the roots, and (6) a combination of all these factors (Khan 1995; Nielsen et al. 2004; Tuheteru et al. 2015; Stenlund and Charvat 1994).

Aerenchyma mediates as an internal gas exchange, maintaining the respiration of the roots and providing oxygen to the roots for colonization and survival of the AMF (Cooke et al. 1993; Brown and Bledsoe 1996; Wang et al. 2010). According to Nielsen et al. (2004), extensive aerenchyma supports the development of the AMF in aquatic plants. The same was reported for the roots of *Jaumea carnosa* (Brown and Bledsoe 1996), *Nyssa sylvatica* (Keeley 1980), and *Casuarina cunninghamiana* (Khan 1993). Some researchers reported that there is no clear relationship with the AMF colonization-to-aerenchyma ratio (Cornwell et al. 2001; Sraj-Krzic et al. 2006). Besides aerenchyma, another strategy is more AMF spores near the roots. Marshall and Pattullo (1981) and Keeley (1980) reported that the AMF abundance decreases with the distance from the root. It is strongly associated with the supply of photosynthesis and C of the redox potential to facilitate colonization of the AMF. Some types of AMF are well adapted and dominant in the aquatic system including *F. mosseae*, *F. geosporus*, *R. fasciculatus*, *R. intraradices*, *Gigaspora margarita*, and *Acaulospora foveata.*

Under inundation condition, the structure of the vesicles is found in large quantities. Vesicles of AMF are the structure that acts as energy storage needed for adaptation to the conditions of inundation (Garcia et al. 2008). Under the condition of inundation, vesicles are plenty on the roots of *Typha* (Stenlund and Charvat 1994), *Lotus tenuis* (Mendoza et al. 2005; Garcia et al. 2008), *Citrus junos* (Wu et al. 2013), *Nauclea orientalis* (Tuheteru et al. 2015), and aquatic species and a semi-aquatic in Yangliao Lake (Wang et al. 2016).

3.4 Factors Influencing Colonization, Abundance, Richness, and Diversity of AMF

The inhibitory effect by flooding on the symbiosis has been reported by several researchers (Miller 2000; Bajwa et al. 2001; Wirsel 2004; Wang et al. 2010; Wang et al. 2011). Earlier studies had reported that AMF diversity was low in wetlands (Nielsen et al. 2004; Wilde et al. 2009), but several other studies reported high diversities (Wirsel 2004; Wang et al. 2011, 2015, 2016). Waterlogging conditions can reduce colonization of roots by the AMF (Miller and Bever 1999; Bajwa et al. 2001; Stevens et al. 2002; Mendoza et al. 2005; Ray and Inouye 2006; Zou et al. 2014;

Orchard et al. 2016; Zhang et al. 2016). Conversely, some studies have shown a relatively high colonization on puddle condition (Bauer et al. 2003; Escudero and Mendoza 2005; Mendoza et al. 2005; Fougnies et al. 2007; Garcia et al. 2008). Thus, AMF colonization and AMF diversity in wetlands are not always low. Here, some factors that affect the AMF symbiosis with aquatic plants are as follows.

The Presence of Oxygen AMF including the aerobic type of fungi that depend on the oxygen concentrates in the roots and around the root zone (Wigand et al. 1998). AMF including aerobic puddles can affect the distribution and effectiveness. Supply of O_2 from the low aerenchyma contributes to colonization (Matsumura et al. 2008). Nielsen et al. (2004) explained that aerenchyma stimulates the development of the AMF on aquatic plants. AMF colonization in wetland trees was higher in aerobic microsite (Cantelmo and Ehrenfeld 1999). Turner and Friese (1998) reported that the increase in soil moisture could reduce AMF activities.

Availability of P and Organic Matter P in high availability causes plants to have a low dependency on the AMF. There is normally a decline in the frequency of *Acaulospora*, *Archaespora*, *Glomus*, and *Paraglomus* at high P conditions (Baar et al. 2011). Increase in the availability of P affects AMF colonization in some species of wetland plants (Wetzel and van der Valk 1996; Wigand et al. 1998; White and Charvat 1999; Sraj-Krzic et al. 2006). The addition of P and organic materials can suppress the colonization of the AMF (Tanner and Clayton 1985a, b; Ipsilantis and Sylvia 2007). There was a decrease in root colonization AMF on *Nauclea orientalis* in wetland construction to acid mine drainage with increasing organic material (Tuheteru 2015). Decrease in AMF community and richness is with the increase of soil P levels (Tanner and Clayton 1985a, b; Wetzel and van der Valk 1996; Kennedy et al. 2002). In mangrove ecosystems, AMF spore richness significantly negatively correlated with the available P (Kumar and Ghose 2008).

Plant Species Plant species is also a contributing factor to the structure and composition of the AMF. Zhao et al. (2016) reported that the structure of plant communities strongly influenced the presence of mycorrhizae. Weishampel and Bedford (2006) reported that AMF colonization in dicot was higher (58%) than monocots (13%). Sraj-Krzic et al. (2006) and Miller (2000) reported that high colonization was found in the roots of plants than a sub-emergent. *Pistia stratiotes* colonized AMF (Radhika and Rodrigues 2007) and was also reported not to be colonized by AMF (Ragupathy et al. 1990; Beck-Nielsen and Madsen 2001; Kai and Zhiwei 2006; de Marins et al. 2009; Seerangan and Thangavelu 2014). Nonetheless, Sraj-Krzic et al. (2006) and de Marins et al. (2009) showed that there was no connection biotype macrospore and colonization of AMF. According to Carvalho et al. (2001), the distribution of mycorrhizal in salt marsh is more affected by stress than the host species.

AMF Species AMF colonization in inundation condition depends on AMF species. Sah et al. (2006) reported that high colonization corresponds to mix *Gigaspora margarita* and *Gigaspora rosea*. *R. intraradices* including AMF aggressive type of colonization where the intensity is high, the production of vesicles and spores as well as utilize the products of photosynthesis (Ruiz-lozano et al. 2001; Fougnies et al. 2007). According to Stevens and Peterson (1996), colonization by AMF (*G. clarum* and *G. versiforme*) is highest at the root of *L. salicaria* in wet conditions, and *Gigaspora margarita* was not found for colonization.

The Depth and Duration of Inundation The depth of water contributes to the distribution of AMF. Colonization decreases with the increase of water depth and duration of inundation (Clayton and Bagyaraj 1984; Stevens and Peterson 1996; Wigand et al. 1998; Miller et al. 1999; Miller and Bever 1999; Miller 2000; Stevens et al. 2002; Mendoza et al. 2005; Sudova et al. 2011). According to Sudova et al. (2011), there was a low colonization by AMF as soil depth due to lower redox potential, higher oxygen consuming organic sediment, and higher phosphate. Wang et al. (2004) showed that the number of AMF spores decreases with soil depth with many spores found between 0 and 40 cm layer as compared to depths >40 cm.

Seasonal Variations Seasonal variations strongly influence the change in the phenology of plant roots to the establishment of a new one (Oliveira et al. 2001). Seasonal variations of AMF had a positive effect on survival during the aerate period (increased oxygen aeration) (Anderson 1984: Miller and Bever 1999). AMF colonization dynamics occurred at three types of salt marsh plants as a reflection of plant phenology (Bohrer et al. 2004). Escudero and Mendoza (2005) found that AMF colonization in the roots of *L. glaber* was higher in summer season and spring than in winter and autumn. Khan (1974) reported that the number of spores is influenced by seasonality; spores increased during the month of October to December and started to decline in March to August. The distribution and number of AMF spores was strongly influenced by the dynamics of the season or in the water table and soil moisture levels (Khan 1974; Bohrer et al. 2004; Wilde et al. 2009). There is no specific influence of species of plants and seasonal changes of the community structure of mycorrhizae (Likar et al. 2009).

3.5 The Role of AMF on Waterlogging

Results related to the role of AMF reviewed for publications demonstrate that the AMF can improve crop tolerance on puddle condition. AMF increase the tolerance of crops in the form of nutrient absorption, growth and biomass of plants, and improvement in the biochemistry and physiology as well as accelerate succession of aquatic vegetation in the area.

3.5.1 Improving Nutrient Uptake

Increased growth of plants by AMF is strongly associated with increased nutrient uptake, specifically P (Smith and Read 2008). The repair of P plant puddle with mycorrhizal inoculation has been reported on several plants. Under tropical peat swamp forest in Central Kalimantan (Indonesia), inoculation with *Glomus clarum* and *Gigaspora decipiens* increased the N and P content of *Dyera polyphylla* (Graham et al. 2013). Inoculating with *Glomus intraradices* increased the P plant *Pterocarpus officinalis* (Fougnies et al. 2007). Similar results were reported on the *Panicum hemitomon* and *Leersia hexandra* plants (Miller and Sharitz 2000; Ipsilantis and Sylvia 2007), *Typha latifolia* (Ipsilantis and Sylvia 2007; Dunham et al. 2003), *Prunus persica* (Ruto et al. 2002), *Vallisneria americana* (Wigand and Stevenson 1997), *Ranunculus* sp. (Tanner and Clayton 1985a, b), and *Oryza sativa* (Secilia and Bagyaraj 1994; Solaiman and Hirata 1996). In addition to P, AMF also increase the N plant *Aster tripolium* (Neto et al. 2006), *T. latifolia* (Ipsilantis and Sylvia 2007; Dunham et al. 2003), *Prunus persica* (Ruto et al. 2002), and *Casuarina equisetifolia* (Osundina 1998).

3.5.2 Enhancing the Growth and Biomass of Plants

Several research reports indicate that plants inoculated with AMF performed better in growth and biomass than plants without mycorrhiza. Inoculation with *G. intraradices* enhanced the growth of *Pterocarpus officinalis* (Fougnies et al. 2007); *Diversispora spurca* increased the plant height of *Citrus junos* (Wu et al. 2013) and *Poncirus trifoliata* (Zou et al. 2014); and colonization by *Gigaspora margarita* increased the growth of *Carex tribuloides*, *Phalaris arundinacea*, *Rumex orbiculatus* (Fraser and Feinstein 2005), *Casuarina equisetifolia* (Osundina 1998), *T. latifolia* (Dunham et al. 2003), rice (Secilia and Bagyaraj 1992, 1994), and *Sclerocarya birrea* (Muok and Ishii 2006). *Gigaspora margarita* too promoted growth and biomass of *Prunus persica* before 12 weeks of stagnancy (Ruto et al. 2002). There was an increase in plant biomass of inoculated *Nyssa sylvatica* (Marshal and Pattullo 1981), *Phragmites australis* (Dolinar and Gaberščk 2010), and rice (Secilia and Bagyaraj 1994). Applications of *Glomus* spp. increase grain number (Secilia and Bagyaraj 1994; Solaiman and Hirata 1997). AMF (*Diversispora spurca*) increased the root system architecture and morphology of *Citrus junos* plants (Wu et al. 2013) and *Poncirus trifoliata* (Zou et al. 2014). In addition to the growth and biomass, mycorrhiza can improve survival rates in conditions of inundation. *Glomus intraradices* promoted the survival of some species under seawater flooding (Camprubi et al. 2012).

3.5.3 Repairing Physio-biochemical Activities

Proline Proline is an organic osmolyte accumulated in plants in response to environmental stress (Ashraf and Foolad 2007). Proline contributes to the regulation of water in the form of stability and integrity of subcellular structures mainly for membrane protein and increases the range of enzyme activity (Ashraf and Foolad 2007;

Szabados and Savoure 2009). AMF can promote the accumulation of plant proline in puddle condition. AMF inoculation (*Glomus geosporum*) improved the osmotic adjustment of *Aster tripolium* (Neto et al. 2006) through the accumulation of sugar solution and proline. Tuo et al. (2015) reported that the inoculation of *F. mosseae* increased the accumulation of proline in peach plants exposed to 12-day inundation. Inoculation with AMF also suppressed the accumulation of toxic products of anaerobic respiration, such as ethanol. Applications of *Gigaspora margarita* suppressed the accumulation of ethanol in *P. persica* (Ruto et al. 2002) and *C. equisetifolia* (Osundina 1998).

Antioxidants Environmental stress commonly causes increased reactive oxygen species (ROS) including hydrogen peroxide and hydroxyl radicals. ROS can cause oxidative damage to cellular components, while plants produce antioxidant enzymes to eliminate and reduce ROS production (Hussain et al. 2008). Inoculation of *Diversispora spurca* increased the activity of antioxidant enzymes (SOD and CAT) in *C. junos* plants (Wu et al. 2013) and *Poncirus trifoliata* plants (Zou et al. 2014) under inundation conditions.

Chlorophyll and Photosynthesis Availability of lower O_2 in a puddle condition can lead to reducing stomatal conductance and root hydraulic conductivity (internal deficit of water) affecting the net photosynthesis. Inoculation of *F. mosseae* increased the concentration of chlorophyll a, b, and a/b in peach plants (Tuo et al. 2015). Increased levels of chlorophyll can support photosynthesis activity. Inoculated *T. latifolia* plants had a higher photosynthesis on puddle condition (Dunham et al. 2003). Increased uptake of water by roots with mycorrhiza had effect on stomatal conductance and increased photosynthesis (Rozema et al. 1986). Under flooded conditions, inoculation of *R. irregularis* improved roots' hydraulic conductivity of tomato plants, and this effect was correlated to a higher expression of the plant aquaporin SlPIP1;1 and of the fungal aquaporin GintAQP1 (Calvo-Polanco et al. 2014).

Root Nodulation *Glomus clarum* colonization is essential for the root nodulation of *Casuarina equisetifolia* on waterlogged soil (Osundina 1998). Mycorrhizal root nodulation was allegedly high due to improved plant nutrient uptake P and reduced accumulation of toxic compounds such as Fe and Mn. However, Fougnies et al. (2007) found no association between colonization AMF with nodulation on the roots of *Pterocarpus officinalis*.

Successional Vegetation and Structure of Plant Communities The AMF presence is essential to support vegetation succession. AMF existence along the river is thought to contribute to help the growth and development of early succession of plants in these conditions (Nakatsubo 1997; Harner et al. 2009, 2010). According to Nakatsubo (1997), colonization by AMF increases the aboveground biomass, nodule weight, leaf N concentration, and legume seed production of *Kummerowia striata* on alluvial soil media. Harner et al. (2010) found that the AMF helped the early succession of *Centaurea stoebe* in alluvial soil along the river/floodplain in Montana, USA. In the salt marsh conditions, AMF colonization on salt marsh plants may be essential for a successful restoration of marshy land (Cooke and Lefor 1990) through increased crop ecological adaptation (Sengupta and Chaudhuri 1990; Zhang et al. 2014). Ecological implications are that the association of AMF and AMF colonization dynamics became the basis for the ecological restoration of wetland functions and important component in the study of ecosystem dynamics (Radhika and Rodrigues 2007).

Potential of Phytoremediation Under constructed wetland acid mine drainage conditions, AMF is potential as phytoremediation (via phytostabilization) of Mn and Zn (Tuheteru et al. 2016).

Thus, the effect of AMF inoculation on plant performance in the wetland is not consistent (Stevens et al. 2011). In addition to improving the tolerance of plants, some studies reported that the AMF caused the growth depression in plants as reported by Stevens et al. (2011) on the *Eclipta prostrata* plant. AMF also lowered *L. salicaria* (Stevens et al. 2002). Dunham et al. (2003) have also reported that the AMF has a negative influence on the growth and weight of *T. latifolia*.

Acknowledgments This study was supported by the Plan in Scientific and Technological Innovation Team of Outstanding Young, Hubei Provincial Department of Education (T201604).

References

Akhtar MS, Siddiqui ZA (2008) Arbuscular mycorrhizal fungi as potential bioprotectants against plant pathogens. In: Siddiqui ZA, Sayeed Akhtar M, Futai K (eds) Mycorrhizae: sustainable agriculture and forestry. Springer, Dordrecht, pp 61–97

Aliasgharzadeh N, Rastin NS, Alizadeh HTA (2001) Occurrence of arbuscular mycorrhizal fungi in saline soils of the Tabriz plain of Iran in relation to some physical and chemical properties of soil. Mycorrhiza 11:119–122

Anderson RC, Liberta AE, Dickman LA (1984) Interaction of vascular plants and vesiculararbuscular mycorrhizal fungi across a soil moisture-nutrient gradient. Oecologia 64:111–117

Ashraf MA (2012) Waterlogging stress in plants: a review. Afr J Agric Res 7(13):1976–1981

Ashraf M, Foolad MR (2007) Roles of glycine betaine and proline in improving plant abiotic stress resistance. Environ Exp Bot 59:206–216

Augé RM (2001) Water relations, drought and vesicular-arbuscular mycorrhizal symbiosis. Mycorrhiza 11:3–42

Baar J, Paradi I, Lucassen ECHE et al (2011) Molecular analysis of AMF diversity in aquatic macrophytes: a comparison of oligotrophic and ultra-oligotrophic lakes. Aquat Bot 94:53–61

Bajwa R, Yaqoob A, Javaid A (2001) Seasonal variation in VAM in wet land plants. Pak J Biol Sci 4:464–470

Bauer CR, Kellogg CH, Bridgham SD et al (2003) Mycorrhizal colonization across hydrologic gradients in restored and reference freshwater wetlands. Wetlands 23(4):961–968

Beauchamp VB, Stromberg JC, Stutz JC (2006) Arbuscular mycorrhizal fungi associated with *Populus salix* stands in a semi-arid riparian ecosystem. New Phytol 170:369–380

Becerra AG, Nouhra ER, Silva MP et al (2009) Ectomycorrhizae, arbuscular mycorrhizae, and dark-septate fungi on *Salix humboldtiana* in two riparian populations from Central Argentina. Mycoscience 50:343–352

Beck-Nielsen D, Madsen TV (2001) Occurrence of vesicular-arbuscular mycorrhiza in aquatic macrophytes from lakes and streams. Aquat Bot 71:141–148

Bohrer KE, Friese CF, Amon JP (2004) Seasonal dynamics of arbuscular mycorrhizal fungi in differing wetland habitats. Mycorrhiza 14:329–337

Borriello R, Berruti A, Lumini E et al (2015) Edaphic factors trigger diverse AM fungal communities associated to exotic camellias in closely located Lake Maggiore (Italy) sites. Mycorrhiza 25:253–265

Brown AM, Bledsoe C (1996) Spatial and temporal dynamics of mycorrhizas in *Jaumea carnosa*, a tidal salt marsh halophyte. J Ecol 84:703–715

Brundrett MC, Ashwath N (2013) Glomeromycotan mycorrhizal fungi from tropical Australia III. Measuring diversity in natural and disturbed habitats. Plant Soil 370:419–433

Calvo-Polanco M, Molina S, Zamarreño AM et al (2014) The symbiosis with the arbuscular mycorrhizal fungus *Rhizophagus irregularis* drives root water transport in flooded tomato plants. Plant Cell Physiol 55:1017–1029

Camprubi A, Abril M, Estaun V et al (2012) Contribution of arbuscular mycorrhizal symbiosis to the survival of psammophilic plants after sea water flooding. Plant Soil 351:97–105

Cantelmo AJ Jr, Ehrenfeld JG (1999) Effects of microtopography on mycorrhizal infection in Atlantic white cedar (*Chamaecyparis thyoides* (L.) Mills.) Mycorrhiza 8:175–180

Carvalho LM, Cacador I, Martins-Loucao MA (2001) Temporal and spatial variation of arbuscular mycorrhizas in salt marsh plants of Tagus estuary (Portugal). Mycorrhiza 11:303–309

Choudhury B, Kalita MC, Azad P (2010) Distribution of arbuscular mycorrhizal fungi in marshy and shoreline vegetation of Deepar Beel Ramsar site of Assam, India. World J Microbiol Biotechnol 26:1965–1971

Clayton JS, Bagyaraj DJ (1984) Vesicular-arbuscular mycorrhizas in submerged aquatic plants of New Zealand. Aquat Plant 19:251–262

Cooke JC, Lefor MW (1990) Comparison of vesicular-arbuscular mycorrhizae in plants from disturbed and adjacent undisturbed regions of a costal salt marsh in Clinton, Connecticut, USA. J Environ Manag 14:131–137

Cooke JC, Butler RH, Madole G (1993) Some observations on the vertical distribution of vesicular arbuscular mycorrhizae in roots of salt marsh grasses growing in saturated soils. Mycologia 85:547–550

Cornwell WK, Bedford BL, Chapin CT (2001) Occurrence of arbuscular mycorrhizal fungi in a phosphorus-poor wetland and mycorrhizal response to phosphorus fertilization. Am J Bot 88:1824–1829

de Marins JF, Carrenho R, Thomaz SM (2009) Occurrence and coexistence of arbuscular mycorrhizal fungi and dark septate fungi in aquatic macrophytes in a tropical river-floodplain system. Aquat Bot 91:13–19

D'Souza J (2016) Arbuscular mycorrhizal diversity from mangroves: a review. In: Pagano MC (ed) Recent advances on mycorrhizal fungi. Springer International Publishing, Cham, pp 109–116

D'Souza J, Rodrigues BF (2013) Biodiversity of arbuscular mycorrhizal (AM) fungi in mangroves of Goa, in West India. J For 24(3):515–523

Dolinar N, Gaberščk A (2010) Mycorrhizal colonization and growth of *Phragmites australis* in an intermittent wetland. Aquat Bot 93:93–98

Dunham RM, Ray AM, Inouye RS (2003) Growth, physiology, and chemistry of mycorrhizal and Nonmycorrhizal *Typha latifolia* seedlings. Wetlands 23(4):890–896

Elzenga JTM, van Veen H (2010) Waterlogging and plant nutrient uptake. In: Stefano M, Shabala S (eds) . Springer, Berlin Heidelberg, pp 23–35

Escudero V, Mendoza R (2005) Seasonal variation of arbuscular mycorrhizal fungi in temperature grassland along a wide hydrologic gradient. Mycorrhiza 15:291–299

Farmer AM (1985) The occurrence of vesicular-arbuscular mycorrhiza in isoetid-type submerged aquatic macrophytes under naturally varying conditions. Aquat Bot 21:245–249

Fougnies L, Renciot S, Muller F et al (2007) Arbuscular mycorrhizal colonization and nodulation improve flooding tolerance in *Pterocarpus officinalis* Jacq. Seedlings. Mycorrhiza 17:159–166

Fraser LH, Feinstein LM (2005) Effects of mycorrhizal inoculant, N:P supply ratio, and water depth on the growth and biomass allocation of three wetland plant species. Can J Bot 83:1117–1125

Fuchs B, Haselwandter K (2004) Red list plants: colonization by arbuscular mycorrhizal fungi and dark septate endophytes. Mycorrhiza 14:277–281

Garcia I, Mendoza R, Pomar MC (2008) Deficit and excess of soil water impact on plant growth of *Lotus tenuis* by affecting nutrient uptake and arbuscular mycorrhizal symbiosis. Plant Soil 304:117–131

Graham LL, Turjaman M, Page SE (2013) *Shorea balangeran* and *Dyera polyphylla* (syn. Dyera Lowii) as tropical peat swamp forest restoration transplant species: effects of mycorrhizae and level of disturbance. Wetl Ecol Manag 21(5):307–321

Gupta N, Basak UC, Das P (2002) Arbuscular mycorrhizal association of mangroves in saline and non saline soils. Mycorrhiza News 13(4):14–19

Gupta N, Bihari KM, Sengupta I (2016) Diversity of arbuscular mycorrhizal fungi in different salinity of mangrove ecosystem of Odisha, India. Adv Plant Agric Res 3(1):85

Harner MJ, Piotrowski JS, Lekberg Y et al (2009) Heterogeneity of mycorrhizal inoculum potential in flood-deposited sediments. Aquat Sci 71:331–337

Harner MJ, Mummey DL, Stanford JA et al (2010) Arbuscular mycorrhizal fungi enhance spotted knapweed growth across a riparian chronosequence. Biol Invasions 12:1481–1490

Harner MJ, Opitz N, Geluso K et al (2011) Arbuscular mycorrhizal fungi on developing islands within a dynamic river floodplain: an investigation across successional gradients and soil depth. Aquat Sci 73:35–42

Hashimoto Y, Higuchi R (2003) Ectomycorrhizal and arbuscular mycorrhizal colonization of two species of floodplain willows. Mycosciense 44:339–343

Hildebrandt U, Janetta K, Ouziad F et al (2001) Arbuscular mycorrhizal colonization of halophytes in central European salt marshes. Mycorrhiza 10:175–183

Hoefnagels MH, Broome SW, Shafer SR (1993) Vesicular-arbuscular mycorrhizae. Estuaries 16(4):851–858

Husna BSWR, Mansur I, Kusmana C (2016) Growth and nutrient status of kayu kuku [*Pericopsis mooniana* (Thw.) Thw] with mycorrhiza in soil media of nickel post mining. Pak J Biol Sci 19:158–170

Hussain TM, Chandrasekhar T, Hazara M et al (2008) Recent advances in salt stress biology-a review (standard review). Biotechnol Mol Biol Rev 3(1):8–13

Ipsilantis I, Sylvia DM (2007) Interactions of assemblages of mycorrhizal fungi with two Florida wetland plants. Appl Soil Ecol 35:261–271

Kai W, Zhiwei Z (2006) Occurrence of arbuscular mycorrhizas and dark septate endophytes in hydrophytes from lakes and streams in Southwest China. Int Rev Hydrobiol 91:29–37

Keeley JE (1980) Endomycorrhizae influence growth of black gum seedlings in flooded soils. Am J Bot 67:6–9

Kennedy LJ, Tiller RL, Stutz JC (2002) Associations between arbuscular mycorrhizal fungi and *Sporobolus wrightii* in riparian habitats in arid South-Western North America. J Arid Environ 50:459–475

Khan AG (1974) The occurrence of mycorrhizas in halophytes, hydrophytes and xerophytes and of *Endogone* spores in adjacent soils. J Gen MicrobioI 81:7–14

Khan AG (1993) Occurrence and importance of mycorrhiza in aquatic trees of new South Wales, Australia. Mycorrhiza 3:31–38

Khan AG (1995) Effects of various soil environmental stresses on the occurrence, distribution and effectiveness of VA Mycorrhizae. Biotropia 8:39–44

Khan AG (2004) Mycotrophy and its significance in wetland ecology and wetland management. In: Wong MH (ed) Wetlands ecosystems in asia: function and management. Elsevier, Amsterdam, pp 95–113

Khan AG, Belik M (1995) Occurrence and ecological significance of mycorrhizal symbiosis in aquatic plants. In: Varma A, Hock B (eds) Mycorrhiza: structure, function, molecular biology and biotechnology. Springer, Heidelberg, pp 627–666

Kothamasi D, Kothamasi S, Bhattacharyya A et al (2006) Arbuscular mycorrhizae and phosphate solubilising bacteria of the rhizosphere of the mangrove ecosystem of great Nicobar island, India. Biol Fertil Soils 42:358–361

Kozlowski TT (1984) In: Kozlowksi TT (ed) Responses of woody plants to flooding. Academic, San Diego

Kozlowski TT, Pallardy SG (1997) Growth control in woody plant. Academic, San Diego

Kumar T, Ghose M (2008) Status of arbuscular mycorrhizal fungi (AMF) in the Sundarbans of India in relation to tidal inundation and chemical properties of soil. Wetl Ecol Manag 16:471–483

Landwehr M, Hildebrandt U, Wilde P et al (2002) The arbuscular mycorrhizal fungus *Glomus geosporum* in European saline, sodic and gypsum soils. Mycorrhiza 12:199–211

Likar M, Regvar M, Mandic-Mulec I et al (2009) Diversity and seasonal variations of mycorrhiza and rhizosphere bacteria in three common plant species at the Slovenian Ljubljana marsh. Biol Fertil Soils 45:573–583

Liu RJ, Jiao H, Li Y et al (2009) Research advances in species diversity of arbuscular mycorrhizal fungi. J Appl Ecol 20:2301–2307

Lodge DJ (1989) The influence of soil moisture and flooding on formation of VA- endo- and ecto-mycorrhizae in Populus and Salix. Plant Soil 117:243–253

Lovelock CE, Andersen K, Morton JB (2003) Arbuscular mycorrhizal communities in tropical forests are affected by host tree species and environment. Oecologia 135:268–279

Marshal PE, Pattullo N (1981) Mycorrhizae occurrence in willows in a northern freshwater wetland. Plant Soil 59:465–471

Mason E (1928) Note on presence of mycorrhiza in the roots of salt marsh plants. New Phytol 27:193–195

Mathur N, Singh J, Bohra S, Vyas A (2007) Arbuscular mycorrhizal status medicinal halophytes in saline areas of Indian Thar Desert. Int J Soil Sci 2(2):119–127

Matsumura A, Horii S, Ishii T (2008) Observation of arbuscular mycorrhizal network system between trifoliate orange and some grasses underwaterlogged conditions. Acta Hortic 773:69–71

Mendoza R, Escudero V, Garcia I (2005) Plant growth, nutrient acquisition and mycorrhizal symbioses of a waterlogging tolerant legume (*Lotus glaber* mill.) in a saline-sodic soil. Plant Soil 275:305–315

Miller SP (2000) Arbuscular mycorrhizal colonization of semiaquatic grasses along a wide hydrologic gradient. New Phytol 145:145–155

Miller SP, Bever JD (1999) Distribution of arbuscular mycorrhizal fungi in stands of wetlands grass *Panicum hemitomon* along a wide hydrologic gradient. Oecologia 119:586–592

Miller SP, Sharitz RR (2000) Manipulation of flooding and arbuscular mycorrhiza formation influences growth and nutrition of two semiaquatic grass species. Funct Ecol 14:738–748

Miller RM, Smith CI, Jastrow JD et al (1999) Mycorrhizal status of the genus *Carex* (Cyperaceae). Am J Bot 86(4):547–553

Møller CL, Kjøller R, Sand-Jensen K (2013) Organic enrichment of sediments reduces arbuscular mycorrhizal fungi in oligotrophic lake plants. Freshw Biol 58:769–779

Moora M, Öpik M, Davison J et al (2016) AM fungal communities inhabiting the roots of submerged aquatic plant *Lobelia dortmanna* are diverse and include a high proportion of novel taxa. Mycorrhiza. doi:10.1007/s00572-016-0709-0

Muok BO, Ishii T (2006) Effect of arbuscular mycorrhizal fungi on tree growth and nutrient uptake of *Sclerocarya birrea* under water stress, salt stress and flooding. J Jpn Soc Hortic Sci 75:26–31

Nakatsubo T (1997) Effects of arbuscular mycorrhizal infection on the growth and reproduction of the annual legume Kummerowia striata growing in a nutrient-poor alluvial soil. Ecol Res 12:231–237

Nakatsubo T, Kaniyu M, Nakagoshi N et al (1994) Distribution of vesicular-arbuscular mycorrhizae in plants growing in a river floodplain. Bull Jap Soc Microbial Ecol 9(3):109–117

Neto D, Carvalho LM, Cruz C et al (2006) How do mycorrhizas affect C and N relationships in flooded *Aster tripolium* plants? Plant Soil 279:51–63
Nielsen KB, Kjøller R, Olsson PA et al (2004) Colonisation and molecular diversity of arbuscular mycorrhizal fungi in the aquatic plants *Littorella uniflora* and *Lobelia dortmanna* in southern Sweden. Mycol Res 108:616–625
Nobis A, Błaszkowski J, Zubek S (2015) Arbuscular mycorrhizal fungi associations of vascular plants confined to river valleys: towards understanding the river corridor plant distribution. J Plant Res 128:127–137
Oliveira RS, Dodd JC, Castro PML (2001) The mycorrhizal status of *Phragmites australis* in several polluted soils and sediments of an industrialised region of northern Portugal. Mycorrhiza 10:241–247
Orchard S, Standish RJ, Nicol D et al (2016) The response of fine root endophyte (*Glomus tenue*) to waterlogging is dependent on host plant species and soil type. Plant Soil 403(1):305–315
Osundina MA (1998) Nodulation and growth of mycorrhizal *Casuarina equisetifolia* J.R. And G. First in response to flooding. Biol Fertil Soils 26:95–99
Othira JO, Omolo JO, Kiruki S et al (2014) Molecular diversity of arbuscular mycorrhizal fungi (AMF) in Lake Victoria Basin of Kenya. J Ecol Nat Environ 6(4):145–152
Radhika KP, Rodrigues BF (2007) Arbuscular mycorrhizae in association with aquatic and marshy plant species in Goa, India. Aquat Bot 86:291–294
Ragupathy S, Mohankumar V, Mhadeven A (1990) Occurrence of vesicular-Arbuscular mycorrhizae in tropical hydrophytes. Aquat Bot 36:287–291
Ray AM, Inouye RS (2006) Effects of water-level fluctuations on the arbuscular mycorrhizal colonization of *Typha latifolia* L. Aquat Bot 84:210–216
Rozema J, Arp W, van Diggelen J et al (1986) Occurrence and ecological significance of vesicular-arbuscular in the salt marsh environment. Acta Bot Neerl 35:457–467
Ruiz-Lozano JM, Collados C, Barea JM et al (2001) Cloning of cDNAs encoding SODs from lettuce plants which show differential regulation by arbuscular mycorrhizal symbiosis and by drought stress. J Exp Bot 52:2241–2242
Ruto KL, Fusao M, Kazumi K (2002) Effect of root-zone flooding on mycorrhizal and non-mycorrhizal peach (*Prunus persica* Batsch) seedlings. Sci Hortic 94:285–295
Sah S, Reed S, Jayachandran K, Dunn C et al (2006) The effect of repeated short-term flooding on mycorrhizal survival in snap bean roots. HortSci 41:598–602
Sannazzaro AI, Ruiz OA, Alberto E et al (2004) Presence of different arbuscular mycorrhizal infection patterns in roots of *Lotus glaber* plants growing in the Salado River basin. Mycorrhiza 14:139–142
Secilia J, Bagyaraj DJ (1992) Selection of efficient vesicular-arbuscular mycorrhizal fungi for wetland rice (*Oryza sativa* L.) BioI Fertil Soils 13:108–111
Secilia J, Bagyaraj DJ (1994) Selection of efficient vesicular-arbuscular mycorrhizal fungi for wetland rice - a preliminary screen. Mycorrhiza 4:265–268
Seerangan K, Thangavelu M (2014) Arbuscular mycorrhizal and dark septate endophyte fungal associations in south indian aquatic and wetland macrophytes. J Bot:1–14
Sengupta A, Chaudhuri S (1990) Vesicular-arbuscular mycorrhiza (VAM) in pioneer salt marsh plants of the Ganges river delta in West Bengal (India). Plant Soil 122:111–113
Sengupta A, Chaudhuri S (2002) Arbuscular mycorrhizal relations of mangrove plant community at the Ganges river estuary in India. Mycorrhiza 12:169–174
Shah MA (2014) Mycorrhizas: novel dimensions in the changing world. Springer, New Delhi
Smith SE, Read DJ (2008) Mycorrhizal symbiosis, Third edn. Academic, New York
Solaiman MZ, Hirata H (1996) Effectiveness of mycorrhizal colonization at nursery-stage on growth and nutrition in wetland rice (*Oryza sativa* L.) after transplanting under different soil fertility and water regimes soil Sci. Plant Nutr 42:561–571
Solaiman MZ, Hirata H (1997) Effect of arbuscular mycorrhizal fungi inoculation of rice seedlings at the nursery stage upon performance in the paddy field and greenhouse. Plant Soil 191:1–12
Søndergaard M, Laegaard S (1977) Vesicular-arbuscular mycorrhiza in some aquatic vascular plants. Nature 268:232–233

Sraj-Krzic N, Pongrac P, Klemenc M et al (2006) Mycorrhizal colonisation in plants from intermittent aquatic habitats. Aquat Bot 85:331–336

Stenlund DL, Charvat ID (1994) Vesicular arbuscular mycorrhizae in floating wetland mat communities dominated by Typha. Mycorrhiza 4:131–137

Stevens KJ, Peterson RL (1996) The effect of a water gradient on the vesicular-arbuscular mycorrhizal status of *Lythrum salicaria* L. (purple loosestrife). Mycorrhiza 6:99–104

Stevens KJ, Spender SW, Peterson RL (2002) Phosphorus, arbuscular mycorrhizal fungi and performance of the wetland plant *Lythrum salicaria* L. under inundated conditions. Mycorrhiza 12:277–283

Stevens KJ, Wellner MR, Acevedo MF (2010) Dark septate endophyte and arbuscular mycorrhizal status of vegetation colonizing a bottomland hardwood forest after a 100 year flood. Aquat Bot 92:105–111

Stevens KJ, Wall CB, Janssen JA (2011) Effects of arbuscular mycorrhizal fungi on seedling growth and development of two wetland plants, *Bidens frondosa* L., and *Eclipta prostrata* L., grown under three levels of water availability. Mycorrhiza 21:279–288

Sudova R, Rydlova J, Ctvrtlikova M et al (2011) The incidence of arbuscular mycorrhiza in two submerged Isoetes species. Aquat Bot 94:183–187

Sudova R, Sykorova Z, Rydlova J et al (2015) *Rhizoglomus melanum*, a new arbuscular mycorrhizal fungal species associated with submerged plants in freshwater lake Avsjoen in Norway. Mycol Prog 14:9

Szabados L, Savoure A (2009) Proline: a multifunctional amino acid. Trends Plant Sci 15(2):89–97

Tanaka K, Masumori M, Yamanoshita T et al (2011) Morphological and anatomical changes of *Melaleuca cajuputi* under submergence. Trees 25:695–704

Tanner CC, Clayton JS (1985a) Vesicular-arbuscular mycorrhiza studies with a submerged aquatic plant. Trans Br Mycol Soc 85:683–688

Tanner CC, Clayton JS (1985b) Effects of vesicular-arbuscular mycorrhizas on growth and nutrition of a submerged aquatic plant. Aquat Bot 22:377–386

Tawaraya K, Takaya Y, Turjaman M et al (2003) Arbuscular mycorrhizal colonization of tree species grown in peat swamp forests of Central Kalimantan, Indonesia. For Ecol Manag 182:381–386

Thormann MN, Currah RS, Bayley SE (1999) The mycorrhizal status of the dominant vegetation along a peatland gradient in southern boreal Alberta, Canada. Wetlands 19(2):438–450

Tuheteru FD (2015) Phytoremediation potential of lonkida (*Nauclea orientalis* L.) in an acid mine drainage artificial wetland. Dissertation. Bogor Agricultural University

Tuheteru FD, Kusmana C, Mansur I et al (2015) Response of lonkida (*Nauclea orientalis* L.) towards mycorrhizal inoculum in waterlogged condition. Biotropia 22(1):61–71

Tuheteru FD, Kusmana C, Mansur I et al (2016) Potential of lonkida (*Nauclea orientalis* L.) for phytoremediation of acid mined drainage at PT. Bukit Asam Tbk (Persero), Indonesia. Res J Bot 11:9–17

Tuo XQ, Li S, Wu QS et al (2015) Alleviation of waterlogged stress in peach seedlings inoculated with *Funneliformis mosseae*: changes in chlorophyll and proline metabolism. Sci Hortic 197:130–134

Turner SD, Friese CF (1998) Plant mycorrhizal community dynamics associated with a moisture gradient within a rehabilitated prairie fen. Restor Ecol 6:44–51

Turner ST, Amon JP, Schnebl RM et al (2000) Mycorrhizal fungi associated with plants in groundwater fed wetlands. Wetlands 20:200–204

Wang B, Qiu YL (2006) Phylogenetic distribution and evolution of mycorrhizas in land plants. Mycorrhiza 16:299–363

Wang FY, Liu RJ, Lin XG et al (2004) Arbuscular mycorrhizal status of wild plants in saline- alkaline soils of the Yellow River Delta. Mycorrhiza 14:133–137

Wang YT, Qui Q, Yang Zy HZJ et al (2010) Arbuscular mycorrhizal fungi in two mangroves in South China. Plant Soil 33:181–191

Wang Y, Huang Y, Qiu Q et al (2011) Flooding greatly affects the diversity of arbuscular mycorrhizal fungi communities in the roots of wetland plants. PLoS One 6(9):1–10

Wang Y, Li T, Li Y et al (2015) Distribution of arbuscular mycorrhizal fungi in four semi-mangrove plant communities. Ann Microbiol 65:603–610

Wang Y, Li Y, Bao X et al (2016) Response differences of arbuscular mycorrhizal fungi communities in the roots of an aquatic and a semiaquatic species to various flooding regimes. Plant Soil 403:361–373

Weishampel PA, Bedfor BL (2006) Wetland dicots and monocots differ in colonization by arbuscular mycorrhizal fungi and dark septate endophytes. Mycorrhiza 16:495–502

Wetzel PR, van der Valk AG (1996) Vesicular-arbuscular mycorrhizae in prairie pothole wetland vegetation in Iowa and North Dakota. Can J Bot 74:883–890

White JA, Charvat I (1999) The mycorrhizal status of an emergent aquatic, *Lythrum salicaria* L., at different levels of phosphorus availability. Mycorrhiza 9(4):191–197

Wigand C, Stevenson JC (1997) Facilitation of phosphate assimilation by aquatic mycorrhizae of *Vallisneria americana* Michx. Hydrobiologia 242:35–41

Wigand C, Andersen FO, Christensen KK et al (1998) Endomycorrhizae of isoëtids along a biogeochemical gradient. Limnol Oceanogr 43:508–515

Wilde P, Manal A, Stodden M et al (2009) Biodiversity of arbuscular mycorrhizal fungi in roots and soils of two salt marshes. Environ Microbiol 11:1548–1561

Wirsel SGR (2004) Homogenous stands of a wetland grass harbor diverse consortia of arbuscular mycorrhizal fungi. FEMS Microbiol Ecol 48:129–138

Wolfe BE, Mummey DL, Rillig MC et al (2007) Small-scale spatial heterogeneity of arbuscular mycorrhizal fungal abundance and community composition in a wetland plant community. Mycorrhiza 17:175–183

Wu QS, Zou YN, He XH (2010) Contribution of arbuscular mycorrhizal fungi to growth, photosynthesis, root morphology and ionic balance of citrus seedlings under salt stress. Acta Physiol Plant 32:297–304

Wu QS, Zou YN, Huang YM (2013) The arbuscular mycorrhizal fungus *Diversispora spurca* ameliorates effects of waterlogging on growth, root system architecture and antioxidant enzyme activities of citrus seedlings. Fungal Ecol 6:37–43

Yang H, Koide RT, Zhang Q (2016) Short-term waterlogging increases arbuscular mycorrhizal fungal species richness and shifts community composition. Plant Soil 404:373–384

Yin D, Zhang Z, Luo H (2012) Anatomical responses to waterlogging in *Chrysanthemum zawadskii*. Sci Hortic 146:86–91

Zhang Q, Sun Q, Koide RT et al (2014) Arbuscular mycorrhizal fungal mediation of plant-plant interactions in a marshland plant community. Sci World J 2014:1–10

Zhang X, Wu S, Ren B et al (2016) Water management, rice varieties and mycorrhizal inoculation influence arsenic concentration and speciation in rice grains. Mycorrhiza 26(4):299–309

Zhao X, Yuan S, Song H et al (2016) Arbuscular mycorrhizal and dark septate fungal associations in riparian plants of the three gorges reservoir region, Southwest China. Aquat Bot 133:28–37

Zou YN, Srivastava AK, Wu QS, Huang YM (2014) Increased tolerance of trifoliate orange (*Poncirus trifoliata*) seedlings to waterlogging after inoculation with arbuscular mycorrhizal fungi. J Anim Plant Sci 24(5):1415–1420

Chapter 4
Arbuscular Mycorrhizal Fungi and Tolerance of Salt Stress in Plants

Bhawna Saxena, Kamlesh Shukla, and Bhoopander Giri

Abstract Soil salinity has become a serious land degradation problem and is increasing steadily in many parts of the world, particularly in arid and semiarid areas. Increased salinization of arable land is expected to have devastating global effects and would lead to 30% land loss within the next 25 years and up to 50% by the middle of the twenty-first century. Plants growing in saline conditions generally last to three distinct stresses, ionic, osmotic, and oxidative. The toxic effects of specific ions such as Na^+ and Cl^-, prevalent in saline soils, disrupt the structure of enzymes and other macromolecules, damage cell organelles, disrupt general metabolic activities, inhibit protein synthesis, and induce ion deficiency. Plants exposed to low water potential face the problem of physiological drought condition. Moreover, salinity creates nutrient imbalance in the plant due to decreased and differential uptake of nutrients and/or their translocation to the shoot and leaf tissues, rendering the plants weak and unproductive. Arbuscular mycorrhizal fungi (AMF) are ubiquitous soil microorganisms that build up symbiotic associations with the majority of higher plants, establish a direct physical link between soil and plant roots, constitute an integral component of the natural ecosystems, and predominantly exist in the saline environment. The extraradical hyphae of AMF run several meters away from the depletion zone, increase root surface area and facilitate nutrient absorption by the plant. Indeed, AMF improve physiological processes and general metabolic activities of the plant and help in the mitigation of physiological drought, which is often imposed under saline conditions. Therefore, the application of AMF could offer a cheaper and cost-effective alternative to counteract the problem of salinity. In this review chapter, we have discussed the factors influencing soil salinization and possible approaches to overcome the problem of salinity stress. The underlying physiological, biochemical and molecular mechanisms by which mycorrhizal plants could improve salt tolerance has also been described.

B. Saxena • B. Giri (✉)
Department of Botany, Swami Shraddhanand College, University of Delhi, Delhi, India
e-mail: bhoopg@yahoo.com

K. Shukla
School of Studies in Biotechnology, Pt. Ravishankar Shukla University, Raipur, Chhattisgarh, India

Q.-S. Wu (ed.), *Arbuscular Mycorrhizas and Stress Tolerance of Plants*,
DOI 10.1007/978-981-10-4115-0_4

Keywords Soil salinization • Reactive oxygen species • Mycorrhizal fungi • Salinity tolerance

4.1 Introduction

Soil solution possesses various important soluble salts and is considered as one of the best media for plant growth and development. Plants through their roots absorb these soluble salts and translocate them to different parts where these salts are required to perform numerous metabolic activities. However, excessive salts in soil reduce plant water and nutrient uptake and disrupt the distribution of ions at both the cellular and the whole-plant levels, thereby inducing osmotic and ionic imbalances. Such drastic changes result in stunted plant growth and development and can lead to death of the plant. Higher accumulation of salts like Na^+ and Cl^- in plant tissues leads to oxidative damage (also considered as secondary stress), affecting integrity of plant membranes (damage to lipids, proteins, and nucleic acids), impairing activities of biocatalysts and functioning of photosynthetic apparatus, which is ascribed to the deleterious effects of the reactive oxygen species (ROS) often generated by salt stress (Zhu 2001; Kumar et al. 2014).

The rhizosphere, an area in the immediate vicinity of the plant root is predominantly affected by the activities of soil microbes. These microbes can be harmful – the plant pathogens, which carry genes for pathogenicity and induce biotic stress in the plants – or can be useful, which alleviate detrimental effects of biotic and abiotic stresses, therefore promoting plant growth and survival. They could be known as stress modulators or potential sources of bio-fertilizers (nitrogen-fixing bacteria, phosphate solubilizers, and mycorrhizae). Among them, arbuscular mycorrhizal fungi (AMF) are considered as one of the most common associations due to their ubiquitous distribution among wide taxa of terrestrial plants. They are the key components of the natural ecosystems and are known to be prevalent in the saline environment (Aliasgharzadeh et al. 2001; Giri et al. 2003). AMF establish a direct physical link between both plant roots and soils and facilitate host plant to acquire mineral nutrients from soils particularly under nutrient stress conditions and modify the environment of rhizosphere, thereby alleviating the adverse effects of salinity stress (Smith and Read 1997; Jahromi et al. 2008; Evelin et al. 2009). AMF have been found to improve salt tolerance in different plant species such as tomato, cucumber, maize, lettuce, clover, fenugreek, sesbania, and acacia (Ruiz-Lozano et al. 1996; Al-Karaki 2000; Feng et al. 2002; Giri et al. 2003, 2007; Giri and Mukerji 2004; Evelin et al. 2012, 2013).

The application of AMF offers a cheaper and cost-effective alternative to counteract the problem of salinity stress. More than hundreds of research papers evidenced that this association diminishes adverse effects of salt stress and improves plant growth under saline condition; however, the direct and exact physiological and biochemical mechanisms involved in the increased salt tolerance in mycorrhizal plants are yet to be documented (Ruiz-Lozano 2003; Evelin et al. 2009; Porcel et al. 2012; Kapoor et al. 2013). Excessive salt in soils drastically limits the availability of phosphorus (P) to plants, because phosphate ions precipitate with Ca^{2+}, Mg^{2+}, and Zn^{2+} ions and become immobile in a saline soil (Azcón-Aguilar et al. 1979). However, AMF strongly influence the uptake of P under saline condition (Ojala et al. 1983; Giri et al. 2003). The contribution of AMF under saline condition is not limited to P uptake; indeed, AMF allow extraradical hyphae to explore more soil volume for better absorption of other nutrients particularly in a stress condition (Ruiz-Lozano and Azcón 2000). Improved nutrient acquisition, particularly P by AMF, could ascribe to be the primary mechanism by which mycorrhizal fungi mitigate the adverse effects of salinity stress on plant growth (Giri et al. 2003, 2007). However, it is tempting to consider that increased nutrient uptake is not the only mechanism by which AM plants increase their tolerance against high levels of soluble salts. The involvement of non-mediated nutritive effects, which could play a more precise role than nutritional effects in the improvement of salt tolerance in AM plants under salinity, has also been suggested (Giri and Mukerji 2004). In fact, AMF modulate several physiological and biochemical processes and regulate expression of salt-related genes to mitigate adverse effects of toxic ions under saline condition (Evelin et al. 2009; Porcel et al. 2012).

In the last decades, researchers focused on identifying the underlying aspects of salt tolerance in mycorrhizal plants to understand what actually conspires between them to initiate certain responses for alleviating deleterious effects of salt stress (Ruiz-Lozano 2003; Evelin et al. 2009; Aroca et al. 2013; Kapoor et al. 2013). In addition to improved mineral nutrition, the involvement of AMF in improving plant photosynthetic capacity, stomatal conductance, root hydraulic conductivity, water use efficiency, accumulation of enzymatic and nonenzymatic antioxidants, compatible organic solutes (help in detoxification of damaging reactive oxygen species), and osmotic adjustment (protect integrity of cell membrane and organelle and stabilize proteins) has been evidenced in AM plants growing under salinity stress (Sharifi et al. 2007; Sheng et al. 2008; Evelin et al. 2009; Abdel Latef and Chaoxing 2011; Porcel et al. 2012; Kumar et al. 2014; Auge et al. 2014). Therefore, the mitigation of adverse effects of salinity stress in AM plants seems to be an integrated approach, instead being nutritive only. In this chapter, the important processes by which AM plants improve their salt stress tolerance have been discussed. Emphasis has been given on the nutrient acquisition, ionic and osmotic homeostasis, photosynthetic protection, antioxidant production and changes in the cell ultrastructures, and expression of salt-responsive genes in AM plants under salinity stress.

4.2 Soil Salinity

Soil salinity, a soil condition characterized by high concentration of soluble salts, is a complex and harmful threat faced by plants due to disruption of ionic, osmotic, and cell-water homeostasis (Bojorquez-Quintal et al. 2014). Soil salinity can be

Table 4.1 Properties of salt-affected soil

Soil property	Saline soil	Sodic soil	Saline-sodic soil
Electric conductivity (dS/m)	>4.0	<4.0	>4.0
Soil pH	<8.5	>8.5	<8.5
Ion present	Sulfates and chlorides of Ca and Mg	Excess levels of Na	Excess levels of Na
Reclamation	Very easy	Slow and costly	Easy

defined as the "electrical conductivity (EC) of a saturated soil solution," which is generally expressed in unit decisiemens per meter (dS/m). Soils are classified as saline when their EC is 4 dS m^{-1} or higher, which is equal to approximately 40 mM NaCl and generates an osmotic pressure of about 0.2 MPa (Munns and Tester 2008; Kapoor et al. 2013). However, Richards (1954) suggested that the electric conductivity of saline soil is always above 4 dS m^{-1} at 25 °C. Salinity has become one of the most severe problems for sustainable agriculture production as it adversely affects growth and productivity of many plant species all over the world. The most common cations and anions in a soil solution associated with salinization of soil are sodium (Na^+), potassium (K^+), calcium (Ca^{+2}), magnesium (Mg^{+2}), chloride (Cl^-), sulfate (SO_4^{2-}), carbonate (HCO_3^-), and nitrates (NO_3^-). In a hypersaline soil or water, the other ion ingredients of salinity may be B, Sr^{2+}, SiO_2, Mo, Ba^{2+}, and Al^{3+} (Hu and Schmidhalter 2002).

On the whole, sodium chloride is the most common salt in a saline soil (Munns and Tester 2008). Depending upon the types of ion and its concentration present in a soil solution, salt-affected soils have been categorized under three different categories: saline, sodic, and saline-sodic soils (Table 4.1). Soil salinity came into existence long before humans and agriculture as salts are present in the soils due to the parental rock's ionic content; however, it has now become a matter of concern as accumulation of these salts in the soil is aggravating by frail agriculture practices and climate change (Zhu 2001; Maathuis et al. 2014). High concentrations of salts in soil, especially Na^+ ions drastically affect the basic structure of soil (Mahajan and Tuteja 2005). The presence of Na^+ ions in the cation exchange complex makes the soil compact and subsequently decreases soil porosity and aeration, which hampers plant growth and hinders their productivity (Garg and Manchanda 2008). The presence of the salts of calcium and magnesium forms a white crust on the soil surface that changes soil water osmotic potential; therefore, plants growing in the saline soils face salt-induced physiological drought conditions. Salinity impairs plant's major processes such as photosynthesis, protein and lipid metabolism, nutrient acquisition, and ion homeostasis. Indeed, water moves out of the plant due to salt-induced osmotic stress, which makes plant dehydrated and eventually ascribes to the death of plant (Evelin et al. 2011).

4.3 Statistics of Salt-Affected Areas

Salinization of soils is one of the most prevalent agricultural problems commonly occurring in the arid and semiarid and low-lying coastal areas of the world (Evelin et al. 2009; Porcel et al. 2012; Kumar et al. 2010). According to the FAO, 7% of the global soil surface is salt affected, out of which 15.7 Mha is in North America, 1.9 Mha in Mexico and Central America, 129.16 Mha in South America, 80.43 Mha in Africa, 85.10 Mha in South and West Asia, 19.98 Mha in Southeast Asia, 211.68 Mha in North and Central Asia, 357.33 Mha in Australia, and 52.08 Mha in Europe (FAO 1999; Ruiz-Lozano et al. 2001). Globally, salinization of soil is increasing due to rise in the sea levels by climate change and also due to wrong irrigation practices of agricultural lands (Bothe 2012; Maathuis et al. 2014). In some areas, it is increasing due to extensive use of salt on roads to prevent frozen glaze in winter (Bothe 2012). The distribution of saline and alkali soils in the world is shown in Fig. 4.1. According to the Central Soil Salinity Research Institute (CSSRI), India, the total salt-affected area in India is 6.74 Mha, with 1.7 Mha under saline soil, 3.7 Mha under sodic soil, and the remaining as coastal soil areas. The distribution pattern of the different soil types in the sixteen states of India is shown in Fig. 4.2. Out of the 15 states and one union territory, saline soil is present in the seven states, and alkaline soil is present in the 11 states and one union territory.

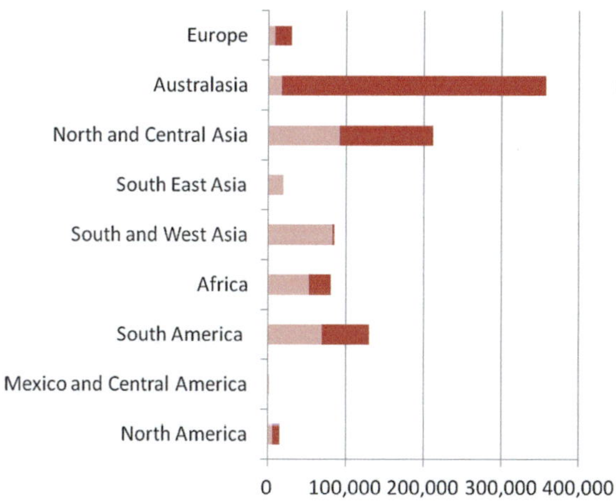

Fig. 4.1 Distribution pattern of saline and alkaline soils in the world (Source: FAO/UNESCO soil map)

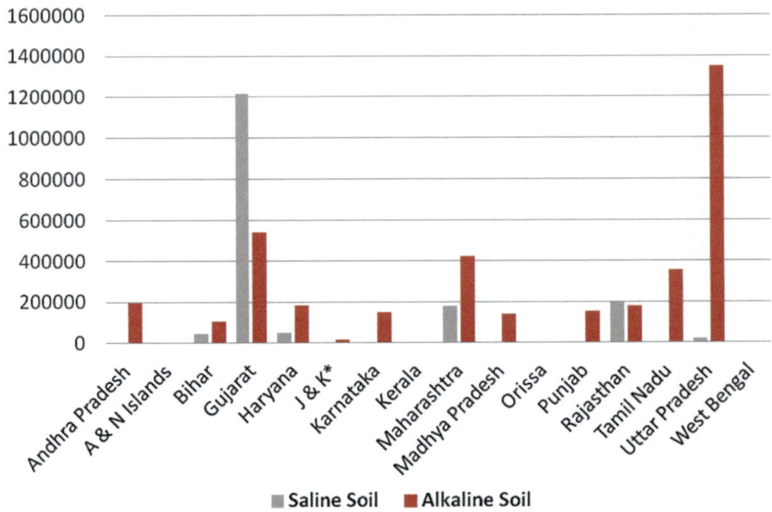

Fig. 4.2 Salt-affected areas in India (Source: CSSRI, India soil salinity database http://www.cssri.org/index.php?option=com_content&view=article&id=122&Itemid=126)

4.4 Genesis of Soil Salinity

The process of addition of salts to the soil is known as salinization. Soil salinization may be the result of natural or man-made factors (Maathuis et al. 2014).

4.4.1 Natural Factors Contributing to Soil Salinity

The natural factors contributing to salinization of soils include weathering of rocks, deposition of oceanic salts, topographical factor, groundwater table fluctuation, salt lake, scald, fallow period, and flood water. However, the major contributors of the soil salinization are weathering of parental rocks, oceanic salt deposition, and groundwater table fluctuation. Rocks are mostly rich in sodium and other salts. Weathering of rocks has been considered to be the primary cause for the presence of salts in the soil. During this process the parent rock is loosen or gets broken down into the salt component and releases salts of calcium, magnesium, and sodium and to some extent salts of carbonates and sulfates. The term weathering means "with no movement"; thus, the salts generated upon the process of weathering become accumulated in the soil. If the weather is humid, then these salts get percolated down to the water table, from where they are transported to the sea. The mineral composition of the parent rock and its porosity determines the soil fertility level and rates of soil formation, respectively.

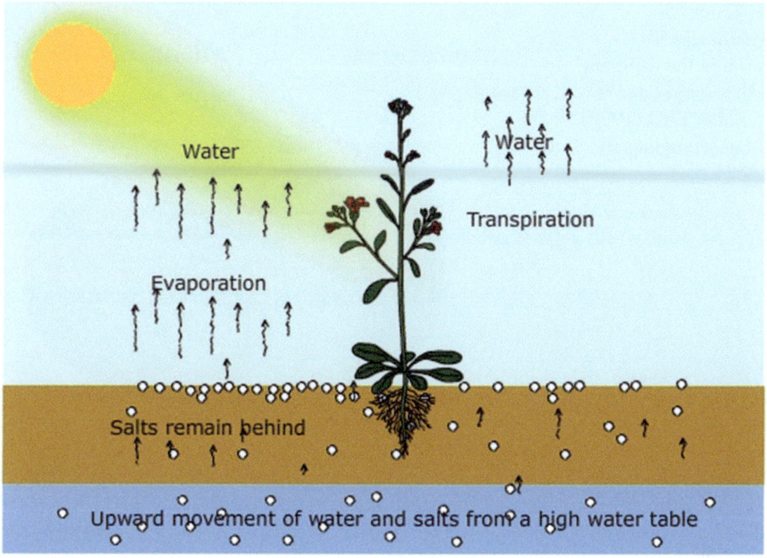

Fig. 4.3 The picture depicts accumulation and deposition of salts in root zone forming a crust on the soil surface thereby inducing salt stress

Agriculture land in the coastal areas gets polluted with the seawater intrusion on the land. Water gets evaporated leaving behind salts. Accumulation of these salts increase salinization of soil of the area. Oceanic salts may also be carried inland by wind and rain. Munns and Tester (2008) demonstrated that rain containing 10 mg Kg^{-1} of NaCl would deposit 10 mg Kg^{-1} of salt for each 100 mm of rainfall/year. Climatic conditions of an area also have a potential role in the soil salinization. The high temperature and low rainfall, in the arid and semiarid areas, result in the upward movement of the soil water. Salts present in the soil solution get accumulated by evaporation of water and cannot be leached downward (Fig. 4.3). Excess rainfall sometimes causes an increase in the level of groundwater; consequently, salts present in the deeper layer move upward and get deposited in root zone. Due to high temperature and precipitation, salts present in the root zone form a crust on the soil surface and increase soil salinity.

4.4.2 *Man-Made Factors Contributing to Soil Salinity*

Man-made causes of soil salinization are usually related to irrigation. Figure 4.4 shows the extent of human-induced salinization in the different parts of the world. Most common man-made factors contributing to salinization are the irrigational water and the use of chemical fertilizers and pesticides. Irrigation is an important factor in the crop productivity. Therefore, the use of high salt-containing water

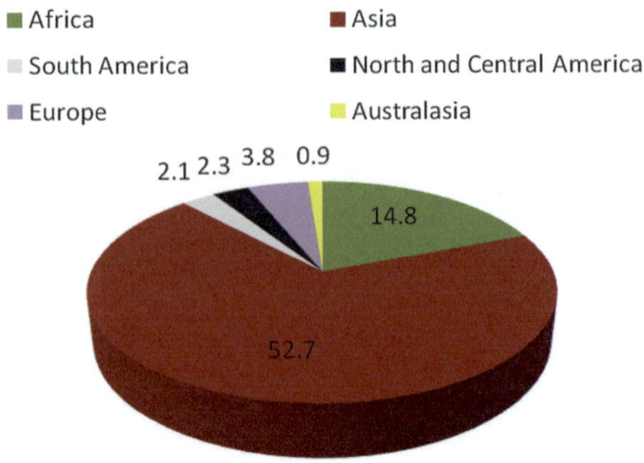

Fig. 4.4 Extent of human-induced soil salinization in the different parts of the world (area in Mha) (Source: FAO (1999) ftp://ftp.fao.org/agl/agll/docs/misc23.pdf)

(Khara pani) for crop irrigation and poor management practices for appropriate leaching of salts lead to the accumulation of salts and deteriorate the crop fields (Zhu 2007). Flood irrigation is another factor responsible for salinization. When a crop field is irrigated, water percolates downward and increases the level of water table that accumulates dissolved salts in the plant root zone. These salts remain present in the soil, as the standing crops absorb root zone water, therefore contributing to soil salinization. The increased temperature often exaggerates this problem because evaporation of water from the soil triggers precipitation of certain ions such as Ca^{+2} and Mg^{+2} and leaves behind Na^{+} in the soil. The increasing world population is putting high pressure on the crop fields to meet the world food demand. The continuous application of chemical fertilizers and pesticides to increase the crop productivity has been a potential source of increasing salt concentrations in soil (Kapoor et al. 2013). In European countries, NaCl is commonly used for the removal of snow and frost from the roadside (Bothe 2012). These salts reach plants either through horizontal flow (by soil) or vertical flow (by air) and reduce plant health and survival. On the other hand, increasing urbanization in both developed and developing countries have led to the problem of deforestation. Estimates indicate that every year 18 million acres of the forest cover is removed. Overgrazing-induced desertification of land and mismanagement of industrial wastewater discharge are other problems contributing to soil salinization.

4.5 Strategies to Overcome the Problem of Salinity

As discussed earlier in this chapter, the earth is a saline planet. High salt concentration is a threat to soil fertility and crop production as well. Estimation indicates that by the middle of the twenty-first century, salinity may affect 50% of the arable land (Wang et al. 2003) that necessitates increase in the food production for the growing world population (Alexandratos and Bruinsma 2012). There are two possible strategies to overcome the problem of salt stress and food insecurity (Epstein et al. 1980). First is to carry out the massive reclamation program, and second is the exploitation and development of salt-tolerant plant species. For the implementation of the second approach, we need to explore germplasm for the salt-tolerant traits/genes and their incorporation through conventional breeding and/or genetic engineering in the cultivated crops. However, the conventional breeding is a time-consuming and labor-intensive approach. Further, this technique has certain drawbacks, like lack of availability of salt-tolerant germplasm, reproductive barrier, transfer of undesirable genes along with the desired genes, etc. (Chinnusamy et al. 2005). Therefore, very limited success has been obtained in developing salt-tolerant plants through conventional breeding program (Ashraf 1994; Flowers 2004).

According to Greenway and Munns (1980), the mechanism of salt tolerance is dependent on two different factors: one is osmotic effect and the other is ionic effect. Success of the salt-tolerant breeding procedure depends on the knowledge of the nature of inheritance pattern and heritability of salt-tolerant genes. The development of salt-tolerant variety requires donor for salt-tolerant genes, knowledge of the inheritance of the trait at a particular development stage, and an effective screening procedure (Ashraf and Foolad 2012). Many wild species of the crop show higher tolerance rate than the cultivated ones. Efforts have been made to develop salt tolerance in wheat and rice in India (Misra et al. 1996; CSSRI 1999) and the International Rice Research Institute (IRRI), Philippines (Dedolph and Hettel 1997; Senadhira et al. 2002). Indeed, tetraploid wheat with low Na^+ accumulation has been demonstrated by Munns et al. (2000).

Though the earlier work on abiotic stress tolerance involved screening of resistance germplasm, nowadays hybridization is a more desirable method for incorporating resistance in the commercial varieties. Lath et al. (2004) transferred the genes for salinity from tetraploid wild rice *Porteresia coarctata* to cultivated variety of rice *Oryza sativa* using the bridge-crossing technique. Similarly, most of the useful characters for salinity tolerance were transferred from the *Aegilops* to cultivated wheat by Cox (1990) and Cox et al. (1991). Salt-tolerant genes were successfully transferred from wheat grass to the cultivated wheat (Dvorak and Ross 1986; Dvorak et al. 1985). *Elytrigia pontica* (wheat grass) showed exclusion of Na^+ and Cl^- uptake under salt stress condition (Shannon 1978).

Therefore, attention has been paid on the transgenic approach as an alternative to convention breeding program. Despite different ethical issues, the cultivation of transgenic plants has been accepted by many countries like the USA, China, Brazil, Argentina, and Canada along with the conventional crop production (ISAAA 2006). For the process of genetic transformation, the first requirement is the identification of the genes involved in salinity tolerance. Different organisms such as yeast, *Mesembryanthemum crystallinum*, *Arabidopsis thaliana* (glycophyte), and tomato have been used to identify the genes for salt tolerance. In yeast, salt tolerance mechanism is similar to plant; therefore, a number of genes have been identified controlling the process (Toone and Jones 1998; Serrano et al. 1999). Different salt-tolerant mutants have also been developed for the identification of genes having role in salt tolerance mechanism (Brewster et al. 1993; Mendoza et al. 1994). Plant protein homologous to the stress signaling pathway in yeast has been identified by different researchers (Popping et al. 1996; Ichimura et al. 1998; Urao et al. 1999; Piao et al. 1999; Lippuner et al. 1996). *Mesembryanthemum crystallinum* being a halophyte has been used as a model plant to study salt tolerance mechanism. *Arabidopsis* has also been used as a model plant for the identification of salt-tolerant genes (Zhu 2000), and a salt overly sensitive (*SOS1*) gene that encodes Na^+/H^+ antiporter has been identified in this plant (Shi et al. 2000). Numerous reports have shown that the genes isolated from these organisms could successfully be used to develop transgenic plants by either transferring these genes to susceptible plant or altering the expression of existing genes. These include genes for osmolytes, oxidative stress, ion homeostasis, transcription factors, DNA helicases, and chaperons. Salt stress is a multigenic trait; hence, genes transferred for salt stress tolerance may tempt tolerance against drought and cold stresses.

The implementation of such approaches may solve the salt stress problem to a larger extent, but such techniques are expensive and beyond the approach of common farmers. Moreover, acceptance of the transgenic products by the society is remaining a matter of concern in various countries, including India. However, an eco-friendly, low-cost, and efficient approach to overcome the problem of salinity stress could be the utilization of beneficial microbes such as arbuscular mycorrhizal fungi to enhance the salt tolerance in crop plants.

4.6 Arbuscular Mycorrhizal Fungi in Alleviation of Salinity Stress

The high concentration of salts in soil may induce three types of stresses, osmotic, ionic, and oxidative, which drastically affect plant growth and productivity. Osmotic stress leads to altered water potential, thereby reducing the water use efficiency of plant due to induction of physiological drought conditions. Whereas ionic stress causes disruption of ion homoeostasis at both cellular and whole-plant levels, oxidative stress elicits release of reactive oxygen species, which inhibit cell growth and plant metabolism. Experiments carried out to understand AMF-salinity interaction revealed that mycorrhizal fungi reduce negative effects of these stresses and promote plant growth (Zuccarini and Okurowska 2008; Evelin et al. 2009; Wu et al. 2006, 2010a, b). It has been widely accepted that AMF improve water use efficiency and nutrient uptake of plant under saline condition, thus helping in reducing negative impact of salt stress. AMF diminish detrimental effects of toxic ions on membrane permeability and cell organelle, maintain the level of compatible organic solutes and increase antioxidant production (both enzymatic and nonenzymatic), and positively control expression of salt-related genes. In the recent reviews, researchers have presented several physiological, biochemical, and molecular approaches by which AM plants could alleviate salt stress (Evelin et al. 2009; Ruiz-Lozano et al. 2012) (Fig. 4.5). These are (i) increased accumulation of osmolytes; (ii) control over ion uptake by roots, compartmentation of ions, and their transport into plant tissues to maintain ion homeostasis; (iii) increased uptake of water and its

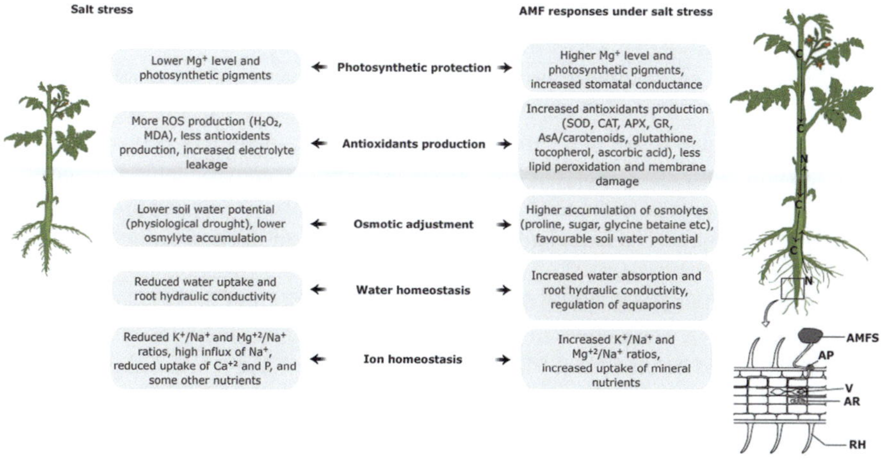

Fig. 4.5 Mycorrhizal and nonmycorrhizal plants response under salt stress. Under salt stress condition, mycorrhizal plant (*right*) shows improved plant growth and yield by improving water and nutrient uptake, mitigating injurious effects of ROS, controlling lipid peroxidation and cell membrane damage, accumulating osmolytes, and maintaining a favorable soil water potential and water and ion homeostasis. On the other hand, nonmycorrhizal plant (*left*) is deprived of basic requirements of water and nutrients, faces the problem of decreased (more negative) soil water potential, accumulates more ROS, and produces less antioxidants, therefore performing poor growth and productivity under salt stress. *AMFS* arbuscular mycorrhizal fungi spores, *AP* appressorium, *V* vesicles, *AR* arbuscules, *RH* root hairs

distribution to plant tissues with the help of aquaporins; (iv) enhanced production of antioxidants, which control oxidative damage; (v) selective buildup or exclusion of salts; (vi) managing adequate rate of photosynthesis for better plant growth, (vii) maintaining membrane structure and integrity; (viii) regulating phytohormone synthesis (Turkan and Demiral 2009); and (ix) controlling ultrastructure damage (Evelin et al. 2013). These strategies seem to develop integrated responses in a concerted manner to improve plant salinity tolerance.

4.6.1 Improved Growth and Nutritional Balance

Increased concentration of Na^+ and Cl^- in the soil solution competes with the uptake of vital ions such as Ca^{+2}, P, K^+, Mg^{+2}, and N and alters the ideal salt ratios in the soil solution, therefore affecting plant nutrient acquisition and restricting plant growth and biomass. In a study, Ullah et al. (1994) reported that irrigation of tomato plants with seawater increased the uptake of sodium chloride and decreased the uptake of P and Fe. The high concentrations of sodium chloride ions in a saline soil lowered the concentration of potassium ions in sugar beet (Ghoulam et al. 2002). It becomes increasingly difficult for transporter proteins to distinguish between Na^+ and K^+, when the ratio of these ions is in the same proportion in the soil solution, indeed affecting the plant's uptake of K^+ from the soil, consequently lowering accumulation of K^+, and affecting cell turgor pressure, membrane potential, and activity of most of the enzymes of plants (Blumwald 2000).

Plant growth and biomass production is an integrative measurement of plant response to the stress conditions; therefore, the symbiotic efficiency of AM fungi has been measured in terms of plant growth or biomass accumulation (Evelin et al. 2009; Ruiz-Lozano et al. 2012; Porcel et al. 2016). Several researchers demonstrated that under salt stress plants colonized by mycorrhizal fungi grow better and produce more biomass than nonmycorrhizal plants (Al-Karaki 2000; Cantrell and Linderman 2001; Giri et al. 2007; Sannazzaro et al. 2007; Zuccarini and Okurowska 2008, Shokri and Maadi 2009; Wu et al. 2010b, Evelin et al. 2011; Abdel Latef and Chaoxing 2011; Porcel et al. 2016). The improved growth of mycorrhizal plants under salt stress has been suggested to be attributed to better nutrient uptake by mycorrhizal plants (Giri et al. 2003; Garg and Manchanda 2009; Evelin et al. 2011; Kumar and Sharma 2011). Besides, mycorrhizal fungi modify morphogenetic characters of roots under salt stress. They alter meristematic activity of root apices and promote formation of lateral roots (Berta et al. 1990; Giri et al. 2003). These modifications in root morphology could influence acquisition of mineral nutrients and water use efficiency of host plant growing in a saline soil. Moreover, AM symbiosis influences various morphological parameters such as plant height, leaf area, root density, and fresh and dry plant weight under saline conditions (Campanelli et al. 2013).

Increased absorption of P is believed to be the major contribution of the mycorrhizal fungi to improve plant growth under salt stress; however, other metabolic processes that are mediated by other nutrients like N, K, and Mg seem to be involved in alleviating deleterious effects of salinity (Giri and Mukerji 2004; Evelin et al. 2009; Evelin et al. 2012). Giri et al. (2007) examined the effect of AMF *Glomus fasciculatum* and salinity on the growth of *Acacia nilotica*, under different salinity levels. They reported that soil salinity decreased root and shoot dry biomass, but regardless of salt concentrations mycorrhizal inoculation of acacia plants showed a positive impact on plant growth and biomass accumulation. Inoculation of rice plants with AM fungal strain *Claroideoglomus etunicatum* obtained from saline soils (Estrada et al. 2013) increased shoot dry biomass production at all levels of salinity (Porcel et al. 2016). Porras-Soriano and coworkers (2009) applied inoculum of *Glomus mosseae, Glomus intraradices*, or *Glomus claroideum* to olive plants grown in the nonsaline as well as saline media. They observed a significant increase in the plant growth and uptake of N, P, and K. Abdel Latef and Chaoxing (2011) reported higher accumulation of dry biomass in case of shoot tissues than root tissues that might be attributed to the greater allocation of a portion of photosynthate to the shoot than root tissues in mycorrhizal plants. Experiments conducted by Khalil et al. (2011) on mycorrhizal fungi and citrus seedlings showed significantly higher plant growth under salinity stress. AMF inoculation significantly increased the uptake of macronutrients P, K, and Mg and micronutrients Cu, Fe, Mn, and Zn in fenugreek plants under different concentrations of NaCl (Giri et al. 2003, 2007; Giri and Mukerji 2004; Evelin et al. 2012). Hajiboland et al. (2010) observed that in spite of negative effects of excessive salts, the inoculation of tomato cultivars with AMF showed considerably higher growth and biomass production that substantiate the findings of previous studies (Al-Karaki 2000, Sannazzaro et al. 2007; Zuccarini and Okurowska 2008).

4.6.2 Maintenance of Water Homeostasis

Under saline condition, plants face the problem of water acquirement, as the salinity lowers water potential; consequently plant growth is declined. Therefore, maintaining water homeostasis and the functioning of photosynthetic apparatus are essential for alleviating the impact of salinity on plant growth and crop production. Soil salinity reduces root hydraulic conductivity of plants (Martínez-Ballesta et al. 2003); however, exact mechanism by which salinity reduces hydraulic conductivity of roots is not clear; someway it is corroborated with the functioning of aquaporins or their abundance in the membrane (Ruiz-Lozano et al. 2012). Under stress conditions, inoculation with mycorrhiza often improves plant water status due to the contribution of extraradical hyphae for absorbing more water, and change in root morphology and soil structure (Augé 2001). In addition, mycorrhizal fungi alter the concentrations of several key compatible organic solutes involving in the improved plant osmotic adjustment (Auge et al. 2014). Mycorrhizal fungi largely alter root hydraulic conductance under salinity stress (Aroca et al. 2007). The colonization by AMF increases active solute transport as a mechanism to continue water flow across the plant roots. Mycorrhizal plants are known to better uptake of K^+ under salt stress as compared to nonmycorrhizal plants, therefore contributing to greater root hydraulic conductivity and improved water status under salinity stress (El-Mesbahi et al. 2012; Auge et al. 2014). The regulation of root hydraulic properties by AMF can be strongly correlated with the regulation of aquaporin abundance and the regulation of PIP genes under salinity stress (Aroca et al. 2007). Mycorrhizal fungi alter the accumulation of numerous plant growth regulators; one of them is ABA, which is involved in regulating leaf gas exchange (Dodd 2003). Although meager information is available on the mechanisms by which AMF regulate root expression of specific aquaporin, it has been suggested that AMF-mediated alteration of abscisic acid (ABA) may be responsible for maintaining water status of plant under saline conditions, because root hydraulic conductivity and aquaporins are both affected by the ABA (Aroca et al. 2006). Therefore, it is proposed that ABA could be the signal molecule that differentially regulates the root hydraulic conductivity and aquaporins in AMF and non-AMF plants under salinity stress (Porcel et al. 2006; Ruiz-Lozano and Aroca 2010; Dodd and Pérez-Alfocea 2012).

4.6.3 Maintenance of Ion Homeostasis and Mitigation of Deleterious Effect of Ion Toxicity

In saline soil, plants are often exposed to the toxicity of specific ions such as sodium and chloride, which damage cell organelles and disrupt metabolism (Porcel et al. 2016). Although the plant cell contains semipermeable membrane, Na^+ and Cl^- largely enter across the membrane under high salt condition. The accumulation of Na^+ in the chloroplast reduces plant photosynthesis; in the cell cytoplasm affects the integrity of cell membrane and dissipates the membrane potential, inhibits the enzymatic activity, and eventually affects overall plant growth (Tuteja 2007). Therefore, Na^+ uptake and distribution within the plant are major determinants for glycophytes. The typical Na^+ ion toxicity involves scorching and burning-like symptoms along the edges of the leaf. Na^+ ion toxicity symptoms appear first along the edges, eventually progress toward the midvein, and finally reach the center of the leaf. Plants have evolved several intrinsic processes to cope with excessive level of salts, especially Na^+, such as the control of the entry of Na^+ into the root, sequestration into the vacuole, and translocation and allocation within the leaf, therefore combating adverse effects of a saline environment (Ruiz-Lozano et al. 2012). Four types of transporters, which are involved in the Na^+ and K^+ homeostasis, have been reported in *Arabidopsis thaliana* (Turkan and Demiral 2009). These are:

(i) HKT-type transporter: It involves in the transport of Na^+ from the root to the shoot tissues and maintains K^+/Na^+ ratios in the root (Munns 2005).
(ii) AKT-type transporter: It provides a channel for K^+ with higher selectivity for K^+ as compared to Na^+ (Munns 2005).
(iii) NHX-type transporter: It is a Na^+/H^+ antiporter located in the vacuole and is expressed in root and leaf tissues and retains Na^+ into the vacuole (Munns 2005).
(iv) SOS1-type transporter: It is a Na^+/H^+ antiporter involved in the Na^+ expulsion from the cell and may also participate in Na^+ extrusion from the root to the external medium (Munns 2005).

In the previous decades, studies have shown that AMF alleviate the detrimental effect of salinity by reducing uptake of toxic ions, favoring Na^+ extrusion from the cytoplasm and sequestration into the vacuole, unloading Na^+ from the xylem, and recirculation of Na^+ from photosynthetic organs to root under saline condition (Giri et al. 2003; Porcel et al. 2016). Porcel et al. (2016) reported increased accumulation of Na^+ in rice root than shoot tissues due to a decreased root-to-shoot distribution of Na^+ and suggested to be one of the key strategies for alleviating salinity stress and maintaining plant growth under salt conditions. These findings corroborate with the results of previous studies (Giri et al. 2003, 2007; Sharifi et al. 2007). The study carried out by Giri et al. (2007) has shown increased accumulation of Na^+ in root and reduced translocation of Na^+ in shoot tissues of AMF-colonized *Acacia nilotica*, regardless of the salinity levels. Further, the reduced accumulation of Na^+ and higher Mg^{+2} has been suggested to maintain increased chlorophyll concentration in mycorrhizal than nonmycorrhizal plants (Giri et al. 2003). Preventing excess uptake of Na^+, mycorrhizal fungi impart a regulatory effect on its translocation to shoot tissues, therefore maintaining a lower Na^+ shoot-to-root ratios in mycorrhizal than nonmycorrhizal plants (Evelin et al. 2009, 2013).

Potassium is one of the most important essential elements for plant growth as it plays a key role in plant metabolism. It represents 2–10% of the plant dry biomass. Potassium participates in various vital processes of plants such as opening and closing of stomata, maintains osmotic balance and plasma membrane polarization, acts as a cofactor for many enzymes, is involved in the biosynthesis of proteins, and participates in the binding of tRNA to the ribosome and adaptation to environmental stresses such as drought and salinity (Blaha et al. 2000; Wang and Wu 2013; Anschütz et al. 2014; Shabala and Pottosin 2014); thus, functions of K^+ are irreplaceable by Na^+, which necessitates maintaining optimal ratios of K^+/Na^+ in plant tissues for the uninterrupted functioning of vital physiological processes (Giri et al. 2007; Evelin et al. 2009; Ruiz-Lozano et al. 2012; Benito and Gonzalez-Guerrero 2014; Shin and Adams 2014). Studies carried out by Giri et al. (2003, 2007) have shown increased accumulation of K^+ and reduced translocation of Na^+ in shoot tissues of AMF-colonized acacia plants grown in a saline soil. A significant decrease in the leaf Na^+ concentration and increase in the leaf K^+ and Mg^{+2} and Ca^{+2} concentrations in AMF-inoculated citrus seedlings have also been reported by Wu et al. (2010a, b). They noted that mycorrhizal fungi (*Glomus mosseae* and *Paraglomus occultum*) improved the ratios of K^+/Na^+, Ca^{+2}/Na^+, and Mg^{+2}/Na^+ in citrus seedlings growing under saline condition. The improved K^+/Na^+ ratios in root and shoot tissues of mycorrhizal plants indicated AMF involvement in protecting disruption of K-mediated enzymatic processes and maintaining ion homoeostasis (Evelin et al. 2012), and increased Mg^{+2}/Na^+ revealed AMF contribution for the better regulation of photosynthesis under salinity stress. The higher ratios of Ca^{2+}/Na^+ and Ca^{2+}/Mg^{2+} in mycorrhizal fenugreek plant confirms mitigation of NaCl-induced ionic imbalance (Evelin et al. (2012). Hammer et al. (2011) observed impact of AMF on the selective uptake of Ca^{2+} and Na^+, which promotes higher plant Ca^{2+}/Na^+ ratios. Besides ion homeostasis, the increased AMF-mediated accumulation of K^+ ions also contribute to greater root hydraulic conductivity and improved water status under salinity stress (El-Mesbahi et al. 2012; Auge et al. 2014).

4.6.4 Regulation of Photosynthesis Process

Soil salinity reduces photosynthetic ability of plant as it affects photosynthetic apparatus and concentration of plant's photosynthetic pigments (Doganlar et al. 2010; Ayala-Astorga and Alcaraz-Melendez 2010; Perveen et al. 2010; Akram and Ashraf 2011), suppresses precursors of chlorophyll biosynthesis (Santos et al. 2001; Santos 2004), and lowers uptake of mineral nutrient like magnesium (Giri and Mukerji 2004; Murkute et al. 2006; Sheng et al. 2008). Photosystem II (PS II) is known to play a central role in photosynthesis; however, salinity alters structure and activity of PS II (Al-Taweel et al. 2007; Abdeshahian et al. 2010; Saleem et al. 2011) and largely affects the reaction center (Ruiz-Lozano et al. 2012). Plant failure to dissipate the excessive energy may distress the photosynthetic machinery (Ruiz-Lozano et al. 2012). Under stress conditions, the photosynthetic activity of the plant is often affected due to the alteration in the level and the activity of various photosynthetic enzymes such as rubisco (Aragao et al. 2005; Valentine et al. 2006), sucrose phosphate synthase and fructose-1,6-bisphosphatase (Seemann and Sharkey 1982; Ghosh et al. 2001), and PEP carboxylase (Abdel-Latif 2008). It is pertinent to corroborate that mycorrhizal plants exhibit higher photosynthetic rates than nonmycorrhizal plants under salinity stress; however, to the best of our knowledge, there is no direct report available about the alteration in the activity of rubisco in AM plants under salinity stress, and therefore this aspect deserves attention of researchers.

The influence of AMF on photosynthesis has been reported in many mycorrhizal plants growing under salinity stress (Sheng et al. 2008; Zuccarini and Okurowska 2008; Hajiboland et al. 2010; Wu et al. 2010a, b). Mycorrhizal fungi (*Rhizophagus intraradices, Claroideoglomus etunicatum*, and *Septoglomus constrictum*) isolated from the rhizosphere of *Asteriscus maritimus* (Cabo de Gata Natural Park, Almería, Spain) show improved performance of photosystem II and higher stomatal conductance when subjected to different levels of salinity, indicating that AMF minimize salinity-induced damage in the photosynthetic machinery and enhanced transpiration rates in maize plants (Estrada et al. 2013), which substantiate the results of previous studies conducted by Sheng et al. (2008), Hajiboland et al. (2010), and Querejeta et al. (2006). While studying salinity-AMF interaction on maize, Sheng et al. (2008) found improved photosynthetic capacity of maize plants by increasing the capacity of gaseous exchange and the efficiency of PS II and regulating the energy flow between photochemical and non-photochemical reactions. Wu et al. (2010a, b) observed increased photosynthetic rates and stomatal conductance in AMF compared to non-AMF plants under salinity stress. The increased rate of photosynthesis in AMF-colonized plants under salinity stress has been correlated with the lower intercellular CO_2 concentration in mycorrhizal plants, since the higher photosynthetic capacity increases water use efficiency for the assimilation of more carbon per unit water transpiration (Nandy et al. 2007; Sheng et al. 2008). The improved net assimilation rates by protecting photochemical machinery of photosystem II and elevating stomatal conductance under salinity stress are believed to be ascribed for salt stress tolerance in AMF plants (Estrada et al. 2013). Plants growing in a saline soil face the problem of the availability of atmospheric CO_2 and its fixation due to the salinity-induced closure of stomata and reduced consumption of NADPH by the Calvin cycle (Ruiz-Lozano et al. 2012).

4.6.5 Maintenance of Osmotic Homeostasis

Under saline conditions, one of the major problems that plants often face is the decreased (more negative) soil water potential. Plant fails to acquire adequate amount of the water from the soil under such conditions. Therefore, it is essential for the plants to adjust their water potential in order to maintain a favorable gradient of water flow from the soil to the plant roots to keep plant away from the cells dryness. To deal with such problems, plants accumulate active organic solutes like amino acids, amides, proteins (nitrogen-containing compounds), betaines (ammonium compounds), and polyamines (small organic cations) (Rabie and Almadini 2005), therefore maintaining osmotic homoeostasis. However, the accumulation of compatible organic solutes in plants gets affected due to the high concentration of salts in soil. AMF modify the composition and abundance of compatible organic solutes such as soluble sugars (Porcel et al. 2003; Sheng et al. 2011), nonreducing disaccharide, trehalose (Ocón et al. 2007), proline (Ruiz-Lozano and Azcón 1995), betaines (Al-Garni 2006), and polyamines (Sannazzaro et al. 2007; Campanelli et al. 2013; Evelin et al. 2013), therefore helping in maintaining osmotic homeostasis under saline condition. Readers may refer to the recently published reviews (Evelin et al. 2009; Ruiz-Lozano et al. 2012; Kapoor et al. 2013) for a detailed discussion on this aspect.

4.6.6 Mitigation of Oxidative Damage/Stress

Oxidative damage occurs due to the production of reactive oxygen species (ROS) and the poor response of plant to counteract or detoxify harmful effects of free radicals through neutralization by antioxidants. These reactive molecules are produced as by-products during the chloroplast and mitochondrial electron transport and have potential to cause a number of deleterious effects (Turrens 2003; Evelin et al. 2009). Free radicals have the tendency to interact with cell components such as DNA, proteins, and lipids and destabilize them in the absence of protective mechanisms (Evelin et al. 2009). Indeed, ROS have a role in cell signaling including apoptosis, gene expression, and the activation of cell signaling cascades (Miller et al. 2010). The production of ROS is generally low under favorable growth conditions; however, their production increases dramatically under salinity or drought stresses (Miller et al. 2010). Elevated levels of ROS under saline conditions may be due to excessive accumulation of salts, which not only affects different pathways in plant metabolism but also the cell membrane integrity and permeability (Evelin et al. 2012), subsequently inducing a variety of reactive oxygen species such as superoxide anion radical (O_2^-), hydrogen peroxide (H_2O_2), and hydroxyl radical ($^\bullet OH$). Mittler et al. (2004) demonstrated that the accumulation of ROS largely depends on the balance between ROS synthesis and ROS scavenging during stress condition; however, an imbalance between accumulation and consumption of assimilatory power ($NADPH_2$ and ATP) triggers ROS production during oxidative stress (Ruiz-Lozano et al. 2012). The overreduction of CO_2 fixation in chloroplasts and electron transport chain in both chloroplast and mitochondria and photorespiration are the main causes of ROS generation during stress condition (Mittler et al. 2004). For the detoxification of ROS both under normal and stress conditions, higher plants develop an efficient antioxidant system as they produce antioxidant molecules and enzymes to protect them from oxidative damage (Wu et al. 2006), but the production rates of these compounds noticeably decreased under salinity stress (Evelin et al. 2009). Antioxidant system includes ROS scavenging enzymes such as superoxide dismutase (SOD), ascorbate peroxidase (APX), catalase (CAT) and glutathione peroxidase (GPX), glutathione reductase (GR) (Alguacil et al. 2003; Ghorbanli et al. 2004; Yamane et al. 2004; Wu et al. 2006), dehydroascorbate reductase (Ghorbanli et al. 2004; Wu et al. 2006), monodehydroascorbate reductase (Ghorbanli et al. 2004), guaiacol peroxidase, and oxidized glutathione (Wu et al. 2006) and the enzymes involved in the ascorbate-glutathione cycle (Alguacil et al. 2003).

In addition, some nonenzymatic antioxidant molecules such as ascorbic acid (AsA) and glutathione (GSH) (Mittler et al. 2004; Miller et al. 2010; Dietz et al. 2006), carotenoids, and tocopherols also scavenge activated oxygen species (Alguacil et al. 2003; Wu et al. 2006; Evelin et al. 2009). AsA plays an important role in antioxidant network and is involved in several developmental processes of plant (cell division, expansion of cell wall) (Pignocchi and Foyer 2003). AsA-GSH cycle is vital for plant resistance under stress conditions (Anjum et al. 2011), and glutathione detoxifies the excessive H_2O_2, thereby controlling ROS level in plants (Rausch et al. 2007).

Several studies have demonstrated that AM association can modulate plant physiological and biochemical processes in order to cope with environmental stresses like salinity and drought. Although available reports showed crash of AMF on enzymatic and nonenzymatic accumulation in host plants, the exact mechanism by which AMF influence the ROS metabolism of host plant is not well known. Researchers demonstrated a correlation between salt stress tolerance and antioxidant capacity in several plant species (Benavides et al. 2000; Nunez et al. 2003), suggesting that plants with higher accumulation of antioxidant molecules have better resistance against osmotic stresses (Jiang and Zhang 2002; Evelin et al. 2009; Ruiz-Lozano et al. 2012). AMF facilitate plant to modulate salinity stress by increasing the activities of antioxidant enzymes such as SOD, CAT, GR, and POX or by increasing the concentration of nonenzymatic antioxidant molecules such as ascorbate and glutathione (Alguacil et al. 2003; Porcel et al. 2003; Porcel and Ruiz-Lozano 2004; Garg and Manchanda 2009; Ruiz-Sanchez et al. 2010; Wu et al. 2010a, b; Talaat and Shawky 2011). Inoculation with AMF has shown higher activities of SOD, peroxidase, and APOX in mycorrhizal soybean (Ghorbanli et al. 2004) and CAT, APOX, and SOD in mycorrhizal *Olea europaea* and *Retama sphaerocarpa*; however, no change was recorded in the activities of CAT and polyphenol peroxidase (Alguacil et al. 2003).

Considering duration of salinity treatments an important factor, Evelin and Kapoor (2014) studied the impact of AMF-salinity interaction on the activity of antioxidant enzymes and the accumulation of nonenzymatic antioxidants after 1 and 14 days of salinity treatment in roots and leaves of fenugreek plants growing in saline condition. The study demonstrated that lipid peroxidation and H_2O_2 accumulation increase with increasing severity and duration of salinity; however, the degree of damage attenuated in AMF as compared to non-AMF plants, which may be attributed to the enhanced activity of antioxidant enzymes and the greater accumulation of antioxidant compounds. Study revealed that the AMF influence on antioxidant production varies among plant tissues, salinity level, and duration of salinity treatment. The impact of AMF on antioxidative enzyme production was more apparent at the initial stage of salinity treatment (1 day) in both plant tissues; however, the accumulation of antioxidant compounds increased at the later stage of salinity treatment (14 days). The study revealed that AMF improve antioxidative defense system of plant, hence accelerating rapid dismutation of reactive oxygen species (O_2^- to H_2O_2) and subsequent prevention of H_2O_2 accumulation through improved activities of antioxidants. Recently, Sarwat et al. (2016) studied salinity-AMF interaction in *Brassica juncea* plants. The study demonstrated that AMF-inoculated plants exhibit higher ROS scavenging capacity that could be attributed to the increased activities of antioxidant enzymes. The increased activity of antioxidant enzymes in AMF plants with less accumulation of MDA content reduced oxidative damage in tomato plant under saline condition (Abdel Latef and Chaoxing 2011). Indeed, mycorrhizal fungi possess various oxidative stress-related genes (*SOD* genes), which upon upregulation enable mycorrhizal plants to tolerate environmental stresses like drought and salinity (Wu et al. 2010a, b; Hajiboland et al. 2012; Ruiz-Lozano et al. 2001; Huang et al. 2014).

4.6.7 Control Over Ultrastructural Changes

High level of soluble salts in the soil may result in the alteration of cell wall thickness, increase in the frequency of plasmodesmata, vacuolization of cytoplasm, shrinkage of protoplasm, widening of apoplastic space between cell wall and cell membrane, disruption of thylakoid and thylakoid membrane, swelling and reduction in the number of thylakoids, disintegration of chloroplast membrane, accumulation of plastoglobules, dilation of cristae and denser matrix in mitochondria and dissipation of mitochondrial membrane potential, aggregation of chromatin in nucleus, and induction of ovule abortion, thereby hampering the plant physiology and metabolism (Yamane et al. 2004; Sun et al. 2004; Hauser et al. 2006; Mahmoodzadeh 2008; Evelin et al. 2009, 2013). Under saline condition, these alterations are probably due to Na^+ and Cl^- toxicity and osmotic imbalance. In fact, excessive salts exert detrimental effect on cell membrane and organelles; unfortunately, there has been scarcity of data understanding the role of AMF in plants in this aspect of salinity stress, except a report that has recently been published by Evelin et al. (2013). Experiments conducted on *Trigonella foenum-graecum* plants with *G. intraradices* showed that salinity stress impairs ultrastructures; however, the extent of salt-induced ultrastructural damage was lesser in mycorrhizal plants. The lower lipid peroxidation and electrolyte leakage in mycorrhizal plants substantiated the fact that AMF prevent membrane damage. It was suggested that the lesser ultrastructural damage in mycorrhizal plants may be due to increased osmolyte and polyamine accumulation and more and bigger plastoglobule formation in mycorrhizal than nonmycorrhizal plants. On the other hand, lesser Na^+ and Cl^- accumulation assures less ionic toxicity, and higher osmolytes and tocopherols ensure osmotic adjustment and better free radical scavenging capacity of mycorrhizal plants under salinity stress. To confirm and better understand the role of AMF in mitigation of injurious effects of salts on cell ultrastructures under saline condition, this aspect needs more experimentation and attention of researchers.

4.6.8 Molecular Approaches for Mitigating Salinity Stress

As discussed previously in this chapter, AMF play a positive role in improving plant salinity stress tolerance. Although salinity is a complex trait and is difficult to manipulate, little success has been obtained in transferring salt-tolerant genes to plant. However, some reports show the successful transfer of salt-tolerant genes to AM-colonized plants. The whole strategy revolves around the synthesis and accumulation of compatible organic solutes (osmolytes), ion homeostasis, and water use efficiency of plant. In a few studies, the role of ABA and late embryogenesis abundant (LEA) protein has also been discussed. Increased salt tolerance has been studied in transgenic tobacco through overexpression of pyrroline-5-carboxylate reductase (P5CR) gene (Kishor et al. 1995). P5CR along with pyrroline-5-carboxylate synthetase (P5CS) has been found to be involved in the synthesis of proline from glutamic acid (Delauney and Verma 1993; Abraham et al. 2003). In contrary, some studies show that the accumulation of proline in AM plant is due to salt stress instead of mycorrhizal association (Wang et al. 2004; Jahromi et al. 2008; Sheng et al. 2011). There are also reports on the accumulation of osmoprotectant, glycine betaine (Robinson and Jones 1986; Evelin et al. 2009), which has a role in protecting oxygen evolving photosystem II and enzyme rubisco at elevated salt concentration (Murata et al. 1992). Glycine betaine is synthesized by the action of two enzymes choline dehydrogenase (COD) and betaine aldehyde dehydrogenase (BADH) (Weigel et al. 1986). The BADH gene has been isolated and characterized in *Atriplex* (Xiao et al. 1995), *Beta vulgaris* (McCue and Hanson 1992), *Sorghum bicolor* (Wood et al. 1996), and *Spinacia oleracea* (Weretilnyk and Hanson 1990). Although several reports indicate the successful transformation of this gene in plants (Jia et al. 2002; Kumar et al. 2004; Holmstrom et al. 2000; Zhang et al. 2011), the transformation studies in AM-colonized plants are yet to be examined.

The adverse effect of salinity on plant growth is mainly due to the uptake of Na^+ which damages the internal organelle and disrupts plant metabolism (Evelin et al. 2013). To mitigate the adverse effects of excessive salts, plants have evolved different transporters to carry out the accumulation of Na^+ in the vacuole, thereby limiting Na^+ entry and its distribution in the underground and areal parts of the plant (Ruiz-Lozano et al. 2012). These transporters include Na^+/H^+ antiporter NHX, plasma membrane Na^+/K^+ antiporter SOS1, and Na^+/K^+ antiporter HKT (Porcel et al. 2016). Genes for these antiporters may serve as a potential candidate to overcome salinity stress in plants (Asins et al. 2013). Moreover, many reports in the literature suggest that under saline condition, AM plants generally show a higher K^+/Na^+ ratio than non-AM plants (Al-Karaki and Hammad 2001; Rabie and Almadini 2005; Sannazzaro et al. 2007); however, the molecular mechanisms for maintaining a favorable K^+/Na^+ ratio in plant under saline condition are still not well evaluated. Indeed, Ouziad et al. (2006) studied the effect of two Na^+/H^+ antiporter genes involved in maintaining ion homeostasis and aquaporin synthesis in mycorrhizal tomato plants. They reported that AM symbiosis did not alter the expression of antiporter genes *LeNHX1* and *LeNHX2*. Nevertheless, the expression of three genes, namely, *ZmAKT2*, *ZmSOS1*, and *ZmSKOR*, maintaining ion homeostasis under saline condition was studied in mycorrhizal maize plants inoculated with three AMF isolated from saline environment (Estrada et al. 2013). *ZmSOS1* gene was upregulated at 100 mM NaCl in AM plants but downregulated with increasing salinity in non-AM plant. They reasoned that the posttranscriptional and posttranslational changes act as the regulators for such genes. This study revealed that the protective effect of AMF on maize plants under saline condition was mediated by improved K^+ retention in the plant tissues due to upregulation of *ZmAKT2* and *ZmSKOR* (Estrada et al. 2013). Recently, Porcel et al. (2016) showed that rice plants inoculated with AM fungus *Claroideoglomus etunicatum* showed a lower distribution of Na^+ from root to shoot tissues. This lower root to shoot distribution of Na^+ ions could be due to increased regulation of rice transporter genes *OsNHX3*, *OsSOS1*, *OsHKT2;1*, and *OsHKT1;5* involved in maintaining ion homeostasis by mycorrhizal fungi. They further suggested that the expression of these genes triggers Na^+ efflux from the cytoplasm and xylem and promotes Na^+ sequestration into the vacuole.

Recently, the cyclic nucleotide-gated ion channel (CNGC) genes have also been suggested to play an important role in the amelioration of various environmental stresses in plant (Clough et al. 2000; Balagué et al. 2003; Gobert et al. 2006; Ma et al. 2006; Yoshioka et al. 2006; Ruiz Lozano et al. 2012; Porcel et al. 2012). These nonselective cation channels are found to be involved in the uptake of Na^+, K^+, and Ca^{++} (Kaplan et al. 2007) and reallocation of Na^+ within the plant tissues. Therefore, it is tempting to speculate that genes for CNGCs could alleviate the toxic effects of excessive salts in plants growing under saline condition (Porcel et al. 2012; Kugler et al. 2009). Although it is evident from the literature that under saline condition, AM plants maintain more favorable K^+/Na^+ ratio than non-AM plant, the molecular mechanisms involved in maintaining favorable ionic balance are not well documented. Therefore, attention is required to identify possible regulation of genes determining HKT, AKT, SOS, and NHX ion transporters and most likely CNGCs in AMF-colonized plant growing under salt stress (Ruiz Lozano et al. 2012).

Besides these ion channels, AMF also enhance the functioning of water channel protein – the aquaporins (Roussel et al. 1997; Krajinski et al. 2000). These proteins belong to the family of major intrinsic protein (MIP) and have a role in passive movement of water molecules following a water potential gradient and maintain the cellular osmoregularity (Maurel et al. 2002; Kjelbom et al. 1999). In addition to water, aquaporins can also transport low molecular weight compounds such as NH_4, urea, glycerol, and CO_2 (Flexas et al. 2006; Katsuhara et al. 2008; Maurel et al. 2008; Maurel et al. 2009). These proteins are subdivided into five distinct subfamilies (Ruiz Lozano et al. 2012):

(i) Plasma membrane intrinsic proteins (PIPs)
(ii) Tonoplast intrinsic proteins (TIPs)
(iii) Small basic intrinsic proteins (SIPs)
(iv) Nodulin 26-like intrinsic proteins (NIPs)
(v) Uncharacterized X intrinsic proteins (XIPs)

Aroca et al. (2007) studied the differential expression of three aquaporin genes. They observed that under saline condition the gene expression was upregulated in mycorrhizal as compared to nonmycorrhizal *Phaseolus vulgaris* plants. The AMF *Glomus mosseae*-colonizing *Glycine max* has been reported to downregulate the PIP gene expression under water stress (Porcel et al. 2006). Jahromi et al. (2008) conducted experiments on *Lactuca sativa* and studied the role of AMF in controlling the expression of aquaporin gene *LsPIP1* under saline condition. They observed that the *LsPIP1* activity enhances under saline condition but suppresses under non-saline condition, which could be ascribed to the complex and multigenic nature of the salinity.

Studies have shown that the ABA-dependent signaling controls the expression of aquaporin genes in plant (Jang et al. 2004; Hose et al. 2000; Ruiz-Lozano et al. 2009). Hose et al. (2000) demonstrated that ABA has an important role in increasing water conductivity and improving water permeability of maize cortical cells. Ruiz-Lozano et al. (2009) demonstrated that the application of exogenous ABA enhanced root hydraulic conductivity (L) in all plants, regardless of water conditions, while mycorrhizal plants showed reduced root hydraulic conductivity than nonmycorrhizal plants. This effect was further correlated with the accumulation pattern of different PIPs in plant. Close (1996) reported another desiccation tolerance plant protein – the late embryogenesis abundant (LEA) protein, which noticeably accumulated under stress conditions. LEA protein is synthesized during the seed stage under normal condition, while under poor water condition it accumulates and serves as a molecular chaperon and maintains the structure of the other protein; however, the expression of LEA in AM plants under saline condition is yet to be determined.

4.7 Concluding Remarks and Future Perspective

Overall, soil salinity is one of the most severe abiotic stresses affecting plant establishment, growth, and production worldwide. Salinity stress induces osmotic, ionic, oxidative, and nutritional disorders, therefore altering the major metabolic processes of plants. Sodium chloride is the most prevalent salt in the soil environment, which lowers water potential of the soil solution around the root, causing the osmotic stress that subsequently induces ionic stress due to the excessive accumulation of salts under saline condition (Deinlein et al. 2014; Gupta and Huang 2014; Wei et al. 2016). Mycorrhizal association has been shown to improve plant growth and yield under salinity stress and seems to be in particular advantageous to host plants subjected to the osmotic stress, induced by excessive accumulation of soluble salts (Auge et al. 2014). Under salinity stress, mycorrhizal fungi have been found to regulate several mechanisms such as the accumulation of osmolytes (compatible organic solutes), control of water and ionic homeostasis, maintenance of photosynthetic processes, reduction in the oxidative damage, and control over ultrastructure alteration; indeed, many aspects of these physiological and biochemical mechanisms remain unknown at molecular level and might be due to the complexity of this trait. Therefore, identification of the salt stress-responsive genes involved in the production of antioxidants and enzymes controlling the synthesis of various osmoregulators and involved in photosynthesis (Rubisco) is a promising field and needs to be addressed in future studies. Know-how of such genes will lead to better understanding of the process and will provide further insights confirming the involvement of this intimate association in improving plant salt stress tolerance.

Acknowledgment The authors duly acknowledge the technical support provided by Mr. Harsh Sharma, Institute of Life Long Learning, University of Delhi, Delhi, India.

References

Abdel Latef AA, Chaoxing H (2011) Effect of arbuscular mycorrhizal fungi on growth, mineral nutrition, antioxidant enzymes activity and fruit yield of tomato grown under salinity stress. Sci Hortic 127:228–233

Abdel-Latif A (2008) Phosphoenolpyruvate carboxylase activity of wheat and maize seedlings subjected to salt stress. Aust J Basic Appl Sci 2:37–41

Abdeshahian M, Nabipour M, Meskarbashee M (2010) Chlorophyll fluorescence as criterion for the diagnosis salt stress in wheat (*Triticum aestivum*) plants. Int J Chem Biol Eng 4:184–186

Abraham E, Rigo G, Szekely G et al (2003) Light dependent induction of proline biosynthesis by abscisic acid and salt stress is inhibited by brassinosteriods in Arabidopsis. Plant Mol Biol 51:363–372

Akram MS, Ashraf M (2011) Exogenous application of potassium dihydrogen phosphate can alleviate the adverse effects of salt stress on sunflower (*Helianthus annuus* L.) J Plant Nutr 34:1041–1057

Alexandratos N, Bruinsma J (2012) World agriculture towards 2030/2050: the 2012 Revision, ESA Working Paper, 12–03. FAO, Rome

Al-Garni SMS (2006) Increasing NaCl-salt tolerance of a halophytic plant *Phragmites australis* by mycorrhizal symbiosis. Am Eurasian J Agric Environ Sci 1:119–126

Alguacil MM, Hernandez JA, Caravaca F et al (2003) Antioxidant enzyme activities in shoots from three mycorrhizal shrub species afforested in a degraded semi-arid soil. Physiol Plant 118:562–570

Aliasgharzadeh N, Saleh Rastin N, Towfighi H et al (2001) Occurrence of arbuscular mycorrhizal fungi in saline soils of the Tabriz plain of Iran in relation to some physical and chemical properties of soil. Mycorrhiza 11:119–122

Al-Karaki GN (2000) Growth of mycorrhizal tomato and mineral acquisition under salt stress. Mycorrhiza 10:51–54

Al-Karaki GN, Hammad R (2001) Mycorrhizal influence on fruit yield and mineral content of tomato grown under salt stress. J Plant Nutr 24:1311–1323

Al-Taweel K, Iwaki T, Yabuta Y et al (2007) A bacterial transgene for catalase protects translation of D1 protein during exposure of salt-stressed tobacco leaves to strong light. Plant Physiol 145:258–265

Anjum SA, Xie XY, Wang LC et al (2011) Morphological, physiological and biochemical responses of plants to drought stress. Afr J Agric Res 6(9):2026–2032

Anschütz U, Becker D, Shabala S (2014) Going beyond nutrition: regulation of potassium homoeostasis as a common denominator of plant adaptive responses to environment. J Plant Physiol 171:670–687

Aragao MEF, Guedes MM, Otoch MLO et al (2005) Differential responses of ribulose-1,5-bisphosphate carboxylase/oxygenase activities of two *Vigna unguiculata* cultivars to salt stress. Braz J Plant Physiol 17:207–212

Aroca R, Ferrante A, Vernieri P et al (2006) Drought, abscisic acid and transpiration rate effects on the regulation of PIP aquaporin gene expression and abundance in *Phaseolus vulgaris* plants. Ann Bot 98(6):1301–1310

Aroca R, Porcel R, Ruiz-Lozano JM (2007) How does arbuscular mycorrhizal symbiosis regulate root hydraulic properties and plasma membrane aquaporins in *Phaseolus vulgaris* under drought, cold or salinity stresses? New Phytol 173:808–816

Aroca R, Ruiz-Lozano JM, Zamarreño AM et al (2013) Arbuscular mycorrhizal symbiosis influences strigolactone production under salinity and alleviates salt stress in lettuce plants. J Plant Physiol 170:47–55

Ashraf M (1994) Breeding for salinity tolerance in plants. Crit Rev Plant Sci 13:17–42

Ashraf M, Foolad MR (2012) Crop breeding for salt tolerance in the era of molecular markers and marker-assisted selection. Plant Breed 132:1–20

Asins MJ, Villalta I, Aly MM et al (2013) Two closely linked tomato HKT coding genes are positional candidates for the major tomato QTL involved in Na+/K+ homeostasis. Plant Cell Environ 36(6):1171–1191

Augé RM (2001) Water relations, drought and VA mycorrhizal symbiosis. Mycorrhiza 11:3–42

Auge RM, Toler HD, Saxton AM et al (2014) Arbuscular mycorrhizal symbiosis and osmotic adjustment in response to NaCl stress: a meta-analysis. Front Plant Sci 5:562–574

Ayala-Astorga GI, Alcaraz-Melendez L (2010) Salinity effects on protein content, lipid peroxidation, pigments, and praline in *Paulownia imperialis* (Siebold and Zaccarini) and *Paulownia fortunei* (Seemann and Hemsley) plant grown in vitro. Electron J Biomed 13:1–15

Azcón-Aguilar C, Azcón R, Barea JM (1979) Endomycorrhizal fungi and *Rhizobium* as biological fertilizers for *Medicago sativa* in normal cultivation. Nature 279:325–327

Balagué C, Lin BQ, Alcon C et al (2003) HLM1, an essential signaling component in the hypersensitive response, is a member of the cyclic nucleotide-gated channel ion channel family. Plant Cell 15:365–379

Benavides MP, Marconi PL, Gallego SM et al (2000) Relationship between antioxidant defence systems and salt tolerance in *Solanum tuberosum*. Aust J Plant Physiol 27:273–278

Benito B, Gonzalez-Guerrero M (2014) Unravelling potassium nutrition in ectomycorrhizal associations. New Phytol 201:707–709

Berta G, Fusconi A, Trotta A et al (1990) Morphogenetic modifications induced by the mycorrhizal fungus *Glomus* strain E3 in the root system of *Allium porrum*. New Phytol 114:207–215

Blaha G, Stelzl U, Spahn CMT et al (2000) Preparation of functional ribosomal complexes and effect of buffer conditions on tRNA positions observed by cryoelectron microscopy. Methods Enzymol 317:292–309

Blumwald E (2000) Sodium transport and salt tolerance in plants. Curr Opin Cell Biol 12:431–434

Bojorquez-Quintal J, Velarde A, Ku A et al (2014) Mechanisms of salt tolerance in habanero pepper plants (*Capsicum chinense* Jacq.): proline accumulation, ions dynamics, root-shoot partition and compartmentation. Front. Plant Sci 5:605

Bothe H (2012) Arbuscular mycorrhiza and salt tolerance of plants. Symbiosis 58:7–16

Brewster JL, de Valoir T, Dwyer ND et al (1993) An osmosensing signal transduction pathway in yeast. Science 259:1760–1763

Campanelli A, Ruta C, DeMastro G et al (2013) The role of arbuscular mycorrhizal fungi in alleviating salt stress in *Medicago sativa* L. car. Icon. Symbiosis 59:65–76

Cantrell IC, Linderman RG (2001) Preinoculation of lettuce and onion with VA mycorrhizal fungi reduces deleterious effects of soil salinity. Plant Soil 233:269–281

Chinnusamy V, Jagendorf A, Zhu JK (2005) Understanding and improving salt tolerance in plants. Crop Sci 45:437–448

Close TJ (1996) Dehydrins: emergence of a biochemical role of a family of plant dehydration proteins. Physiol Plant 97:795–803

Clough SJ, Fengler KA, Yu IC et al (2000) The Arabidopsis dnd1 "defense, no death" gene encodes a mutated cyclic nucleotide-gated ion channel. Proc Natl Acad Sci U S A 97:9323–9328

Cox JA (1990) The role of intracellular calcium binding proteins. In: Dedman JR, Smith VL (eds) Stimulus response coupling. CRC Press, Boca Raton, pp 266–269

Cox TS, Harrel LG, Chen P et al (1991) Reproductive behaviour of hexaploid/diploid wheat hybrids. Plant Breed 107:105–118

CSSRI (1999) Central Soil Salinity Research Institute, Karnal, India. Annual report. pp 41–52

Dedolph C, Hettel G (eds) (1997) Rice varieties boost yield and improve saline soils. Partners making a difference. IRRI, Manila, p 37
Deinlein U, Stephan AB, Horie T et al (2014) Plant salt-tolerance mechanisms. Trends Plant Sci 19:371–379
Delauney AJ, Verma DPS (1993) Proline biosynthesis and osmoregulation in plants. Plant J 4:215–223
Dietz KJ, Jacob S, Oelze ML et al (2006) The function of peroxiredoxins in plant organelle redox metabolism. J Exp Bot 57:1697–1709
Dodd IC (2003) Hormonal interactions and stomatal responses. J Plant Growth Regul 22(1):32–46
Dodd IC, Pérez-Alfocea F (2012) Microbial amelioration of crop salinity stress. J Exp Bot 63(9):3415–3428
Doganlar ZB, Demir K, Basak H et al (2010) Effects of salt stress on pigment and total soluble protein contents of three different tomato cultivars. Afr J Agric Res 5:2056–2065
Dvorak J, Ross K (1986) Expression of tolerance of Na^+, K^+, Mg^+, Cl^- and SO_4^+ ions and sea water in the amphiploid of *Triticum aestivum*×*Elytrigia elongata*. Crop Sci 26:658–660
Dvorak J, Kathleen R, Mendlinger S (1985) Transfer of salt tolerance from *Elytrigia pontica* to wheat by the addition of an incomplete Elytrigia genome. Crop Sci 25:306–309
El-Mesbahi MN, Azcón R, Ruiz-Lozano JM et al (2012) Plant potassium content modifies the effects of arbuscular mycorrhizal symbiosis on root hydraulic properties in maize plants. Mycorrhiza 22(7):555–564
Epstein E, Norlyn JD, Rush DW et al (1980) Saline culture of crops: a genetic approach. Science 210:399–404
Estrada B, Aroca MFJM et al (2013) Arbuscular mycorrhizal fungi native from a Mediterranean saline area enhance maize tolerance to salinity through improved ion homeostasis. Plant Cell Environ 36:1771–1782
Evelin H, Kapoor R (2014) Arbuscular mycorrhizal symbiosis modulates antioxidant response in salt stressed *Trigonella foenum-graecum* plants. Mycorrhiza 24:197–208
Evelin H, Kapoor R, Giri B (2009) Arbuscular mycorrhizal fungi in alleviation of salt stress: a review. Ann Bot 104:1263–1280
Evelin H, Giri B, Kapoor R (2011) Contribution of Glomus intraradices inoculation to nutrient acquisition and mitigation of ionic imbalance in NaCl-stressed *Trigonella foenum-graecum*. Mycorrhiza 22:1–15
Evelin H, Giri B, Kapoor R (2012) Contribution of *Glomus intraradices* inoculation to nutrient acquisition and mitigation of ionic imbalance in NaCl-stressed *Trigonella foenum-graecum*. Mycorrhiza 22:203–217
Evelin H, Giri B, Kapoor R (2013) Ultrastructural evidence for AMF mediated salt stress mitigation in *Trigonella foenum-graecum*. Mycorrhiza 23(1):71–86
FAO (1999) Integrated soil management for sustainable agriculture and food security in southern and east Africa. In: Nabhan H, Mashali AM, Mermut AR (eds) Proceedings of the expert consultation Harare, Zimbabwe, 8–12 December 1997
Feng G, Zhang FS, XI L et al (2002) Improved tolerance of maize plants to salt stress by arbuscular mycorrhiza is related to higher accumulation of soluble sugars in roots. Mycorrhiza 12:185–190
Flexas J, Ribas Carbó M, Hanson DT et al (2006) Tobacco aquaporin NtAQP1 is involved in mesophyll conductance to CO_2 *in vivo*. Plant J 48:427–439
Flowers TJ (2004) Improving crop salt tolerance. J Exp Bot 55:307–319
Garg N, Manchanda G (2008) Effect of arbuscular mycorrhizal inoculation of salt-induced nodule senescence in *Cajanus cajan* (pigeonpea). J Plant Growth Regul 27:115–124
Garg N, Manchanda G (2009) Role of arbuscular mycorrhizae in the alleviation of ionic, osmotic and oxidative stresses induced by salinity in *Cajanus cajan* (L.) millsp. (pigeonpea). J Agron Crop Sci 195:110–123

Ghorbanli M, Ebrahimzadeh H, Sharifi M (2004) Effects of NaCl and mycorrhizal fungi on antioxidative enzymes in soybean. Biol Plant 48:575–581

Ghosh S, Bagchi S, Majumder AL (2001) Chloroplast fructose- 1,6-bisphosphatase from *Oryza* differs in salt tolerance property from the *Porteresia* enzyme and is protected by osmolytes. Plant Sci 160:1171–1181

Ghoulam C, Foursy A, Fares K (2002) Effects of salt stress on growth, inorganic ions and proline accumulation in relation to osmotic adjustment in five sugar beet cultivars. Environ Exp Bot 47:39–50

Giri B, Mukerji KG (2004) Mycorrhizal inoculant alleviates salt stress in *Sesbania aegyptiaca* and *Sesbania grandiflora* under field conditions: evidence for reduced sodium and improved magnesium uptake. Mycorrhiza 14:307–312

Giri B, Kapoor R, Mukerji KG (2003) Influence of arbuscular mycorrhizal fungi and salinity on growth, biomass and mineral nutrition of *Acacia auriculiformis*. Biol Fertil Soils 38:170–175

Giri B, Kapoor R, Mukerji KG (2007) Improved tolerance of *Acacia nilotica* to salt stress by arbuscular mycorrhiza, *Glomus fasciculatum*, may be partly related to elevated K+/Na+ ratios in root and shoot tissues. Microb Ecol 54:753–760

Gobert A, Park G, Amtmann A et al (2006) *Arabidopsis thaliana* Cyclic nucleotide-gated channel 3 forms a non-selective ion transporter involved in germination and cation transport. J Exp Bot 57:791–800

Greenway H, Munns R (1980) Mechanisms of salt tolerance in nonhalophytes. Annu Rev Plant Physiol 31:149–190

Gupta B, Huang B (2014) Mechanism of salinity tolerance in plants: physiological, biochemical, and molecular characterization. Int J Genom 2014:1–18

Hajiboland R, Aliasgharzadeh N, Laiegh SF et al (2010) Colonization with arbuscular mycorrhizal fungi improves salinity tolerance of tomato (*Solanum lycopersicum* L.) plants. Plant Soil 331:313–327

Hajiboland R, Aliasgharzadeh N, Laiegh SF et al (2012) Colonization with arbuscular mycorrhizal fungi improves salinity tolerance of tomato (*Solanum lycopersicon* L.) plants. Plant Soil 331:313–327

Hammer EC, Nasr H, Pallon J et al (2011) Elemental composition of arbuscular mycorrhizal fungi at high salinity. Mycorrhiza 21:117–129

Hauser BA, Oppenheimer KSDG, Sage TL (2006) Changes in mitochondrial membrane potential and accumulation of reactive oxygen species precede ultra structural changes during ovule abortion. Planta 223:492–499

Holmstrom KO, Somersalo S, Mandal A et al (2000) Improved tolerance to salinity and low temperature in transgenic tobacco producing glycine betaine. J Exp Bot 51:177–185

Hose E, Steudle E, Hartung W (2000) Abscisic acid and hydraulic conductivity of maize roots: a study using cell- and root-pressure probes. Planta 211:874–882

Hu Y, Schmidhalter U (2002) Limitation of salt stress to plant growth. In: Hock B, Elstner CF (eds) Plant toxicology. Marcel Dekker Inc, New York, pp 91–224

Huang YM, Srivastava AK, Zou YN et al (2014) Mycorrhizal-induced calmodulin mediated changes in antioxidant enzymes and growth response of drought-stressed trifoliate orange. Front Microbiol 5:682. doi:10.3389/fmicb.2014.00682

Ichimura K, Mizoguchi T, Irie K et al (1998) Isolation of ATMEKK1 (a MAP kinase kinase kinase)-interacting proteins and analysis of a MAP kinase cascade in *Arabidopsis*. Biochem Biophys Res Commun 253:532–543

ISAAA (2006) International service for the acquisition of agri-biotech applications. Global Status of Commercialized Biotech/GM Crops

Jahromi F, Aroca R, Porcel R et al (2008) Influence of salinity on the *in vitro* development of *Glomus intraradices* and on the *in vivo* physiological and molecular responses of mycorrhizal lettuce plants. Microb Ecol 55:45–53

Jang JY, Kim DG, Kim YO et al (2004) An expression analysis of a gene family encoding plasma membrane aquaporins in response to abiotic stresses in *Arabidopsis thaliana*. Plant Mol Biol 54:713–725

Jia GX, Zhu ZQ, Chang FQ et al (2002) Transformation of tomato with the BADH gene from *Atriplex* improves salt tolerance. Plant Cell Rep 21:141–146

Jiang M, Zhang J (2002) Water stress-induced abscisic acid accumulation triggers the increased generation of reactive oxygen species and up-regulates the activities of antioxidant enzymes in maize-leaves. J Exp Bot 53:2401–2410

Kaplan ST, Fromm H et al (2007) Cyclic nucleotide-gated channels in plants. FEBS Lett 581:2237–2246

Kapoor R, Evelin H, Mathur P et al (2013) Arbuscular mycorrhiza: approaches for abiotic stress tolerance in crop plants for sustainable agriculture. In: Tuteja N, Gill SS (eds) Plant acclimation to environmental stress. Springer, New York, pp 359–401

Katsuhara M, Hanba YT, Shiratake K et al (2008) Expanding roles of plant aquaporins in plasma membranes and cells organelles. Funct Plant Biol 35:1–14

Khalil HA, Eissa AM, El-Shazly SM et al (2011) Improved growth of salinity-stressed citrus after inoculation with mycorrhizal fungi. Sci Hortic 130(3):624–632

Kishor PB, Hong Z, Miao GH et al (1995) Overexpression of D1-pyrroline-5-carboxilate synthetase increases proline production and confers osmotolerance in transgenic plants. Plant Physiol 108:1387–1394

Kjelbom P, Larsson C, Johansson I, Karlsson M, Johanson U (1999) Aquaporins and water homeostasis in plants. Trends Plant Sci 4:308–314

Krajinski F, Biela A, Schubert D et al (2000) Arbuscular mycorrhiza development regulates the mRNA abundance of Mtaqp1 encoding a mercury-insensitive aquaporin of *Medicago truncatula*. Planta 211:85–90

Kugler A, Köhler B, Palme K et al (2009) Salt-dependent regulation of a CNG channel subfamily in Arabidopsis. BMC Plant Biol 9:140

Kumar A, Sharma S (2011) Non-edible oil seeds as biodiesel feedstock for meeting energy demands in India. Renew Sust Energ Rev 15:1791–1800

Kumar S, Dhingra A, Daniell H (2004) Plastid expressed b*etaine aldehyde dehydrogenase* gene in carrot cultured cells, roots, and leaves confers enhanced salt tolerance. Plant Physiol 136(1):2843–2854

Kumar A, Sharma S, Mishra S (2010) Influence of arbuscular mycorrhizal (AM) fungi and salinity on seedling growth, solute accumulation and mycorrhizal dependency of *Jatropha curcas* L. J Plant Growth Regul 29:297–306

Kumar A, Dames JF, Gupta A (2014) Current developments in arbuscular mycorrhizal fungi research and its role in salinity stress alleviation: a biotechnological perspective. Crit Rev Biotechnol:1–14

Lath R, Srinivas Rao C, Subramaniam HMS et al (2004) Approaches to breeding for salinity tolerance-a case study on *Porteresia coarctata*. Ann Appl Biol 144:177–184

Lippuner V, Cyert MS, Gasser CS (1996) Two classes of plant cDNA clones differentially complement yeast calcineurin mutants and increase salt tolerance of wild-type yeast. J Biol Chem 271:12859–12866

Ma W, Ali R, Berkowitz GA (2006) Characterization of plant phenotypes associated with loss-of-function of AtCNGC1, a plant cyclic nucleotide gated cation channel. Plant Physiol Biochem 44:494–505

Maathuis FJM, Ahmad I, Patishtan J (2014) Regulation of Na^+ fluxes in plants. Front Plant Sci 5:467

Mahajan S, Tuteja N (2005) Cold, salinity and drought stresses: an overview. Arch Biochem Biophys 444(2):139–158

Mahmoodzadeh H (2008) Ultrastructural changes in shoot apical meristem of canola (*Brassica napus* cv. Symbol) treated with sodium chloride. Pak J Biol Sci 11:1161–1164

Martínez-Ballesta MC, Aparicio F, Pallás V et al (2003) Influence of saline stress on root hydraulic conductance and PIP expression in *Arabidopsis*. J Plant Physiol 160:689–697

Maurel C, Javot H, Lauvergeat V et al (2002) Molecular physiology of aquaporins in plants. Int Rev Cytol 215:105–148

Maurel C, Verdoucq L, Luu DT et al (2008) Plant aquaporins: membrane channels with multiple integrated functions. Annu Rev Plant Biol 59:595–624

Maurel C, Santoni V, Luu DT et al (2009) The cellular dynamics of plant aquaporin expression and functions. Curr Opin Plant Biol 12:690–698

McCue KF, Hanson AD (1992) Salt-inducible betaine aldehyde dehydrogenase from sugar beet: cDNA cloning and expression. Plant Mol Biol 18:1–11

Mendoza I, Rubio F, Rodriguez-Navarro A et al (1994) The protein phosphatase calcineurin is essential for NaCl tolerance of *Saccharomyces cerevisiae*. J Biol Chem 269:8792–8796

Miller G, Suzuki N, Ciftci-Yilmaz S et al (2010) Reactive oxygen species homeostasis and signalling during drought and salinity stresses. Plant Cell Environ:33453–33467

Misra AN, Murmu B, Singh P et al (1996) Growth and proline accumulation in mung bean seedlings as affected by sodium chloride. Biol Plant 38:531–536

Mittler R, Vanderauwera S, Gollery M et al (2004) The reactive oxygen gene network in plants. Trends Plant Sci 9:490–498

Munns R (2005) Genes and salt tolerance: bringing them together. New Phytol 167:645–663

Munns R, Tester M (2008) Mechanisms of salinity tolerance. Annu Rev Plant Biol 59:651–681

Munns R, Guo J, Passioura JB et al (2000) Leaf water status controls day-time but not daily rates of leaf expansion in salt-treated barley. Aust J Plant Physiol 27:949–957

Murata N, Mohanty PS, Hayashi H et al (1992) Glycinebetaine stabilizes the association of extrinsic proteins with the photosynthetic oxygen-evolving complex. FEBS Lett 296:187–189

Murkute AA, Sharma S, Singh SK (2006) Studies on salt stress tolerance of citrus rootstock genotypes with arbuscular mycorrhizal fungi. Hortic Sci 33:70–76

Nandy P, Das S, Ghose M (2007) Effects of salinity on photosynthesis, leaf anatomy, ion accumulation and photosynthetic nitrogen use efficiency in five Indian mangroves. Wetl Ecol Manag 15:347–357

Nunez M, Mazzafera P, Mazorra LM et al (2003) Influence of a brassinosteroid analogue on antioxidant enzymes in rice grown in culture medium with NaCl. Biol Plant 47:67–70

Ocón A, Hampp R, Requena N (2007) Trehalose turnover during abiotic stress in arbuscular mycorrhizal fungi. New Phytol 174:879–891

Ojala JC, Jarrell WM, Menge JA et al (1983) Influence of mycorrhizal fungi on the mineral nutrition and yield of onion in saline soil. Agron J 75:255–259

Ouziad F, Wilde P, Schmelzer E et al (2006) Analysis of expression of aquaporins and Na+/H+ transporters in tomato colonized by arbuscular mycorrhizal fungi and affected by salt stress. Environ Exp Bot 57:177–186

Perveen S, Shahbaz M, Ashraf M (2010) Regulation in gas exchange and quantum yield of photosystem II (PSII) in salt-stressed and non-stressed wheat plants raised from seed treated with triacontanol. Pak J Bot 42:3073–3081

Piao HL, Pih KT, Lim JH et al (1999) An *Arabidopsis GSK3/shaggy*-like gene that complements yeast salt stress-sensitive mutants is induced by NaCl and abscisic acid. Plant Physiol 119:1527–1534

Pignocchi C, Foyer CH (2003) Apoplastic ascorbate metabolism and its role in the regulation of cell signalling. COPLBI 6:379–389

Popping B, Gibbons T, Watson MD (1996) The *Pisum sativum* MAP kinase homologue (*PsMAPK*) rescues the *Saccharomyces cerevisiae* hog1 deletion mutant under conditions of high osmotic stress. Plant Mol Biol 31:355–363

Porcel R, Ruiz-Lozano JM (2004) Arbuscular mycorrhizal influence on leaf water potential, solute accumulation and oxidative stress in soybean plants subjected to drought stress. J Exp Bot 55:1743–1750

Porcel R, Barea JM, Ruiz-Lozano JM (2003) Antioxidative activities in mycorrhizal soybean plants under drought stress and their possible relationship to the process of nodule senescence. New Phytol 157:135–143

Porcel R, Aroca R, Azcon R et al (2006) PIP aquaporin gene expression in arbuscular mycorrhizal *Glycine max* and *Lactuca sativa* plants in relation to drought stress tolerance. Plant Mol Biol 60(3):389–404

Porcel R, Aroca R, Ruíz-Lozano JM (2012) Salinity stress alleviation using arbuscular mycorrhizal fungi – a review. Agron Sustain Dev 32:181–200

Porcel R, Aroca R, Azcon R et al (2016) Regulation of cation transporter genes by the arbuscular mycorrhizal symbiosis in rice plants subjected to salinity suggests improved salt tolerance due to reduced Na^+ root-to-shoot distribution. Mycorrhiza 26:673

Porras-Soriano A, Soriano-Martin ML, Porras-Piedra A et al (2009) Arbuscular mycorrhizal fungi increased growth, nutrient uptake and tolerance to salinity in olive trees under nursery conditions. J Plant Physiol 166(13):1350–1359

Querejeta JI, Allen MF, Caravaca FA et al (2006) Differential modulation of host plant $\delta^{13}C$ and $\delta^{18}O$ by native and nonnative arbuscular mycorrhizal fungi in a semiarid environment. New Phytol 169:379–387

Rabie GH, Almadini AM (2005) Role of bioinoculants in development of salt-tolerance of *Vicia faba* plants under salinity stress. Afr J Biotechnol 4:210–222

Rausch T, Gromes R, Liedschulte V et al (2007) Novel insight into the regulation of GSH biosynthesis in higher plants. Plant Biol 9:565–572

Richards LA (1954) Diagnosis and improvement of saline and alkaline soils. United States Salinity Laboratory Staff. Agricultural handbook no 60. United States Department of Agriculture, 160 p

Robinson SP, Jones GP (1986) Accumulation of glycinebetaine in chloroplasts provides osmotic adjustment during salt stress. Aust J Plant Physiol 13:659–668

Roussel H, Bruns S, Gianinazzi-Pearson V et al (1997) Induction of a membrane intrinsic protein-encoding mRNA in arbuscular mycorrhiza and elicitor stimulated cell suspension cultures of parsley. Plant Sci 126:203–210

Ruiz-Lozano JM (2003) Arbuscular mycorrhizal symbiosis and alleviation of osmotic stress: new perspectives for molecular studies. Mycorrhiza 13:309–317

Ruiz-Lozano JM, Aroca R (2010) Modulation of aquaporin genes by the arbuscular mycorrhizal symbiosis in relation to osmotic stress tolerance. In: Symbioses and stress. Springer, Dordrecht, pp 357–374

Ruiz-Lozano JM, Azcón R (1995) Hyphal contribution to water uptake in mycorrhizal plants as affected by the fungal species and water status. Physiol Plant 95:472–478

Ruiz-Lozano JM, Azcón R (2000) Symbiotic efficiency and infectivity of an autochthonous arbuscular mycorrhizal *Glomus* sp from saline soils and *Glomus deserticola* under salinity. Mycorrhiza 10:137–143

Ruiz-Lozano JM, Azcón R, Goméz M (1996) Alleviation of salt stress by arbuscular-mycorrhizal *Glomus* species in *Lactuca sativa* plants. Physiol Plant 98:767–772

Ruiz-Lozano JM, Collados C, Barea JM et al (2001) Arbuscular mycorrhizal symbiosis can alleviate drought-induced nodule senescence in soybean plants. New Phytol 151:493–502

Ruiz-Lozano JM, Alguacil MM, Barzana G et al (2009) Exogenous ABA accentuates the differences in root hydraulic properties between mycorrhizal and non mycorrhizal maize plants through regulation of PIP aquaporins. Plant Mol Biol 70:565–579

Ruiz-Lozano JM, Porcel R, Azcón C, Aroca R (2012) Regulation by arbuscular mycorrhizae of the integrated physiological response to salinity in plants: new challenges in physiological and molecular studies. J Exp Bot 63(11):4033–4044

Ruiz-Sanchez M, Aroca R, Munoz Y et al (2010) The arbuscular mycorrhizal symbiosis enhances the photosynthetic efficiency and the antioxidative response of rice plants subjected to drought stress. J Plant Physiol 167:862–869

Saleem A, Ashraf M, Akram NA (2011) Salt (NaCl) induced modulation in some key physio-biochemical attributes in okra (*Abelmoschus esculentus* L.) J Agron Crop Sci 197:202–213

Sannazzaro AI, Echeverria M, Albertó EO, Ruiz OA, Menéndez AB (2007) Modulation of polyamine balance in Lotus glaber by salinity and arbuscular mycorrhiza. Plant Physiol Biochem 45:39–46

Santos CV (2004) Regulation of chlorophyll biosynthesis and degradation by salt stress in sunflower leaves. Sci Hortic 103:93–99
Santos C, Azevedo H, Caldeira G (2001) *In situ* and *in vitro* senescence induced by KCl stress: nutritional imbalance, lipid peroxidation and antioxidant metabolism. J Exp Bot 52:351–360
Sarwat M, Hashem A, Ahanger MA et al (2016) Mitigation of NaCl stress by arbuscular mycorrhizal fungi through the modulation of osmolytes, antioxidants and secondary metabolites in mustard (*Brassica juncea* L.) plants. Front Plant Sci 7:869–873
Seemann JR, Sharkey TD (1982) The effect of abscisic acid and other inhibitors on photosynthetic capacity and the biochemistry of CO2 assimilation. Plant Physiol 84:696–700
Senadhira D, Zapata-Arias FJ, Gregorio GB et al (2002) Development of the first salt-tolerant rice cultivar through indica/indica anther culture. Field Crop Res 76:103–110
Serrano R, Culiañz-Maciá A, Moreno V (1999) Genetic engineering of salt and drought tolerance with yeast regulatory genes. Sci Hortic 78:261–269
Shabala S, Pottosin I (2014) Regulation of potassium transport in plants under hostile conditions: implications for abiotic and biotic stress tolerance. Physiol Plant 151:257–279
Shannon MC (1978) Testing salt tolerance variability among tall wheatgrass populations. Agron J 70:719–722
Sharifi M, Ghorbanli M, Ebrahimzadeh H (2007) Improved growth of salinity-stressed soybean after inoculation with pre-treated mycorrhizal fungi. J Plant Physiol 164:1144–1151
Sheng M, Tang M, Chan H et al (2008) Influence of arbuscular mycorrhizae on photosynthesis and water status of maize plants under salt stress. Mycorrhiza 18:287–296
Sheng M, Tang M, Zhang FF et al (2011) Influence of arbuscular mycorrhiza on organic solutes in maize leaves under salt stress. Mycorrhiza 21:423–430
Shi H, Ishitani M, Kim C et al (2000) The *Arabidopsis thaliana* salt tolerance gene *SOS1* encodes a putative Na^+/H^+ antiporter. Proc Natl Acad Sci U S A 97(12):6896–6901
Shin R, Adams E (2014) Transport, signalling and homeostasis of potassium and sodium in plants. J Integr Plant Biol 56:231–249
Shokri S, Maadi B (2009) Effects of arbuscular mycorrhizal fungus on the mineral nutrition and yield of *Trifolium alexandrinum* plants under salinity stress. J Agron 8:79–83
Smith SE, Read DJ (1997) Mycorrhizal symbiosis. Academic Press, San Diego
Sun K, Hunt K, Hauser BA (2004) Ovule abortion in Arabidopsis triggered by stress. Plant Physiol 135:2358–2367
Talaat NB, Shawky BT (2011) Influence of arbuscular mycorrhizae on yield, nutrients, organic solutes, and antioxidant enzymes of two wheat cultivars under salt stress. J Plant Nutr Soil Sci 174:283–291
Toone WM, Jones N (1998) Stress-activated signaling pathways in yeast. Genes Cells 3:485–498
Turkan I, Demiral T (2009) Recent developments in understanding salinity tolerance. Environ Exp Bot 67:2–9
Turrens JF (2003) Mitochondrial formation of reactive oxygen species. J Physiol 552:335–344
Tuteja N (2007) Mechanisms of high salinity tolerance in plants. Methods Enzymol 428:420–436
Ullah SM, Gerzabek MH, Soja G (1994) Effect of seawater and soil salinity on ion uptake, yield and quality of tomato (fruit). Die Bodenkultur 45:227–237
Urao T, Yakubov B, Satoh R et al (1999) A transmembrane hybrid type histidine kinase in *Arabidopsis* functions as an osmosensor. Plant Cell 11:1743–1754
Valentine AJ, Mortimer PE, Lintnaar M et al (2006) Drought responses of arbuscular mycorrhizal grapevines. Symbiosis 41(3):127–133
Wang Y, Wu WH (2013) Potassium transport and signaling in higher plants. Annu Rev Plant Biol 64(4):1–4
Wang W, Vinocur B, Altman A (2003) Plant responses to drought, salinity and extreme temperatures: towards genetic engineering for stress tolerance. Planta 218:1–14
Wang FY, Liu RJ, Lin XG et al (2004) Arbuscular mycorrhizal status of wild plants in saline-alkaline soils of the yellow river delta. Mycorrhiza 14:133–137
Wei P, Wang L, Liu A et al (2016) GmCLC1 confers enhanced salt tolerance through regulating chloride accumulation in soybean. Front Plant Sci 7:1082–1091

Weigel P, Weretilnyk EA, Hanson AD (1986) Betaine aldehyde oxidation by spinach chloroplasts. Plant Physiol 82:753–759

Weretilnyk EA, Hanson AD (1990) Molecular cloning of a plant betaine aldehyde dehydrogenase, an enzyme implicated in adaption to salinity and drought. Proc Natl Acad Sci U S A 87:2745–2749

Wood AJ, Saneola H, Rhides D et al (1996) Betaine aldehyde dehydrogenase in Sorghum. Plant Physiol 110:1301–1308

Wu DC, Ré DB, Nagai M et al (2006) The inflammatory NADPH oxidase enzyme modulates motor neuron degeneration in amyotrophic lateral sclerosis mice. Proc Natl Acad Sci U S A 103:12132–12137

Wu QS, Zon YN, Liu W et al (2010a) Alleviation of salt stress in citrus seedlings inoculated with mycorrhiza: changes in leaf antioxidant defense systems. Plant Soil Environ 56:470–475

Wu QS, Zou YN, He XH (2010b) Contributions of arbuscular mycorrhizal fungi to growth, photosynthesis, root morphology and ionic balance of citrus seedlings under salt stress. Acta Physiol Plant 32:297–304

Xiao G, Zhang GY, Liu FH et al (1995) Study on BADH gene from *Atriplex hortensis* L. Chin Sci Bull 40:741–745

Yamane K, Rahman Md S, Kawasaki M et al (2004) Pretreatment with a low concentration of methyl viologen decreases the effects of salt stress on chloroplast ultrastructure in rice leaves (*Oryza sativa* L.) Plant Prot Sci 7:435–441

Yoshioka K, Moeder W, Kang HG et al (2006) The chimeric Arabidopsis cyclic nucleotide-gated ion channel11/12 activates multiple pathogen resistance responses. Plant Cell 18:747–763

Zhang N, Si HJ, Wen G et al (2011) Enhanced drought and salinity tolerance in transgenic potato plants with a BADH gene from spinach. Plant Biotechnol Rep 5:71–77

Zhu JK (2000) Genetic analysis of plant salt tolerance using *Arabidopsis*. Plant Physiol 124(3):941–948

Zhu JK (2001) Plant salt tolerance. Trends Plant Sci 6(2):66–71

Zhu JK (2007) Plant salt stress. John Wiley and Sons Ltd

Zuccarini P, Okurowska P (2008) Effects of mycorrhizal colonization and fertilization on growth and photosynthesis of sweet basil under salt stress. J Plant Nutr 31:497–513

Chapter 5
Arbuscular Mycorrhizal Fungi and Adaption of P Stress in Plants

Bo Shu and Qiang-Sheng Wu

Abstract Phosphorus (P) is an essential macronutrient for plant growth. In general, soil P deficiency is a major constraint for plant production because of the excited forms and slow rate of diffusion of P. Several strategies have been systemically applied for plant sensing and adapting P deficiency, especially mycorrhization in most land plants. This chapter reviewed P stress in plant (stress sensing and signaling molecule) and the influence of arbuscular mycorrhizal fungi on host plant P deficiency adaptation.

Keywords P deficiency • Phosphate transporter • Signaling molecule

5.1 Introduction

Phosphorus (P) is a macronutrient that is a vital component of nucleic acids, phospholipids, and sugar phosphates in all living organisms. P is abundant in the environment, with soil P existing as organic phosphate (Po; 50%–80% of the total) or in immobilized form (50%–80% of the total) (Turner et al. 2002). However, P cannot be assimilated by plants unless it is hydrolyzed into inorganic phosphate (Pi) (Raghothama 1999). Pi concentration in soil seldom exceeds 10 μM (Bieleski 1973), and P deficiency is a major constraint to plant growth.

B. Shu (✉)
South Subtropical Crops Research Institute, Chinese Academy of Tropical Agricultural Science, Key Laboratory of Tropical Fruit Biology, Ministry of Agriculture, Zhanjiang 524091, China
e-mail: bshbest@163.com

Q.-S. Wu
College of Horticulture and Gardening, Yangtze University, Jingzhou, Hubei 434025, China

Institute of Root Biology, Yangtze University, Jingzhou, Hubei 434025, China

Department of Chemistry, Faculty of Science, University of Hradec Kralove, Hradec Kralove 50003, Czech Republic
e-mail: wuqiangsh@163.com

Q.-S. Wu (ed.), *Arbuscular Mycorrhizas and Stress Tolerance of Plants*,
DOI 10.1007/978-981-10-4115-0_5

Plants have evolved complex responsive and adaptive mechanisms for stress sensing and Pi acquisition, especially symbiosis with arbuscular mycorrhiza (AM) fungi, which represents widespread mutualistic fungi belonging to phylum *Glomeromycota*. AM fungi can allocate up to 80% phosphate of the host plant and in return obtain 30% of plant photosynthetic carbohydrates, which may be considered "fair trade" (Kiers et al. 2011; Fellbaum et al. 2012). Spatiotemporal molecular, biochemical, and physiological Pi deficiency responses developed by plants result from local and systemic sensing and signaling pathways involving Pi itself, hormones, miRNAs, and sugar that serve to coordinate Pi deficiency responses. The response of AM fungi and host plant to Pi deficiency is coevolution. In this review, we evaluate recent advances toward developing a comprehensive understanding of molecular events underlying Pi deficiency responding and adapting. Cross talk related to plant and AM fungi (stress sensing and signaling molecules, stress application) was also identified.

5.2 Phosphate Stress Sensing

The whole plant Pi level needs to be integrated via systemic sensing and signaling, but the external Pi status is sensed locally in root tips first. Roots perceive fluctuations in extracellular nutrient levels and send signals to the shoot, via the xylem, as a warning of impending limitation in the supply of the particular nutrient. Shoots sense the root-derived nutrient signals and send signals both to the shoot and roots, via the phloem, to adjust developmental processes and nutrient uptake (Lough and Lucas 2006; Liu et al. 2009; Lucas et al. 2013).

It has been proposed that Pi is sensed by root-localized mechanisms, and currently, there are two ways by which plants are thought to sense Pi availability in the rhizosphere: external Pi concentration changes are sensed by a root cell membrane-localized sensor, or internal nutrient status is sensed by an intracellular sensor (Forde and Lorenzo 2001; Chiou and Lin 2011; Nagarajan and Smith 2012). The root tips are the site where Pi deficiency is sensed (Svistoonoff et al. 2007). When the root tip encounters a region of low Pi, a primary signal or stimulus (most likely Pi concentration in the apoplasm of the root tip) is perceived by a plasma membrane-localized sensor. Alternatively, an internal Pi deficiency signal (such as Pi concentration in the cytoplasm) is perceived by internal sensors in root tip cells. To date, neither external nor internal Pi-stress sensors have been identified. Based on recent progress in the area of nutrient sensing and signaling, an emerging concept has been established in which plasma membrane-localized nutrient transporters and related proteins can function as nutrient sensors (Giots et al. 2003; Holsbeeks et al. 2004; Ho et al. 2009; Popova et al. 2010). Based on these findings, the possibility exists that a member of the phosphate transporter 1 (*Pht1*) family of Pi transporters may sense the external Pi status and function as the external Pi sensor. The question is, which sensors respond to Pi stress in the root whether Pi deficiency is sensed externally, intracellularly, or through a combination (Fig. 5.1A)?

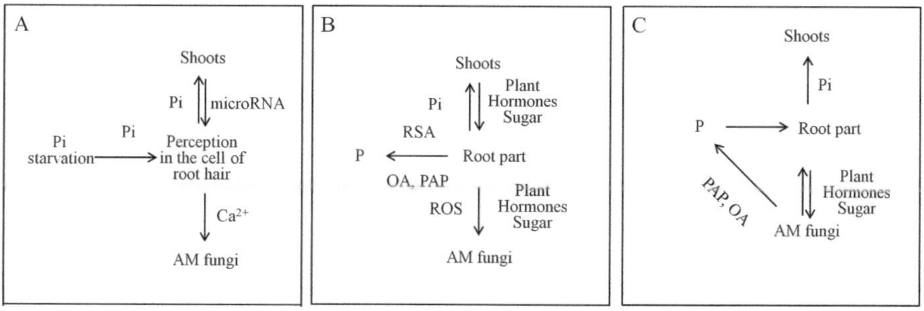

Fig. 5.1 Hypothetical models of mycorrhizal roots sensing and adapting P stress. (**A**) Perception of P stress in host root cell and caution signaling to host shoots. (**B**) Host shoots feedback signaling to roots and AM fungi, and the root modified for P stress (**C**) AM fungi modified for host plant and P resources under P stress

5.3 Signaling Molecular of Phosphate Stress

What are the signaling molecules derived after changes of external Pi concentrations? These are Pi, calcium, reactive oxygen species (ROS), miRNAs, and phytohormones, along with sugars that have all been implicated in the signaling pathways (Chiou and Lin 2011). These signals may act locally and/or serve as systemic signals to elicit responses at a distance.

5.3.1 P

Increased intracellular Pi concentrations could repress Pi starvation responses (PSR). In this context, Pi is straight forwardly rationalized as a signal. However, because Pi is a nutrient, its suppression of PSR can be explained as an effect of adequate nutrition rather than signaling of Pi. Compelling evidence obtained from the results of $H_2PO_3^-$ or HPO_3^{2-}application supports the notion that Pi serves as a signal. $H_2PO_3^-$ or HPO_3^{2-} is taken up by plants through Pi transporters; it cannot be oxidized to Pi or further metabolized once inside cells (Carswell et al. 1996, 1997). Under low Pi conditions, exogenous application of $H_2PO_3^-$ or HPO_3^{2-} attenuates a wide range of PSR, including a reduction in the root-to-shoot biomass ratio, root hair elongation, thocyanin accumulation, lipid remodeling, and expression of many Pi starvation-induced (PSI) genes (Carswell et al. 1996, 1997; Kobayashi et al. 2006; Ticconi et al. 2001; Varadarajan et al. 2002). The interference of gene expression by $H_2PO_3^-$ or HPO_3^{2-} is specific to PSR and is an early event occurring at the level of transcription. The Pi signaling machinery is unable to discriminate $H_2PO_3^-$ or HPO_3^{2-} from Pi given the structural similarity between $H_2PO_3^-$ or HPO_3^{2-} and Pi. In addition, $H_2PO_3^-$ or HPO_3^{2-} inhibits the Pi influx in a competitive manner and accumulates specifically in the cytoplasm (Danova-Alt et al. 2008; Pratt et al. 2009).

It is very likely that elevated intracellular $H_2PO_3^-$ or HPO_3^{2-} concentrations mimic Pi sufficiency, thus interfering with Pi signal transduction pathways, even though plants are starving for Pi. Collectively, these findings demonstrate that Pi is capable of acting as an initial signal. The use of $H_2PO_3^-$ or HPO_3^{2-} shows promise as a tool to dissect the P signaling machinery in the future. Split-root experiments led to the conclusion that several PSR and the expression of PSI genes are regulated according to the whole plant Pi status, which is determined by the translocation and mobilization of Pi within plants (Burleigh and Harrison 1999; Franco-Zorrilla et al. 2007; Liu et al. 1998). Upregulation of PSI genes in one portion of roots grown in Pi-depleted medium is repressed likely by a systemic suppressor that has been transported from another portion of roots grown in Pi-replete medium. Pi has been considered as a potential candidate for the systemic suppressor because of its mobile nature. However, Pi being a signal is argued by the observation that the reduced induction of the *Medicago Mt4* gene, a PSI gene, in the Pi-depleted compartment initiated prior to a rise in Pi concentration and reduction in Pi flow had no effect on the systemic suppression (Burleigh and Harrison 1999). A recent study by the modulation of *PHO1* (*PHOSPHATE1*) expression also supports the existence of an additional signal, independent of Pi, because the PSR could be uncoupled from the low Pi concentration in the shoot (Rouached et al. 2010). Pi-related compounds, such as ATP, were instead suspected to carry out such a role. This seems unlikely because the nonassimilated $H_2PO_3^-$ and HPO_3^{2-} were still subjected in a similar suppression (Ticconi et al. 2001). Dissection of primary causes from subsequent secondary effects is often difficult (Carbonnel and Gutjahr 2014) because Pi serves both as a nutrient and a signal.

5.3.2 Ca

As universal secondary messengers in many signal transduction pathways, Ca^{2+} has been considered as players in Pi sensing and signaling (Chiou and Lin 2011). The level of the plasma membrane Ca^{2+}-ATPase, which is important for Ca^{2+} transport, significantly increased when tomato roots were subjected to Pi starvation conditions (Muchhal et al. 1997), suggesting a possible involvement of Ca^{2+} in Pi-stress signaling. Upon Pi deficiency perception by the root sensing system, downstream adaptive signaling pathways become activated in the Pi-stress sensing cells to generate both cell-autonomous and systemic signals that amplify the primary Pi deficiency signals (Chiou and Lin 2011; Lucas et al. 2013).

5.3.3 ROS

The concentration of ROS in roots increases rapidly (within a few hours) after deprivation of P (Schachtman and Shin 2007; Shin and Schachtman 2004; Shin et al. 2005) and triggers the expression of several nutrient starvation responsive genes. ROS have been shown to undergo an increase in P-deprived roots (Shin and Schachtman 2004; Shin et al. 2005). At high P, the elongation zone and the primary root meristem were the main sites of ROS production; however, at low P, ROS were not detected in the elongation zone, but were present in the proximal part of the lateral root meristem.

5.3.4 miRNA

miR399 is specifically and highly upregulated in Pi-depleted tissues (Aung et al. 2006; Bari et al. 2006). It directs the cleavage of *PHO2* mRNA encoding an ubiquitin-conjugating E2 enzyme. When deficient, miR399 is upregulated and *PHO2* is suppressed, leading to activation of Pi uptake in plant. Promoter–reporter analyses suggest that miR399 and *PHO2* are expressed predominantly in vascular tissues (Aung et al. 2006). Consistent with these observations, miR399 was detected in the phloem sap of rapeseed and pumpkin (Buhtz et al. 2008; Pant et al. 2008). Results from reciprocal grafting between wild-type and miR399-overexpressing plants led to the conclusion that *PHO2* in roots can be suppressed by a shoot-to-root movement of mature miR399 (Lin et al. 2008; Pant et al. 2008). The biological significance of such movement, however, has been questioned because miR399 can be expressed in both shoots and roots. Detailed time-course analysis showed that induction of miR399 expression by Pi deficiency occurred earlier in shoots than in roots (Lin et al. 2008). Estimation of the phloem transport rates also supports rapid movement of miR399 (Pant et al. 2008). It was proposed that the systemic movement of miR399 is an early response to Pi deficiency (Lin et al. 2008). Movement of miR399 serves as a systemic signal for activating Pi uptake in roots and for communicating Pi status in shoots to Pi uptake activity in roots. Induction of miR399 under Pi starvation is positively regulated by the transcription factor PHR1 (phosphate starvation response1) and by the availability of photosynthates; however, its activity on the cleavage of *PHO2* mRNA is suppressed by *AT4/IPS1* RNAs (Bari et al. 2006; Lin et al. 2008; Liu et al. 2010; Rubio et al. 2001; Valdés-LÓpez et al. 2008; Zhou et al. 2008b). Expression of miR399 is reduced in Pi-deprived *phr1* mutants, and a subset of genes is misregulated in both *phr1* and *pho2* mutants, suggesting that miR399 and PHO2 form a branch of the Pi-signaling network downstream of PHR1 (Bari et al. 2006). Surprisingly, sequence analysis of *AT4/IPS1* RNAs revealed a conserved 22-nucleotide sequence that is partially complementary to miR399 (Shin 2011). Such partial complementarily is not sufficient for its cleavage by miR399; instead, *AT4/IPS1* RNAs function as riboregulators, interfering with miR399 targeting of *PHO2* mRNA, an action termed target mimicry (Franco-Zorrilla et al. 2007).

5.3.5 Plant Hormones

Many hormones including auxin, cytokinin, ethylene, abscisic acid, gibberellins, and strigolactones have been associated in Pi signaling. Auxin signaling has been suggested to associate closely with the modification of root system architecture (RSA) caused by Pi deprivation. Treatment with exogenous auxin triggers localized alterations in RSA as seen in Pi-deprived plants (Gilbert et al. 2000; López-Bucio et al. 2003). Pi-deprived plants are more sensitive to exogenous auxin than Pi-replete plants with respect to the arrest of primary root growth and induced formation of lateral roots (Gilbert et al. 2000; López-Bucio et al. 2003). Although results from different experiments using auxin transport inhibitors or auxin-responsive mutants are somewhat controversial (Jain et al. 2007; López-Bucio et al. 2003; Schmidt and Schikora 2001), it becomes clear that an auxin-dependent pathway involved in auxin transport or sensitivity and an auxin-independent pathway may coexist to modulate Pi starvation-induced changes of RSA (López-Bucio et al. 2003). Inhibition of primary root growth and stimulation of root hair growth may be independent of auxin signaling, whereas auxin is required to stimulate lateral root primordium emergence under Pi deficiency (López-Bucio et al. 2003; Pérez-Torres et al. 2008). A recent report showed that enhanced auxin sensitivity due to an increased expression of the TIR1 (transport inhibitor response1) auxin receptor rather than an increased accumulation of free auxin is responsible for lateral root development in Pi-deprived *Arabidopsis* seedlings (Pérez-Torres et al. 2008). The increased level of TIR1 accelerates degradation of Aux/IAA (auxin/indole-3-acetic acid) proteins, thereby liberating ARF (auxin response factor) transcription factors (e.g., ARF19) that modulate expression of genes involved in activating pericycle cell division and lateral root formation (Pérez-Torres et al. 2008). Additionally, auxin and cytokinin signaling are associated in membrane lipid remodeling during Pi starvation by regulating the expression of genes in these processes (Kobayashi et al. 2006; Narise et al. 2010). This response was directly related to the findings that the expression of the *Arabidopsis* auxin receptor TIR1 (transport inhibitor response 1) is higher in Pi-deficient plants than in plants grown under Pi-sufficient conditions and that TIR1 knockout mutants are impaired in the lateral root response to low Pi (Shin et al. 2006). The increased lateral root formation mediated by TIR1 in response to Pi deficiency requires the presence of the ARF7 (auxin response factor 7) and ARF19 transcription factors, which transduced this enhanced auxin response into the formation of new lateral roots (Shin et al. 2006).

The involvement of cytokinin signaling in regulating PSR is well documented. Pi starvation represses the action of cytokinin by reducing its concentrations (Kuiper et al. 1988) and by decreasing the expression of CRE1, a cytokinin receptor (Franco-

Zorrilla et al. 2007). On the other hand, cytokinin negatively regulates a number of PSI genes (Franco-Zorrilla et al. 2007; Wang et al. 2006). The negative regulation by cytokinin is attenuated when *CRE1*- and/or *AHK3*-encoding cytokinin receptors are mutated, thus indicating the presence of a two-component signaling circuit in cytokinin-mediated PSR (Franco-Zorrilla et al. 2007). A decrease in the endogenous level of cytokinin or in its action could ensure a full PSR upon Pi deficiency. A bidirectional interaction between cytokinin and Pi signaling was proposed (Franco-Zorrilla et al. 2007). Transcriptional profiling by microarray analysis in rice revealed global suppression of PSI genes by cytokinin (Wang et al. 2006). Such negative regulation can be explained in part by elevated intracellular Pi concentrations in both shoots and roots, which are probably caused by the release of Pi from internal sources and a pause in growth after cytokinin application. By contrast, Pi concentrations were increased in roots but decreased in shoots upon cytokinin treatment in *Arabidopsis* (Lai et al. 2007). This discrepancy has to be reexamined. Given the mobile nature of cytokinin and the broad range of PSI genes it affects, cytokinin was assumed to be involved in the systemic repression of Pi signaling (Sakakibara 2006). However, in split-root experiments, systemic repression of several PSI genes was not affected by the mutations of CRE1 and AHK3; moreover, repression of PSI genes was confined to the local compartment of roots treated with cytokinins (Franco-Zorrilla et al. 2007). This argues against a role for cytokinin in systemic repression of PSR.

Ethylene has also been implicated in changing the RSA in response to Pi deficiency, as increased levels of ethylene have been detected under these conditions and mimicry phenotypes of Pi-starved roots were obtained after exogenous application of ethylene (Borch et al. 1999; Gilbert et al. 2000; Ma et al. 2003). Analyses of ethylene signaling mutants and application of ethylene precursors or inhibitors of ethylene biosynthesis or action resulted in various effects on root growth. During Pi starvation, ethylene is important for inhibition of primary root growth and for promotion of lateral root elongation but is not required for lateral root initiation (López-Bucio et al. 2003; Ma et al. 2003). With regard to Pi starvation that stimulated root hair development, studies by Schmidt and Schikora (2001) suggested an ethylene-independent signaling pathway. However, results from other studies supported the involvement of ethylene in root hair responses to Pi deficiency (He et al. 2005; Zhang et al. 2003). Apart from its effects on root growth, ethylene regulates Pi recycling during petal senescence (Chapin and Jones 2009).

It has been speculated that abscisic acid (ABA) signaling is involved in PSR because there are some similarities in growth patterns, such as increased root-to-shoot ratio and root hair density, between plants subjected to Pi starvation and those treated with ABA (Ciereszko and Kleczkowski 2002; Trull et al. 1997). However, a direct relation between ABA signaling and PSR has not been established. In Pi-deficient caster bean, xylem transport of ABA was stimulated, but the content of ABA in tissues was not affected (Jeschke et al. 1997). The accumulation of anthocyanin at low Pi concentrations was reduced in the ABA-deficient *aba1* mutant, whereas the phosphatase activity and the root-to-shoot ratio were not altered in *aba1* and ABA-insensitive *abi2-1* mutants under these conditions (Trull et al. 1997).

Normally, GA controls growth and developmental adaptations to low Pi via a DELLA-dependent mechanism (Jiang et al. 2007). Pi starvation reduces the level of bioactive GA, leading to accumulation of DELLA proteins. Pi starvation-induced changes in RSA, such as reduced primary root and increased lateral root growth, and anthocyanin accumulation are repressed by exogenous GA or in DELLA-deficient mutants. However, the P content and the transcript levels of several PSI genes are not affected (Jiang et al. 2007).

Pi starvation induces the production of strigolactones, a specific group of terpenoid lactones (López-Ráez et al. 2008; Yoneyama et al. 2007). Strigolactones serve as a rhizosphere signal for the stimulation of hyphal branching of AM fungi and as a root-derived hormone for optimization of shoot branching by inhibiting axillary bud outgrowth during Pi starvation (Umehara et al. 2010).

Many studies have advocated the importance of sugar signaling in regulating PSR, including increased expression of PSI genes and changes in RSA (Hammond and White 2008; Jain et al. 2007; Karthikeyan et al. 2007; Liu et al. 2005; Zhou et al. 2008a). Moreover, Pi starvation can activate expression of several sugar-responsive genes (Ciereszko and Kleczkowski 2002). Limitation of Pi results in decreased photosynthesis and an increased level of sugars and starch in Pi-deprived leaves (Morcuende et al. 2007). The built-up sugars are translocated to roots via enhanced loading of sucrose into the phloem. The root-to-shoot biomass ratio increases as a consequence of this resource allocation. Analyses of the transcriptome, proteome, and metabolome in response to Pi deficiency revealed a close association between increased gene expression and enzymatic activities relating to carbohydrate biosynthesis and PSR (Hernández et al. 2007; Li et al. 2008; Morcuende et al. 2007; Wasaki et al. 2006). Both hexokinase-dependent and hexokinase-independent signaling pathways have been suggested to be involved in interactions between sugar sensing and PSR (Karthikeyan et al. 2007; Zhang et al. 2014).

5.3.6 Sugar

Sugars demonstrate clear temporal and spatial control of PSR. Increased sucrose concentrations in roots precede the induction of PSR (Hammond and White 2008). Inhibition of sucrose biosynthesis or translocation by reduced photosynthesis, dark treatment, or stem girdling diminishes expression of root PSI genes upon Pi deprivation (Liu et al. 2005). In validation, expression of root PSI genes is impaired in a *pho3* mutant defective in phloem loading of sucrose (Lloyd and Zakhleniuk 2004). Exogenous application of sugars magnifies PSR (Karthikeyan et al. 2007; Liu et al. 2005). It was proposed that carbon assimilation and partitioning are checkpoints for the onset of Pi deficiency and sugars, mainly sucrose, are candidates for the shoot-derived systemic signal, which contributes to the regulation of PSR in roots. Signaling pathways mediated by sugar and different hormones are interconnected (Gibson 2004). Under Pi starvation, sucrose may promote auxin transport and increase the sensitivity of the root system to auxin (Jain et al. 2007; Karthikeyan et al. 2007). Moreover, sugars and cytokinins act antagonistically to regulate the expression of PSI genes (Franco-Zorrilla et al. 2007). It was postulated that cell cycle activity determines Pi demand and specifies the magnitude of PSI gene expression (Lai et al. 2007). Inhibition of cell division reduced PSI gene expression, and manipulation of cell cycle activity dominated over the effects of sugar or cytokinin. The coverage of the regulation requires further investigation.

5.4 The Signaling of Molecules Regulated Mycorrhizal Development in P Deficiency

5.4.1 P

Symbiotic P_i uptake occurs in cortex cells that are colonized by arbuscules (Javot et al. 2007a). Arbuscules are surrounded by a plant-derived periarbuscular membrane that hosts a specific set of membrane proteins (Pumplin and Harrison 2009). Importantly, it contains symbiotic phosphate transporters (PT4/PT11), which import phosphate ions that are released by the arbuscule into the plant cell (Javot et al. 2007b; Yang et al. 2012). *Medicago pt4* and rice *pt11* mutants revealed that PT4/PT11 is not only essential for AM-mediated phosphate uptake but also for arbuscule maintenance (Javot et al. 2007b; Yang et al. 2012) demonstrating that P_i import is crucial for wild-type arbuscule dynamics. It has been suggested that the P_i ion itself could act as a local, cell-autonomous signal that triggers accommodation and maintenance of the arbuscule by the host cell (Javot et al. 2007a; Yang and Paszkowski 2011). This notion is supported by the rice *pt13* mutant, which is deficient in a second AM-induced phosphate transporter called PT13. It is not impaired in symbiotic phosphate uptake, but in arbuscule maintenance. Thus, OsPT13 might act as a P_i sensor rather than a transporter (Yang et al. 2012). Analysis of the promoter regions of symbiotic PTs in different species revealed two conserved cis-elements called MYCS (or CTTC) and P1BS that are often located close to each other (Karandashov et al. 2004; Chen et al. 2011). Deletion of each of these elements from PT promoters driving a GUS reporter gene showed that they are both essential for colonization-responsive promoter activation (Chen et al. 2011), suggesting that at least two transcription factors (TFs) co-regulate the expression of symbiotic PTs. The MYCS element is overrepresented in mycorrhiza-regulated genes, and four repeats of the MYCS element alone are sufficient to drive GUS expression in colonized areas of the root (Lota et al. 2013). Thus, the P1BS element is dispensable when MYCS is taken out of context. The P1BS motif is common to many promoters of P_i starvation-induced genes and is targeted by central regulators of P_i starvation responses, the MYB transcription factor PHR1, and its homologs (Bustos et al. 2010). Thus, promoter induction of symbiotic phosphate transporters and other mycorrhiza-responsive genes likely requires simultaneous activation by symbiosis signaling and

P_i starvation signaling. Repressed expression of a transgene containing MYCS-GUS after fertilization of colonized transgenic roots of *Lotus japonicus* with high P_i for 2 weeks seems to contradict this hypothesis (Lota et al. 2013). However, in *Petunia* it has been shown earlier that such a long period of phosphate replenishment leads to decreased root colonization, while expression of a symbiotic PT gene is already suppressed after 2–4 days of high P_i supply (Breuillin et al. 2010). Therefore, 2 weeks after P_i replenishment, MYCS activation is probably indirectly affected due to fungal senescence and cessation of symbiotic signaling. Nevertheless, the important role of symbiotic PTs in AM symbiosis maintenance (Javot et al. 2007b; Yang et al. 2012) and their transcriptional regulation by P_i conditions (Nagy et al. 2009; Breuillin et al. 2010) makes them possible targets of AM developmental control by nutrients (Carbonnel and Gutjahr 2014).

5.4.2 Ca

Plant endosymbiosis (AM and root nodule symbiosis) development requires a common set of genes called common SYM genes. Their protein products belong to a signal transduction cascade that is triggered by perception of fungal signals (Myc factors) through receptor-like kinases (Gough and Cullimore 2011). Myc factor perception induces nuclear Ca^{2+} spiking that is decoded by a nuclear-localized calcium–calmodulin kinase (CCaMK) and leads to transcriptional activation of symbiosis-related genes by the transcription factor CYCLOPS (Genre et al. 2013; Singh et al. 2014). One potential explanation for suppression of AM development at high P_i could be the repression of common SYM signaling. However, the expression of downstream common SYM genes involved in the generation and interpretation of Ca^{2+} spiking is not affected (Breuillin et al. 2010), and high P_i did not reduce the ability of *Medicago* rhizodermis cells to trigger nuclear Ca^{2+} spiking in response to rare hyphopodia or germinating spore exudates (Balzergue et al. 2013). Variations in intracellular free Ca^{2+} concentration occur as one of the initial steps in signaling pathways activated in plants when they encounter pathogens, fungal biocontrol agents, and nitrogen-fixing bacteria. Molecules secreted by microorganisms, after binding to specific receptors, trigger in plant cells transient changes in cytosolic Ca^{2+} level, due to the influx of the ion from the extracellular environment and/or the release from internal Ca^{2+} storage compartments. Ca^{2+} messages delivered to plant cells are at least partly deciphered on the basis of their spatial and temporal features. Legumes are able to engage in a dual symbiotic interaction, with rhizobia and AM fungi. Components of the Ca^{2+}-mediated signaling pathway are shared by the two symbioses. In the mycorrhizal signal transduction pathway, the involvement of Ca^{2+} has long been speculated, based on the observed similarities with symbiotic nitrogen fixation. It seems that AM fungi announce their presence to the plant through the constitutive release of a chemical signal, even before experiencing the proximity of the plant or its AM symbiotic signals. The notion that the secreted fungal molecules herald, through Ca^{2+}, a beneficial message which can be acknowledged only by competent receivers is supported by the lack of defense response induction and the upregulation of some genes essential for the AM symbiosis initiation in host plant cells and the unresponsiveness of cultured cells from the nonhost plant *Arabidopsis thaliana.* Ca^{2+}-mediated perception of both AM fungal and rhizobial signals by plant cells unifies the signaling pathways activated in the two symbioses. However, the actual occurrence of Ca^{2+} spiking in AM symbiosis remains to be ascertained when applied to an asynchronous cell population due to limitations of the recombinant aequorin method. Contribution of internal Ca^{2+} stores, in particular the nucleus, to the observed Ca^{2+} changes will be a future research goal to be achieved through a pharmacological approach and/or targeting of Ca^{2+} indicators to intracellular compartments.

5.4.3 ROS

From the very early contact times, the oxidative burst (O_2^{-}• generation) and the related enzymes (SOD and POX) measured in the apoplast of the intact roots in contact with the pathogenic fungus (PF) presented values much greater than those in control roots and in those in contact with AMF. The PF-treated root values were even greater than those in roots treated with MeJA, the defense reactions of phytohormone. The phenolic compounds measured in root homogenates also showed the same response. The defense reactions evoked in roots by AMF were, indeed, either similar to the control or just a little stronger given that the mycorrhizal fungus attenuated the defense reactions of roots in order to facilitate the establishment of the mycorrhiza. This was in contrast to the strong defense response evoked by the pathogenic fungus, PF, described above (Espinosa et al. 2014; Kulik et al. 2015). All the enzymatic activities measured in the apoplast of these in vivo roots showed homeostatic oscillations, with SOD and POX roughly in opposition to O_2^{-}• generation, since the excess of O_2^{-}• would induce SOD and POX activities that would keep O_2^{-}• and H_2O_2 accumulation in the apoplast under physiological control. The same was exhibited for the controls, and the MeJA and AMF contact treatments, but not for PF contact. In this last case, strong, steady (without peaks or oscillations) O_2^{-}• generation in the root apoplast was observed throughout the time of the experiments, although SOD and POX showed clear oscillations apparently unrelated to the O_2 generation (Espinosa et al. 2015).

5.4.4 Plant Hormones

Researchers have used molecular techniques for the identification of signaling molecules and plant hormones produced by the host plant affecting AM symbiosis. Akiyama and Hayashi (2006) identified the signaling molecules called strigolactones (5-deoxystrigol), a new class of phytohormones (Umehara et al. 2008) affecting AM plant symbiosis. These plant hormones can also affect the activities of the parasitic weeds, *Striga* and *Orobanche*, and shoot branching in plants. The functional characteristics of strigolactones are determined by their structural properties (Akiyama et al. 2010; Kretzschmar et al. 2012). Strigolactones are sesquiterpene lactones produced in the carotenoid pathway (Matusova et al. 2005; García-Garrido et al. 2009) with little specificity. A total of 10–13 identified forms of strigolactones exist. They include molecules such as strigol, sorgolactone, orbanchyl acetate (alectrol), orobanchol, 2′-epiorbanchol, solanacol, 5-deoxistrigol, and sorgomol, while their production is determined by the strategies used by plants to absorb nutrients (Steinkellner et al. 2007; Yoneyama et al. 2008, 2011). They are able to induce hyphal branching in AM fungi by affecting different molecular and cellular processes and induce the seed germination of the parasitic plants (Brachmann and Parniske 2006; Gomez-Roldan et al. 2007; Rochange 2010). These signaling molecules are produced by a wide range of plants including monocots and dicots (Akiyama et al. 2005). Strigolactones are effective at minute concentrations, indicating the presence of very strong and sensitive receptors in the AM fungal cell membranes. Strigolactones affect the fungal respiration and hence activities by influencing mitochondrial activity and density as well as lipid catabolism in AM fungi. This eventually increases the production of Myc factors (Besserer et al. 2006). Accordingly, strigolactones can significantly affect the process of AM symbiosis and hence ecosystem efficiency. Germinating spores produce exudates in the absence of the host plant. Signaling molecules, such as strigolactones and flavonoids, can also be produced by plant roots even in the absence of mycorrhizal fungi (Akiyama et al. 2005; Besserer et al. 2006; Steinkellner et al. 2007). This indicates that the interactions between the fungi and the host plant can influence the nature of exudates produced by the two symbionts. The exudates of germinating spores are able to activate the symbiotic genes of *Medicago truncatula* by the activation of multiple signaling pathways. However, such active molecules (Myc factors) must be exactly recognized (Mukherjee and Ané 2011). Lendzemo et al. (2009) examined the effects of sorghum (as a host to AM fungi and *Striga* plant) root exudates on the germination of *Striga* seeds. They also investigated such effects using plants that were nonhost to *Striga* but host and nonhost to AM fungi. According to their results, the seeds germinated only in the presence of strigolactones. Although this hormone may also result in hyphal branching, not all root exudates of AM host plants resulted in the germination of seeds. In addition, the exudates of nonhost plants did not stimulate the germination of seeds. This indicates that being a host to AM fungi is not a necessity for the germination of *Striga* seeds. Moreover, strigolactones affect seed germination and hyphal branching through completely different molecular pathways, which are also a function of exudate levels. To assess whether strigolactones can affect the susceptibility of nonhost plants to AM symbiosis, Illana et al. (2011) inoculated four nonhost plant species using synthetic strigolactones and *Glomus intraradices*, which did not result in the colonization of plant roots. In addition, the pea mutants (deficient in strigolactones), which were not colonized by AM fungi, were able to establish symbiotic association after the application of synthetic strigolactones (Illana et al. 2011; Miransari et al. 2014).

The accumulation of cytokinins in shoots and roots is higher in mycorrhizal plants than in non-mycorrhizal plants (Hause et al. 2007). This alteration is usually related to the late stages of symbiosis (Danneberg et al. 1992). However, high levels of cytokinins may have adverse effects on the proliferation of fungal hyphae in *Glomus fistulosum* (Gryndler et al. 1998). AM symbiosis inhibited the expression of two pathogenesis genes but enhanced the levels of zeatin riboside-like cytokinins (Ginzberg et al. 1998). This indicates some of the interesting effects of AM symbiosis in controlling pathogens. Researchers have indicated that cytokinins can act as a chitinase gene inhibitor (Hause et al. 2007). The enhanced level of hormones can increase the host plant P uptake and decrease the expression of pathogenesis-related (PR) genes and some chitinase and glucan genes (Song et al. 2011). However, whether the plant or fungal productions are involved in the alteration of cytokinins remains to be investigated. Enhanced P uptake by mycorrhizal plants is suggested to increase cytokinin production in the roots. Increased concentrations of cytokinins may enhance shoot growth but not root growth (Hause et al. 2007). Plant hormones affect P uptake by affecting root growth and sugar signaling (Rouached et al. 2010). In addition to the increased root-to-shoot ratio, the levels of cytokinins may also decrease under P-deficient conditions. Plant response to P deficiency is regulated by soil P levels and mycorrhizal symbiosis through alteration in the expression of the related genes. The binding of transcription factors for the promotion of the P responsive genes can affect plant response to P stress. Such effects are influenced by the related P pathways, allocation, and homeostasis regulation in plant cells (Sobkowiak et al. 2012). This may also further elucidate the role of cytokinins during P stress in plant.

Ethylene is a stress hormone that controls plant growth interactively with another plant hormone, ABA. Under stress, the amount of ethylene in the plant increases, adversely affecting plant growth. Moreover, ethylene can also significantly decrease the rate of AM colonization, especially under P-deficient conditions (Zsögön et al. 2008; Li and Dong 2015). Using of products like aminoethoxyvinylglycine, which inhibits the production of ethylene, can positively affect mycorrhizal symbiosis. Ethylene can also decrease the entry of fungi into the root cells during symbiosis (Oldroyd et al. 2001; Nukui et al. 2004). The biosynthesis of ethylene is reduced because of the production of phenolic products in the roots of *Solanum tuberosum*, resulting in the development of symbiosis (McArthur and Knowles 1992).

ABA is necessary for the colonization of plant host roots by AM fungi and controls the formation of arbuscules. Application of ABA can result in the formation of functional arbuscules by the fungi as indicated by measuring fungal alkaline phosphatase (ALPase) activity (Herrera-Medina et al. 2007). AM fungi and ABA induced the activities of genes necessary for the production of leghemoglobin. The latter can detoxify the unfavorable effects of nitric oxide by binding to it in mycorrhizal roots and symbiotic nodules under nonsymbiotic conditions. Furthermore, nitric oxide, which is necessary for the signaling pathway, results in the production of ABA under stress. Accordingly, leghemoglobin can control the mechanisms necessary for the regulation of ABA production (Ruan et al. 2004; Vieweg et al. 2005; Grunwald et al. 2009). However, Zsögön et al. (2008) indicated that ABA did not influence the colonization process. ABA production increased in the roots of mycorrhizal maize (*Zea mays* L.) (Danneberg et al. 1992; Bothe et al. 1994) and soybean (*Glycine max* L.) (Meixner et al. 2005).

Alterations in the properties of mycorrhizal roots, including increased number of lateral/fine roots, during the early period of plant growth are similar to the effects of IAA on plant roots. Hence, IAA has been suggested to have a role in mycorrhizal symbiosis (Ludwig-Muller 2000; Ludwig-Müller 2010, 2011). In addition, Gryndler et al. (1998) found that IAA is able to enhance the growth of *G. fistulosum*. Strigolactones can regulate auxin transport in the plant by affecting auxin carriers (Koltai et al. 2010) or the secondary messengers, which are active in auxin signaling (Ferguson and Beveridge 2009; Ludwig-Müller and Güther 2007). This suggests that auxin may indirectly affect the symbiotic level because it is influenced by strigolactones. Accordingly, the interactive effects of strigolactones and auxin may affect plant growth by influencing the symbiont activities. Among different factors affecting symbiotic efficiency are root growth and development, which are affected by plant hormone strigolactones and their interaction with other plant hormones, including auxin (Teale et al. 2008).

Although JA and SA were not the most significant hormones in plant Pi deficiency, they play important roles in AM root development for Pi deficiency adaptation. JA and its derivatives (signals derived from lipid molecules) are plant hormones regulating plant growth as well as plant response under stress. JA and ethylene, which are produced in the transduction pathway, are parts of plant systemic acquired resistance and adjust plant interactions with microorganisms (Pozo et al. 2004; Mabood and Smith 2005; Van Wees et al. 2008; Van der Ent et al. 2009). Hence, JA regulate the expression of genes involved in the deactivation of proteinase, production of phytoalexin and proteins, vegetative growth and storage, and production of thionins, which are responsible for plant resistance to different stresses (Lorenzo and Solano 2005). The level of JA increased in the mycorrhizal roots of plants, including *Hordeum vulgare* (Hause et al. 2002), *Cucumis sativus* (Vierheilig and Piché 2002), *M. truncatula* (Stumpe et al. 2005), and *G. max* (Meixner et al. 2005). The levels of JA increased in mycorrhizal plants and tended to vary with plant species. For example, the increase in *M. truncatula* was between two- and threefold relative to non-mycorrhizal plants (Stumpe et al. 2005), while the relative increase in *H. vulgare* and *C. sativus* was up to 5- and 14-fold, respectively (Hause et al. 2002; Vierheilig and Piché 2002). Such increases may positively affect the process of AM symbiosis. Recent research has also indicated that a higher amount of JA is found in mycorrhizal roots relative to non-mycorrhizal roots (Hause et al. 2002; Meixner et al. 2005). Given that a localized measurement of JA has yet to be conducted, the expression of genes responsible for JA production may be a good indicator of JA produced in plants, such as *H. vulgare* and *M. truncatula* (Maucher et al. 2000; Hause et al. 2002; Stenzel et al. 2003; Isayenkov et al. 2005). The reason for the increased level of JA in mycorrhizal *H. vulgare* has been attributed to the osmotic stress resulting from the increased level of carbohydrates transferred from the shoots to the roots of mycorrhizal plants. This is because mycorrhizal roots, as a sink, have higher nutritional demands than non-mycorrhizal roots (Hause et al. 2007). The increased translocation of carbohydrates from the shoots to the roots of mycorrhizal plants (Alkan et al. 2006; Nakano-Hylander and Olsson 2007; Miransari et al. 2007, 2008; Miransari 2010a, b) and expression of the related genes may indicate the following strategies: source tissues provide carbohydrates, which are transferred to the roots (sink tissues) increasing osmotic stress; the carbohydrates may be able to induce the expression of genes, which are responsible for the production of enzymes producing JA; and the produced JA, which can affect the production of arbuscules in mycorrhizal roots, can also enhance their sink capacity (Hause et al. 2007; Hause and Schaarschmidt 2009).

In addition to the expression of a set of PR genes, the signaling molecule, salicylic acid, is also necessary for plant systematic acquired resistance (Lian et al. 2000) and mycorrhization (Herrera-Medina et al. 2003). Given that plant SA is involved in systemic acquired resistance in the presence of pathogens (Lian et al. 2000), it may also have a role in the initiation of AM plant symbioses (Hause et al. 2007). Exogenous addition of SA to rice (*Oryza sativa* L.) inoculated with AM fungi decreased the colonization rate at the beginning of symbiosis, but the formation of appressoria remained unaffected. The inhibitory effects of salicylic acid on the growth of AM fungi have also been indicated (Hause et al. 2007). Analysis of the gene, *NahG*, which activates the bacterial enzyme salicylate hydroxylase with producing and inhibiting activities of salicylic acid (Gaffney et al. 1993), indicates that salicylic acid has inhibitory effects on root colonization by *G. intraradices* and *Glomus mosseae* (Herrera-Medina et al. 2003). Researchers have also indicated the role of salicylic acid on microbial symbiosis. In addition, during the early stages of AM symbiosis establishment, the mechanisms related to the systematic acquired resistance are activated but are eventually inhibited with the progress of the symbiosis (Herrera-Medina et al. 2003). During pathogenic presence, the genes regulating the plant resistance system are activated, hence enabling the plant to resist the pathogen by activating different mechanisms, including the production of antipathogenic products (Lian et al. 2000). Shekoofeh et al. (2012) evaluated the effects of mycorrhization (*G. mosseae* and *G. intraradices*) and SA (0.2 mM) on the salinity tolerance of green basil (*Ocimum basilicum* L.). SA decreased the preoxidation of lipids and increased the concentrations of proline and proteins in the plant root and shoot.

The related signaling molecules of the host plant that regulated AM development were very complex when the symbiosis underwent Pi deficiency. However, circumstantial evidence suggested that control occurred at multiple levels, which included nutrient, miRNA, phytohormone, and sugar. Although AM colonization increased more rapidly under Pi deficiency, the pathway of signal transduction was still largely unknown. Normally, P starvation, which was attributed to Pi concentration variation, was sensed in the root. Then, the system's signaling molecules, such as microRNA and plant hormones, were induced, causing a systemic reaction in the entire plant. Finally, the systemic signal affects the mycorrhizal root, which induces the promotion of AM. Strictly speaking, identifying the function of signaling molecules triggered by Pi starvation is difficult because of the cross talk of these molecules and their complex effects on AM development. Recently, very important connections between ROS and Ca^{2+} both within and across different cells have been reported. From the earliest contacts, symbiotic interactions have been reported to evoke Ca^{2+} oscillations in the cytosol of epidermal root cells. These oscillations act as a signal to generate ROS in the oxidative burst and induce signaling cascades resulting in the biosynthesis of the compounds involved in mycorrhization (Shu et al. 2016a, b).

Good candidates for such downstream or parallel mechanisms are phytohormone signaling modules, because they steer developmental responses to the nutrient environment (Rubio et al. 2009) and regulate AM formation (Foo et al. 2013; Bucher et al. 2014; Gutjahr 2014). For example, strigolactones are exuded into the rhizosphere and stimulate germination, hyphal branching, and metabolic activity of AM fungi (Akiyama et al. 2005; Besserer et al. 2006, 2008), which increases AM colonization (Gomez-Roldan et al. 2008; Foo et al. 2012; Gutjahr et al. 2012; Kohlen et al. 2012; Kretzschmar et al. 2012; Yoshida et al. 2012). Gibberellins (GAs) negatively regulate AM development. In *Medicago* and pea, exogenous GA treatment in the roots blocked arbuscule formation but did not affect the colonization with intraradical hyphae (Floss et al. 2013; Foo et al. 2013). Moreover, it generally reduced intraradical colonization in rice (Yu et al. 2014). Consistently, DELLA proteins, which are repressors of GA signaling, are required for AM development (Floss et al. 2013; Foo et al. 2013; Yu et al. 2014). Thus, the GA signaling module has the potential to regulate AM development and, in particular, arbuscule formation according to the plant phosphate status (Floss et al. 2013). Interestingly, arbuscule formation of the *L. japonicus* common SYM mutant *cyclops* can be restored by overexpression of a resistant DELLA version. The DELLA/GA module is therefore a good candidate regulator of arbuscule development either downstream of, or in parallel with, common SYM signaling (Floss et al. 2013; Gutjahr 2014). However, whether transgenic expression of a resistant DELLA can counteract the negative impact of high Pi supply on AM symbiosis remains to be tested. Cross talk among plant hormones also exists. The concentrations and structural features of the strigolactones secreted by the plant are important for AM fungal development (Ruyter-Spira et al. 2013), while fungal chitin oligomers, stimulated by strigolactones, trigger the symbiotic programming in the root (Gutjahr 2014). SA, ethylene, and cytokinins have negative effects during the fungal penetration or root colonization steps (Foo et al. 2013). At later stages, phytohormones also regulate arbuscule development and lifespan (Gutjahr and Parniske 2013). Biologically active GAs suppress arbuscule development and, accordingly, the DELLA repressors are essential for their formation (Floss et al. 2013; Foo et al. 2013; Martín-Rodríguez et al. 2014). In contrast, ABA and auxins positively regulate arbuscule development and functionality (Martín-Rodríguez et al. 2011; Etemadi et al. 2014), while positive and negative effects have been described for jasmonates (Wasternack and Hause 2013). As in other plant processes, the impact of hormones on mycorrhizas depends on pathway cross talk. The mainly antagonistic interactions of ABA–ethylene and ABA–GAs regulate AM development and arbuscule formation, respectively (Martín-Rodríguez et al. 2011; Gutjahr 2014; Miransari et al. 2014). Elements of sugar signaling also interact with jasmonates and brassinosteroids to finely modulate mycorrhizas (Bitterlich et al. 2014). Thus, the molecular hubs in hormone cross talk likely integrate external and developmental cues with symbiotic programs during mycorrhiza formation (Gutjahr 2014) (Fig. 5.1B).

5.5 Phosphate Stress Adaptation of Host Roots

5.5.1 Pi Transporters of the PHT1 Family in Response to Low Phosphate Availability

Plants can acquire Pi directly from the soil via transport systems in their roots, or it is alternatively delivered to the root cortical cells by their AM fungal symbionts. Both Pi uptake pathways need phosphate transporters. The *Pht* (*Phosphate Transporter*) genes that encode Pi transporters have been normally grouped into four families, *Pht1*, *Pht2*, *Pht3*, and *Pht4* (Raghothama and Karthikeyan 2005), which transfer Pi across the different cell membranes. Pi transporters encoded by members of the *Pht1* gene family, which are predominantly expressed in epidermal cells and in the outer cortex of the root, have been identified as mediators of Pi uptake at the root–soil interface when Pi is limited (Schünmann et al. 2004; Shu et al. 2012; Zhang et al. 2016). These proteins are part of the so-called direct Pi uptake pathway and transport P as Pi anions, mainly $H_2PO_4^-$ and HPO_4^{2-}, against a concentration gradient between the soil solution (0.1–10 μM Pi) and the cytoplasm of the root epidermal cell (5–10 mM Pi) (Raghothama and Karthikeyan 2005). The number of the *Pht1* transporters relies on plant species. In soybean, 14 members of the *Pht1* family encode high-affinity Pi transporters predominantly expressed in roots under low Pi conditions (Fan et al. 2013). In *M. truncatula*, five transporters encoded by *Pht1* genes are induced by low Pi, of which *M. truncatula* phosphate transporter 1 (MtPT1) and MtPT2 play an important role in Pi uptake. MtPT4 plays a role in Pi translocation from mycorrhizal fungi into the roots (Harrison et al. 2002; Liu et al. 2008; Xiao et al. 2006). In rice, the *Pht1* gene family is composed of 13 members (*OsPT1–13*). Four of these (*OsPT2*, *OsPT3*, *OsPT6*, and *OsPT7*) are highly expressed in Pi-deprived roots, but several (such as *OsPT6*) apparently play a dual role in Pi uptake from the soil and Pi translocation inside the plant (Ai et al. 2009; Goff et al. 2002; Paszkowski et al. 2002). In maize, five *Pht1* genes (*ZmPht1;1–5*) are induced by Pi limitation in roots (Nagy et al. 2006). Therefore, at least some members of the *Pht1* gene family play an important role in Pi uptake from the soil solution, particularly when this nutrient is present in limited amounts.

5.5.2 Root System Modifications in Response to Low Phosphate Availability

RSA modifications induced in response to low Pi stress enhance topsoil foraging and allow the exploitation of these Pi reserves. Significant progress has been made in dissecting the underlying mechanisms that lead to changes in the RSA in response to Pi deficiency. Pi availability alters root traits by modulating the developmental programs that control lateral root primordium initiation and emergence, primary and lateral root growth, the angle of lateral root growth, and the density and elongation rate of root hairs (Pérez-Torres et al. 2008). RSA responses to Pi availability vary significantly among and within species. For instance, in white lupin (*Lupinus albus*), the formation of clusters of lateral roots, known as proteoid roots, is the most evident response to low Pi. However, in common bean (*Phaseolus vulgaris*), changes in the angle of lateral root emergence are the typical response, where perpendicular lateral root growth predominates over downward growth and topsoil foraging is promoted (Lambers et al. 2011; Lynch 2011). RSA traits that promote topsoil foraging in common bean are root shallowness, adventitious root formation, and increased dispersion of lateral branching from the basal roots (Basu et al. 2007; López-Bucio et al. 2003; Liao et al. 2004; Pant et al. 2008; Ramaekers et al. 2010). These characteristics imply that plants grown in soils with low Pi availability have a shallower and broader root system that increases the capacity of the plant to explore the upper layers of the soil, in which organic and inorganic Pi-rich patches are most commonly present. Common bean plants with Pi-efficient genotypes develop root systems with differential growth angles of lateral roots and/or increased adventitious root formation (López-Bucio et al. 2003; Lynch and Ho 2005). The formation of greater root biomass through adventitious roots is less expensive for the plant because adventitious roots are metabolically less demanding per unit Pi (Lynch and Ho 2005; Pant et al. 2008; Nielsen et al. 2001). These root traits are controlled by a low Pi-sensing system and are influenced by ethylene, permitting the plant to explore a larger volume of the upper layers of the soil (Basu et al. 2007; Pant et al. 2008).

5.5.3 Organic Acid (OA) Exudation to Low Phosphate Availability

The production and root exudation of OAs, such as citrate, malate, and oxalate, enhance Pi availability in highly Pi-fixing soils. OAs release Pi bonded to Al^{3+}, Fe^{3+}, and Ca^{2+}, which are common in the upper layers of the soil, where Pi-rich soil patches are most frequently found. OA exudation is controlled at the transcriptional level in monocots and dicots and is highly induced under Pi deficiency in wild species that naturally adapt to Pi-deficient soils, such as white lupin. In crops such as rapeseed (*Brassica napus*), rice, alfalfa (*Medicago sativa*), chickpea (*Cicer arietinum*), maize, wheat (*Triticum aestivum*), soybean, triticale, and rye (*Secale cereale*), OA exudation increases in response to Pi deficiency (Ryan and Delhaize 2001). Interestingly, although OA exudation occurs mainly at the root tip, it is restricted to the low Pi-inducible proteoid roots in white lupin and species of the proteaceae family. These structures not only increase the root surface area by more than 100-fold to aid in Pi exploration but also secrete OAs in millimolar concentrations, mainly as citrate and malate, into the rhizosphere (Lin et al. 2008; Lambers et al. 2013; Vance et al. 2003). The metabolic investment in root OA exudation can reach 20% of the total carbon fixed by photosynthesis, highlighting the importance of OA in enhancing Pi uptake (Lynch and Ho 2005). In white lupin, OA exudation from proteoid roots has been associated with enhanced activity of some metabolic enzymes, such as phosphoenolpyruvate carboxylase, malate dehydrogenase, and citrate synthase (O'Rourke et al. 2013; Uhde-Stone et al. 2003). The quantity and type of OAs secreted by roots determine the level of Pi solubilization in the rhizosphere. The synthesis and exudation of piscidic acid by the roots of some plants, such as pigeon pea (*Cajanus cajan*), have been reported to be highly effective in releasing Pi from iron–Pi and aluminum–Pi complexes (Ishikawa et al. 2002). Pigeon pea has been used as an intercrop because of its capacity to enhance Pi availability. Piscidic acid exudation has not been reported in crops, and the mechanisms of its synthesis and exudation have not been elucidated.

5.5.4 Production and Secretion of Phosphatases During Low Phosphate Availability

Between 30% and 70% of the total P in agricultural and natural soils is present in organic forms. Therefore, the production and secretion of different types of enzymes, such as acid phosphatases (ACPases) and nucleases, into the rhizosphere contribute to the release of Pi from these organic sources (Brinch-Pedersen et al. 2002). For instance, purple ACPases can act on a wide range of organic molecules, cleaving Pi from ester linkage sites, and are relatively stable over wide intervals of pH (4.0–7.6) and temperature (22–60 °F). The importance of purple ACPases in plant Pi nutrition has been clearly shown through the study of the *pup* (*phosphatase under-producer*). Purple ACPase synthesis and secretion have been important in developing low Pi tolerance in soybean and common bean (Liang et al. 2010; Wang et al. 2009; Shu et al. 2014, 2015) (Fig. 5.1C).

5.6 The Response of Mycorrhizae to Pi Deficiency

Symbiosis with AM fungi is one of the most important strategies for host plant adaptation to Pi deficiency. The AM fungi can enhance the availability of mineral nutrients by the small diameter of extraradical hyphae, activating soil nutrients through the release of organic acids and soil enzymes, ensuring the efficiency of soil nutrient transport into extraradical hyphae by lowering the value of the Km nutrient transporter on extraradical hyphae, ensuring the rate of nutrition transport in intraradical hyphae by converting nutrient ions into suitable forms, and promoting the efficiency of the nutrient transport into the host plant by inducing symbiotic plant nutrient transporters. All these aspects effect the phosphate stress adaptation of the host itself.

5.6.1 Inducing pht1 Family Transporters in AM Symbiosis

In general, the Pht1 transporters are divided into three different expression patterns, specially induced, repressed, and constituted by AM fungi. While the Pht1 transporter genes are expressed exclusively during AM symbiosis, their expression is also upregulated substantially in mycorrhizal roots relative to non-colonized control roots (Rausch et al. 2001; Harrison et al. 2002; Paszkowski et al. 2002; Karandashov et al. 2004; Glassop et al. 2005; Nagy et al. 2005, 2006; Maeda et al. 2006; Wegmüller et al. 2008; Grace et al. 2009; Shu et al. 2012; Zhang et al. 2016). Consequently, these transporters are prime candidates for the role of symbiotic Pi transporters. Phylogenetic analyses indicate that AM-induced Pi transporters are dispersed throughout the Pht1 family, which suggests that they are detected in different ways in the different plant families. However, one exception exists. One clade within the Pht1 family, referred to as Pht1, subfamily I (Nagy et al. 2005), contains only AM-induced transporters. Most plants have a single Pi transporter gene in subfamily I except for members of the Solanaceae, in which a duplication event has caused two paralogs. In mycorrhizal roots, the clade within Pht1 is expressed only in cells with arbuscules, while the protein is localized to the periarbuscular membrane, specifically to the regions around the fine branches of the arbuscules (Harrison et al. 2002; Pumplin and Harrison 2009). The Pht1 transporter induced by AM fungi is also highly expressed in Pi-deprived roots, which suggests that the AM Pi absorption pathway plays an important role in Pi deficiency adaptation of mycorrhizal plants (Koltai and Kapulnik 2010).

5.6.2 Promoting Phosphatases in AM Roots

The AM fungi systemically regulated the phosphatase activities, while the mycorrhizal root showed higher activities of phosphatase under Pi deficiency conditions, which was widely accepted. The upregulation had been attributed to two causes. The first was phosphatase upregulation by the fungi itself, and the other was phosphatase upregulation of the plant induced by fungi colonization. Localization of phosphatase activities in the separated intraradical hyphae could easily be observed (Saito 1995). Localization of phosphatase activities in the separated intraradical hyphae of three AM fungus species, *G. mosseae*, *Glomus etunicatum*, and *Gigaspora rosea*, and inhibition studies with the phosphatases indicated that different types of phosphatases may be found in these species even though the localization patterns of their phosphatase activity were the same (Ezawa et al. 1995). Cyto-/histochemical studies have shown the localization of ALPase in arbuscular hyphae (Ezawa et al. 1995; Gianinazzi et al. 1979; Saito 1995). Staining for ALPase activity has been used to estimate the physiological activity of mycorrhizas under various conditions (Jabaji-Hare et al. 1990). The enzyme is localized in vacuoles in intraradical hyphae. The number of ALPase-active arbuscules correlates with the increase in phosphorus uptake by mycorrhizal plants, suggesting that arbuscular ALPase plays a significant role in phosphorus transfer from the AM fungus to the host plant. The activity of ACPase in the crude protein from intraradical hyphae is comparable to, or much greater than, that of ALPase (Saito 1995). However, because this enzyme is mostly soluble, its localization is not well identified. It has been found in relatively young hyphae, suggesting its involvement in hyphal synthesis (Gianinazzi et al. 1979). A similar tendency in neutral phosphatase localization also was found (Jeanmaire et al. 1985). ACPase was found in vacuoles of intraradical hyphae separated from the roots (Ezawa et al. 1995; Saito 1995). In contrast to ALPase, most ACPases have a wide range of substrate specificity, and their physiological roles are not well defined (Duff et al. 1994). Mycorrhiza-specific phosphatases have been found by conventional biochemical methods. A mycorrhiza-specific soluble ALPase was first found by polyacrylamide gel electrophoresis (PAGE) of crude mycorrhizal root extracts (Gianinazzi-Pearson and Gianinazzi 1976, 1978). This ALPase showed characteristics of alkaline phosphomonoesterase, and its activity was closely related to the development of fungal colonization. However, its function in the symbiosis has not yet been clarified. Other mycorrhiza-specific phosphatases were also separated by PAGE (Fabig et al. 1989; Kojima et al. 1998). Phosphatases specific to mycorrhizas found by Fabig et al. (1989) seemed to be of hyphal origin because the activities remained after the digestion of host tissues by cellulase. Kojima et al. (1998) compared the relative mobility of a mycorrhiza-specific phosphatase extracted from roots vs that extracted from intraradical hyphae, which were separated from mycorrhizal roots by enzymatic digestion. They concluded that the phosphatase band was of hyphal origin. One mycorrhiza-specific phosphatase has been partially purified from the *Tagetes patula* (marigold)–*G. etunicatum* symbiosis. The partially purified phosphatase showed characteristics of a nonspecific ACPase in inhibition and substrate specificity tests. Its sequence showed high homology to secreted purple ACPases from *Phaseolus* and *Arabidopsis*, indicating that this enzyme is of host plant origin. Liu et al. (1998) and Shu et al. (2014, 2015) found that a homologous gene of the purple ACPase, induced by phosphate starvation, is downregulated in mycorrhizal roots. This contrasts with the ACPase found by Ezawa and Yoshida (1994), which was increased by mycorrhizal colonization (López-Arredondo et al. 2014) (Fig. 5.1c).

5.7 Future Direction

Nutrient exchange is a basic function of AM. Although information has accumulated through the use of new methods, only fragmentary pieces of the whole picture have been found. Therefore, further work in several areas is essential. The primary need is to clarify the function of the molecules involved in mycorrhizal symbiosis for sensing and adapting to Pi deficiency. Another need is to identify the cross talk of the molecules from host plants and AM fungi for sensing and adapting to Pi deficiency. Third, the network of molecules involved in sensing and adapting to Pi deficiency should be characterized. The "fair trade" of phosphates for carbohydrates between host plants and AM fungi has been widely accepted. Photosynthesis of the host was repressed by Pi deficiency, but the colonization of AM increased under that condition. Identifying how these balances occur between the two partners can be used to improve the strategies of host Pi deficiency adaptation.

Acknowledgments This work was supported by the National Natural Science Fund of China (Project No. 31401818), the Fund on Basic Scientific Research project of Nonprofit Central Research Institutions (No. SSCRI-1630062014006), Germplasm Resource Protection of Ministry of Agriculture (2016NWB053), China Litchi and Longan Industry Technology Research System (Project No.CARS-33-11), and the Plan in Scientific and Technological Innovation Team of Outstanding Young, Hubei Provincial Department of Education (T201604).

References

Ai P, Sun S, Zhao J et al (2009) Two rice phosphate transporters, OsPht1;2 and OsPht1;6, have different functions and kinetic properties in uptake and translocation. Plant J 57:798–809

Akiyama K, Hayashi H (2006) Chemical identification of strigolactones as a host-recognition signal for arbuscular mycorrhizal fungi. Regul Plant Growth Dev 41:141–149

Akiyama K, Matsuzaki K, Hayashi H (2005) Plant sesquiterpenes induce hyphal branching in arbuscular mycorrhizal fungi. Nature 435:824–827

Akiyama K, Ogasawara S, Ito S, Hayashi H (2010) Structural requirements of strigolactones for hyphal branching in AM fungi. Plant Cell Physiol 51:1104–1117

Alkan N, Gadkar V, Yarden O et al (2006) Analysis of quantitative interactions between two species of arbuscular mycorrhizal fungi, Glomus mosseae and G. intraradices, by real-time PCR. Appl Environ Microbiol 72:4192–4199

Aung K, Lin SI, Wu CC et al (2006) *pho2*, a phosphate overaccumulator, is caused by a nonsense mutation in a microRNA399 target gene. Plant Physiol 141:1000–1011

Balzergue C, Chabaud M, Barker DG et al (2013) High phosphate reduces host ability to develop arbuscular mycorrhizal symbiosis without affecting root calcium spiking responses to the fungus. Front Plant Sci 4:426

Bari R, Pant BD, Stitt M et al (2006) PHO2, microRNA399, and PHR1 define a phosphate-signaling pathway in plants. Plant Physiol 141:988–999

Basu P, Zhang YJ, Lynch JP et al (2007) Ethylene modulates genetic, positional, and nutritional regulation of root plagiogravitropism. Funct Plant Biol 34:41–51

Besserer A, Puech-Pagés V, Kiefer P et al (2006) Strigolactones stimulate arbuscular mycorrhizal fungi by activating mitochondria. PLoS Biol 4:e226

Besserer A, Becard G, Jauneau A et al (2008) GR24, a synthetic analog of strigolactones, stimulates the mitosis and growth of the arbuscular mycorrhizal fungus *Gigaspora rosea* by boosting its energy metabolism. Plant Physiol 148:402–413

Bieleski RL (1973) Phosphate pools, phosphate transport and phosphate availability. Annu Rev Plant Physiol 24:225–252

Bitterlich M, Krügel U, Boldt-Burisch K et al (2014) The sucrose transporter SlSUT2 from tomato interacts with brassinosteroid functioning and affects arbuscular mycorrhiza formation. Plant J 78:877–889

Borch K, Bouma TJ, Lynch JP et al (1999) Ethylene: a regulator of root architectural responses to soil phosphorus availability. Plant Cell Environ 22:425–431

Bothe H, Klingner A, Kaldorf M et al (1994) Biochemical approaches to the study of plant–fungal interactions in arbuscular mycorrhiza. Experientia 50:919–925

Brachmann A, Parniske M (2006) The most widespread symbiosis on earth. PL Biol 4:1111–1112

Breuillin F, Schramm J, Hajirezaei M et al (2010) Phosphate systemically inhibits development of arbuscular mycorrhiza in *Petunia hybrida* and represses genes involved in mycorrhizal functioning. Plant J 64:1002–1017

Brinch-Pedersen H, Sørensen LD, Holm PB (2002) Engineering crop plants: getting a handle on phosphate. Trends Plant Sci 7:118–125

Bucher M, Hause B, Krajinski F et al (2014) Through the doors of perception to function in arbuscular mycorrhizal symbioses. New Phytol 204:833–840

Buhtz A, Springer F, Chappell L et al (2008) Identification and characterization of small RNAs from the phloem of *Brassica napus*. Plant J 53:739–749

Burleigh SH, Harrison MJ (1999) The down-regulation of *Mt4*-like genes by phosphate fertilization occurs systemically and involves phosphate translocation to the shoots. Plant Physiol 119:241–248

Bustos R, Castrillo G, Linhares F et al (2010) A central regulatory system largely controls transcriptional activation and repression responses to phosphate starvation in *Arabidopsis*. PLoS Genet 6:e1001102

Carbonnel S, Gutjahr C (2014) Control of arbuscular mycorrhiza development by nutrient signals. Front Plant Sci 5:462

Carswell C, Grant BR, Theodorou ME et al (1996) The fungicide phosphonate disrupts the phosphate-starvation response in *Brassica nigra* seedlings. Plant Physiol 110:105–110

Carswell MC, Grant BR, Plaxton WC (1997) Disruption of the phosphate-starvation response of oilseed rape suspension cells by the fungicide phosphonate. Planta 203:67–74

Chapin LJ, Jones ML (2009) Ethylene regulates phosphorus remobilization and expression of a phosphate transporter (PhPT1) during petunia corolla senescence. J Exp Bot 60:2179–2190

Chen A, Gu M, Sun S et al (2011) Identification of two conserved cis-acting elements, MYCS and P1BS, involved in the regulation of mycorrhiza-activated phosphate transporters in eudicot species. New Phytol 189:1157–1169

Chiou TJ, Lin SI (2011) Signaling network in sensing phosphate availability in plants. Annu Rev Plant Biol 62:185–206

Ciereszko I, Kleczkowski LA (2002) Effects of phosphate deficiency and sugars on expression of rab18 in Arabidopsis: hexokinase-dependent and okadaic acid-sensitive transduction of the sugar signal. Biochim Biophys Acta 1579:43–49

Danneberg G, Latus C, Zimmer W et al (1992) Influence of vesicular–arbuscular mycorrhiza on phytohormone balances in maize (*Zea mays* L.) J Plant Physiol 141:33–39

Danova-Alt R, Dijkema C, DeWaard P et al (2008) Transport and compartmentation of phosphate in higher plant cells-kinetic and ^{31}P nuclear magnetic resonance studies. Plant Cell Environ 31:1510–1521

Duff SMG, Sarath G, Plaxton WC (1994) The role of acid phosphatases in plant phosphorus metabolism. Physiol Plant 90:791–800

Espinosa F, Garrido I, Ortega A et al (2014) ROS, redox activities and NO production by axenically cultured olive seedling roots after interaction with mycorrhizal and pathogenic fungi. PLoS One 9:e100132

Espinosa F, Garrido I, Álvarez-Tinaut MC (2015) Reflexions on some aspects of the interactions among ROS, RNS, and Ca^{2+} in response to a mycorrhizal or a pathogenic fungus. Plant Signal Behav 10:e1049789

Etemadi M, Gutjahr C, Couzigou JM et al (2014) Auxin perception is required for arbuscule development in arbuscular mycorrhizal symbiosis. Plant Physiol 166:281–292

Ezawa T, Yoshida T (1994) Characterization of phosphatase in marigold roots infected with vesicular-arbuscular mycorrhizal fungi. Soil Sci Plant Nutr 40:255–264

Ezawa T, Saito M, Yoshida T (1995) Comparison of phosphatase localization in the intraradical hyphae of arbuscular mycorrhizal fungi, *Glomus* spp. and *Gigaspora* spp. Plant Soil 176:57–63

Fabig B, Vielhauer K, Moawad A et al (1989) Gas-chromatographic separation of organic acids and electrophoretic determination of phosphatases from VA mycorrhizal roots. Z Pflanzenernahr Bodenk 152:261–265

Fan C, Wang X, Hu R et al (2013) The pattern of phosphate transporter 1 genes evolutionary divergence in *Glycine max* L. BMC Plant Biol 3:48

Fellbaum CR, Gachomo EW, Beesetty Y et al (2012) Carbon availability triggers fungal nitrogen uptake and transport in arbuscular mycorrhizal symbiosis. Proc Natl Acad Sci U S A 109:2666–2671

Ferguson BJ, Beveridge CA (2009) Roles for auxin, cytokinin, and strigolactone in regulating shoot branching. Plant Physiol 149:1929–1944

Floss DS, Levy JG, Levesque-Tremblay V et al (2013) DELLA proteins regulate arbuscule formation in arbuscular mycorrhizal symbiosis. Proc Natl Acad Sci U S A 110:E5025–E5034

Foo E, Yoneyama K, Hugill C et al (2012) Strigolactones and the regulation of pea symbioses in response to nitrate and phosphate deficiency. Mol Plant 6:76–87

Foo E, Ross JJ, Jones WT et al (2013) Plant hormones in arbuscular mycorrhizal symbioses: an emerging role for gibberellins. Ann Bot 111:769–779

Forde B, Lorenzo H (2001) The nutritional control of root development. Plant Soil 232:51–68

Franco-Zorrilla JM, Valli A, Todesco M et al (2007) Target mimicry provides a new mechanism for regulation of microRNA activity. Nat Genet 39:1033–1037

Gaffney T, Friedrich L, Vernooij B et al (1993) Requirement of salicylic acid for the induction of systemic acquired resistance. Science 261:754–756

García-Garrido JM, Lendzemo V, Castellanos-Morales V et al (2009) Strigolactones, signals for parasitic plants and arbuscular mycorrhizal fungi. Mycorrhiza 19:449–459

Genre A, Chabaud M, Balzergue C et al (2013) Short-chain chitin oligomers from arbuscular mycorrhizal fungi trigger nuclear Ca^{2+} spiking in *Medicago truncatula* roots and their production is enhanced by strigolactone. New Phytol 198:190–202

Gianinazzi S, Gianinazzi-Pearson V, Dexheimer J (1979) Enzymatic studies on the metabolism of vesicular -arbuscular mycorrhiza. 3. Ultrastructural localization of acid and alkaline phosphatase in onion roots infected by *Glomus mosseae* (Nicol.&Gerd). New Phytol 82:127–132

Gianinazzi-Pearson V, Gianinazzi S (1976) Enzymatic studies on the metabolism of vesicular-arbuscular mycorrhiza. I effect of mycorrhiza formation and phosphorus nutrition on soluble phosphatase activities. Physiol Veg 14:833–841

Gianinazzi-Pearson V, Gianinazzi S (1978) Enzymatic studies on the metabolism of vesicular -arbuscular mycorrhiza. IT. Soluble alkaline phosphatase specific to mycorrhizal infection in onion roots. Physiol. Plant Pathol 12:45–53

Gibson SI (2004) Sugar and phytohormone response pathways: navigating a signalling network. J Exp Bot 55:253–264

Gilbert GA, Knight JD, Vance CP et al (2000) Proteoid root development of phosphorus deficient lupin is mimicked by auxin and phosphonate. Ann Bot 85:921–928

Ginzberg I, David R, Shaul O et al (1998) Glomus intraradices colonization regulates gene expression in tobacco roots. Symbiosis 25:145–157

Giots F, Donaton MCV, Thevelein JM (2003) Inorganic phosphate is sensed by specific phosphate carriers and acts in concert with glucose as a nutrient signal for activation of the protein kinase a pathway in the yeast *Saccharomyces cerevisiae*. Mol Microbiol 47:1163–1181

Glassop D, Smith SE, Smith FW (2005) Cereal phosphate transporters associated with the mycorrhizal pathway of phosphate uptake into roots. Planta 222:688–698

Goff SA, Ricke D, Lan TH et al (2002) A draft sequence of the rice genome (*Oryza sativa* L. ssp. *japonica*). Science 296:92–100

Gomez-Roldan V, Roux C, Girard D et al (2007) Strigolactones: promising plant signals. Plant Signal Behav 2:163–164

Gomez-Roldan V, Fermas S, Brewer PB et al (2008) Strigolactone inhibition of shoot branching. Nature 455:189–194

Gough C, Cullimore J (2011) Lipo-chitooligosaccharide signaling in endosymbiotic plant-microbe interactions. Mol Plant-Microbe Interact 24:867–878
Grace EJ, Cotsaftis O, Tester M et al (2009) Arbuscular mycorrhizal inhibition of growth in barley cannot be attributed to extent of colonization, fungal phosphorus uptake or effects on expression of plant phosphate transporter genes. New Phytol 181:938–949
Grunwald U, Guo W, Fischer K et al (2009) Overlapping expression patterns and differential transcript levels of phosphate transporter genes in arbuscular mycorrhizal, Pi-fertilised and phytohormone-treated *Medicago truncatula* roots. Planta 229:1023–1034
Gryndler M, Hrselova H, Chvatalova I et al (1998) The effect of selected plant hormones on in vitro proliferation of hyphae of Glomus Fistulosum. Biol Plant 41:255–263
Gutjahr C (2014) Phytohormone signaling in arbuscular mycorhiza development. Curr Opin Plant Biol 20:26–34
Gutjahr C, Parniske M (2013) Cell and developmental biology of arbuscular mycorrhiza symbiosis. Annu Rev Cell Dev Biol 29:593–617
Gutjahr C, Radovanovic D, Geoffroy J et al (2012) The half-size ABC transporters STR1 and STR2 are indispensable for mycorrhizal arbuscule formation in rice. Plant J 69:906–920
Hammond JP, White PJ (2008) Sucrose transport in the phloem: integrating root responses to phosphorus starvation. J Exp Bot 59:93–109
Harrison MJ, Dewbre GR, Liu J (2002) A phosphate transporter from *Medicago truncatula* involved in the acquisition of phosphate released by arbuscular mycorrhizal fungi. Plant Cell 14:2413–2429
Hause B, Schaarschmidt S (2009) The role of jasmonates in mutualistic symbioses between plants and soil-born microorganisms. Phytochemistry 70:1589–1599
Hause B, Maier W, Miersch O et al (2002) Induction of jasmonate biosynthesis in arbuscular mycorrhizal barley roots. Plant Physiol 130:1213–1220
Hause B, Mrosk C, Isayenkov S et al (2007) Jasmonates in arbuscular mycorrhizal interactions. Phytochemistry 68:101–110
He Z, Ma Z, Brown KM et al (2005) Assessment of inequality of root hair density in *Arabidopsis thaliana* using the Gini coefficient: a close look at the effect of phosphorus and its interaction with ethylene. Ann Bot 95:287–293
Hernández G, Ramirez M, Valdés-López O et al (2007) Phosphorus stress in common bean: root transcript and metabolic responses. Plant Physiol 144:752–767
Herrera-Medina M, Gagnon H, Piche Y et al (2003) Root colonization by arbuscular mycorrhizal fungi is affected by the salicylic acid content of the plant. Plant Sci 164:993–998
Herrera-Medina MJ, Steinkellner S, Vierheilig H et al (2007) Abscisic acid determines arbuscule development and functionality in the tomato arbuscular mycorrhiza. New Phytol 175:554–564
Ho CH, Lin SH, Hu HC et al (2009) CHL1 functions as a nitrate sensor in plants. Cell 138:1184–1194
Holsbeeks I, Lagatie O, Van Nuland A et al (2004) The eukaryotic plasma membrane as a nutrient-sensing device. Trends Biochem Sci 29:556–564
Illana A, Garcia-Garrido J, Sampedro I et al (2011) Strigolactones seem not to be involved in the nonsusceptibilty of arbuscular mycorrhizal (AM) nonhost plants to AM fungi. Botany 89:285–288
Isayenkov S, Mrosk C, Stenzel I et al (2005) Suppression of allene oxide cyclase in hairy roots of *Medicago truncatula* reduces jasmonate levels and the degree of mycorrhization with *Glomus intraradices*. Plant Physiol 139:1401–1410
Ishikawa S, Adu-Gyamfi JJ, Nakamura T et al (2002) Genotypic variability in phosphorus solubilizing activity of root exudates by pigeonpea grown in low-nutrient environments. Plant Soil 245:71–81
Jabaji-Hare SH, Therien J, Charest PM (1990) High resolution cytochemical study of the vesicular-arbuscular mycorrhizal association, *Glomus clarum -Allium porrum*. New Phytol 114:481–496

Jain A, Poling MD, Karthikeyan AS et al (2007) Differential effects of sucrose and auxin on localized phosphate deficiency-induced modulation of different traits of root system architecture in Arabidopsis. Plant Physiol 144:232–247

Javot H, Pumplin N, Harrison M (2007a) Phosphate in the arbuscular mycorrhizal symbiosis: transport properties and regulatory roles. Plant Cell Environ 30:310–322

Javot H, Varma Penmetsa R, Terzaghi N et al (2007b) A *Medicago truncatula* phosphate transporter indispensible for the arbuscular mycorrhizal symbiosis. Proc Natl Acad Sci U S A 104:1720–1725

Jeanmaire C, Dexheimer J, Marx C et al (1985) Effect of vesicular -arbuscular mycorrhizal infection on the distribution of neutral phosphatase activities in root cortical cells. J. Plant Physiol 119:285–293

Jeschke WD, Peuke AD, Pate JS et al (1997) Transport, synthesis and catabolism of abscisic acid (ABA) in intact plants of castor bean (*Ricinus communis* L.) under phosphate deficiency and moderate salinity. J Exp Bot 48:1737–1747

Jiang C, Gao X, Liao L et al (2007) Phosphate starvation root architecture and anthocyanin accumulation responses are modulated by the gibberellin-DELLA signaling pathway in *Arabidopsis*. Plant Physiol 145:1460–1470

Karandashov V, Nagy R, Wegmuller S et al (2004) Evolutionary conservation of a phosphate transporter in the arbuscular mycorrhizal symbiosis. Proc Natl Acad Sci U S A 101:6285–6290

Karthikeyan A, Varadarajan D, Jain A et al (2007) Phosphate starvation responses are mediated by sugar signaling in Arabidopsis. Planta 225:907–918

Kiers ET, Duhamel M, Beesetty Y et al (2011) Reciprocal rewards stabilize cooperation in the mycorrhizal symbiosis. Science 333:880–882

Kobayashi K, Masuda T, Ki T et al (2006) Membrane lipid alteration during phosphate starvation is regulated by phosphate signaling and auxin/cytokinin cross-talk. Plant J 47:238–248

Kohlen W, Charnikhova T, Lammers M et al (2012) The tomato CAROTENOID CLEAVAGE DIOXYGENASE8 (SlCCD8) regulates rhizosphere signaling, plant architecture and affects reproductive development through strigolactone biosynthesis. New Phytol 196:535–547

Kojima T, Hayatsu M, Saito M (1998) Intraradical hyphae phosphatase of the arbuscular mycorrhizal fungus, *Gigaspora margarita*. Biol Fertil Soils 26:331–335

Koltai H, Kapulnik Y (eds) (2010) Arbuscular mycorrhizas: physiology and function. Springer, Dordrecht/New York

Koltai H, Dor E, Hershenhorn J et al (2010) Strigolactones' effect on root growth and root-hair elongation may be mediated by auxin-efflux carriers. J Plant Growth Regul 29:129–136

Kretzschmar T, Kohlen W, Sasse J et al (2012) A petunia ABC protein controls strigolactone-dependent symbiotic signalling and branching. Nature 483:341–344

Kuiper D, Schuit J, Kuiper P (1988) Effects of internal and external cytokinin concentrations on root growth and shoot to root ratio of *Plantago major ssp. pleiosperma* at different nutrient conditions. Plant Soil 111:231–236

Kulik A, Noirot E, Grandperret V et al (2015) Interplays between nitric oxide and reactive oxygen species in cryptogein signalling. Plant Cell Environ 38:331–348

Lai F, Thacker J, Li Y et al (2007) Cell division activity determines the magnitude of phosphate starvation responses in Arabidopsis. Plant J 50:545–556

Lambers H, Finnegan PM, Laliberte E et al (2011) Phosphorus nutrition of proteaceae in severely phosphorus-impoverished soils: are there lessons to be learned for future crops? Plant Physiol 156:1058–1066

Lambers H, Clements JC, Nelson MN (2013) How a phosphorus-acquisition strategy based on carboxylate exudation powers the success and agronomic potential of lupines (Lupinus, Fabaceae). Am J Bot 100:263–288

Lendzemo V, Kuyper TW, Urban A et al (2009) The arbuscular mycorrhizal host status of plants cannot be linked with the striga seed-germination-activity of plant root exudates. J Plant Dis Protect 11:86–89

Li K, Xu C, Li Z et al (2008) Comparative proteome analyses of phosphorus responses in maize (*Zea mays* L.) roots of wild-type and a low-P-tolerant mutant reveal root characteristics associated with phosphorus efficiency. Plant J 55:927–939

Li S, Dong L (2015) Ethylene and plant responses to phosphate deficiency. Front Plant Sci 6:796

Lian B, Zhou X, Miransari M et al (2000) Effects of salicylic acid on the development and root nodulation of soybean seedlings. J Agron Crop Sci 185:187–192

Liang C, Tian J, Lam HM et al (2010) Biochemical and molecular characterization of PvPAP3, a novel purple acid phosphatase isolated from common bean enhancing extracellular ATP utilization. Plant Physiol 152:854–865

Liao H, Yan X, Rubio G et al (2004) Genetic mapping of basal root gravitropism and phosphorus acquisition efficiency in common bean. Funct Plant Biol 31:959–970

Lin SI, Chiang SF, Lin WY et al (2008) Regulatory network of microRNA399 and *PHO2* by systemic signaling. Plant Physiol 147:732–746

Liu H, Trieu AT, Blaylock LA et al (1998) Cloning and characterization of two phosphate transporters from *Medicago truncatula* roots: regulation in response to phosphate and to colonization by arbuscular mycorrhizal (AM) fungi. Mol Plant-Microbe Interact 11:14–22

Liu J, Samac DA, Bucciarelli B et al (2005) Signaling of phosphorus deficiency-induced gene expression in white lupin requires sugar and phloem transport. Plant J 41:257–268

Liu J, Versaw WK, Pumplin N et al (2008) Closely related members of the *Medicago truncatula* PHT1 phosphate transporter gene family encode phosphate transporters with distinct biochemical activities. J Biol Chem 283:24673–24681

Liu TY, Chang CY, Chiou TJ (2009) The long-distance signaling of mineral macronutrients. Curr Opin Plant Biol 12:312–319

Liu JQ, Allan DL, Vance CP (2010) Systemic signaling and local sensing of phosphate in common bean: cross-talk between photosynthate and microRNA399. Mol Plant 3:428–437

Lloyd JC, Zakhleniuk OV (2004) Responses of primary and secondary metabolism to sugar accumulation revealed by microarray expression analysis of the Arabidopsis mutant, pho3. J Exp Bot 55:1221–1230

López-Arredondo DL, Leyva-González MA, González-Morales SI et al (2014) Phosphate nutrition: improving low-phosphate tolerance in crops. Annu Rev Plant Biol 65:95–123

López-Bucio J, Cruz-Ramírez A, Herrera-Estrella L (2003) The role of nutrient availability in regulating root architecture. Curr Opin Plant Biol 6:280–287

López-Ráez JA, Charnikhova T, Gómez-Roldán V et al (2008) Tomato strigolactones are derived from carotenoids and their biosynthesis is promoted by phosphate starvation. New Phytol 178:863–874

Lorenzo O, Solano R (2005) Molecular players regulating the jasmonate signalling network. Curr Opin Plant Biol 8:532–540

Lota F, Wegmüller S, Buer B et al (2013) The cis-acting CTTC-P1BS module is indicative for gene function of LjVTI12, a Qb-SNARE protein gene, required for arbuscule formation in *Lotus japonicus*. Plant J 74:280–293

Lough TJ, Lucas WJ (2006) Integrative plant biology: role of phloem long – distance macromolecular trafficking. Annu Rev Plant Biol 57:203–232

Lucas WJ, Groover A, Lichtenberger R et al (2013) The plant vascular system: evolution, development and functions. J Integr Plant Biol 55:294–388

Ludwig-Muller J (2000) Hormonal balance in plants during colonization by mycorrhizal fungi. In: Kapulnik Y, Douds D (eds) Arbuscular mycorrhizas: physiology and function. Kluwer Academic Publishers, Amsterdam, pp 263–283

Ludwig-Muller J (2010) Hormonal responses in host plants triggered by arbuscular mycorrhizal fungi. In: Kapulnik Y, Koltai H (eds) Arbuscular mycorrhizas: physiology and function, 2nd edn. Springer, Dordrecht, p 169

Ludwig-Müller J (2011) Auxin conjugates: their role for plant development and in the evolution of land plants. J Exp Bot 62:1757–1773

Ludwig-Müller J, Güther M (2007) Auxins as signals in arbuscular mycorrhiza formation. Plant Signal Behav 2:194–196
Lynch JP (2011) Root phenes for enhanced soil exploration and phosphorus acquisition: tools for future crops. Plant Physiol 156:1041–1049
Lynch JP, Ho MD (2005) Rhizoeconomics: carbon costs of phosphorus acquisition. Plant Soil 269:45–56
Ma Z, Baskin TI, Brown KM, Lynch JP (2003) Regulation of root elongation under phosphorus stress involves changes in ethylene responsiveness. Plant Physiol 131:1381–1390
Mabood F, Smith DL (2005) Pre-incubation of Bradyrhizobium japonicum with jasmonates accelerates nodulation and nitrogen fixation in soybean (*Glycine max*) at optimal and suboptimal root zone temperatures. Physiol Plant 125:311–323
Maeda D, Ashida K, Iguchi K et al (2006) Knockdown of an arbuscular mycorrhiza-inducible phosphate transporter gene of *Lotus japonicus* suppresses mutualistic symbiosis. Plant Cell Physiol 47:807–817
Martín-Rodríguez JÁ, León-Morcillo R, Vierheilig H et al (2011) Ethylene-dependent/ethylene-independent ABA regulation of tomato plants colonized by arbuscular mycorrhiza fungi. New Phytol 190:193–205
Martín-Rodríguez JÁ, Ocampo JA, Molinero-Rosales N et al (2014) Role of gibberellins during arbuscular mycorrhizal formation in tomato: new insights revealed by endogenous quantification and genetic analysis of their metabolism in mycorrhizal roots. Physiol Plant 154:66–81
Matusova R, Rani K, Verstappen FW et al (2005) The strigolactone germination stimulants of the plant-parasitic striga and Orobanche spp. are derived from the carotenoid pathway. Plant Physiol 139:920–934
Maucher H, Hause B, Feussner I et al (2000) Allene oxide synthases of barley (*Hordeum vulgare* cv. Salome): tissue specific regulation in seedling development. Plant J 21:199–213
McArthur DA, Knowles NR (1992) Resistance responses of potato to vesicular-arbuscular mycorrhizal fungi under varying abiotic phosphorus levels. Plant Physiol 100:341–351
Meixner C, Ludwig-Müller J, Miersch O et al (2005) Lack of mycorrhizal autoregulation and phytohormonal changes in the supernodulating soybean mutantnts1007. Planta 222:709–715
Miransari M (2010a) Contribution of arbuscular mycorrhizal symbiosis to plant growth under different types of soil stress. Plant Biol 12:563–569
Miransari M (2010b) Arbuscular mycorrhiza and soil microbes. In: Thangadurai D, Busso CA, Hijri M (eds) Mycorrhizal biotechnology. Science Publishers/CRC Press/Taylor & Francis Group, Enfield/Boca Raton/USA, p 226
Miransari M, Bahrami HA, Rejali F et al (2007) Using arbuscular mycorrhiza to reduce the stressful effects of soil compaction on corn (*Zea mays* L.) growth. Soil Biol Biochem 39:2014–2026
Miransari M, Bahrami HA, Rejali F et al (2008) Using arbuscular mycorrhiza to reduce the stressful effects of soil compaction on wheat (*Triticum aestivum* L.) growth. Soil Biol Biochem 40:1197–1206
Miransari M, Abrishamchi A, Khoshbakht K et al (2014) Plant hormones as signals in arbuscular mycorrhizal symbiosis. Crit Rev Biotechnol 34:123–133
Morcuende R, Bari R, Gibon Y et al (2007) Genome-wide reprogramming of metabolism and regulatory networks of Arabidopsis in response to phosphorus. Plant Cell Environ 30:85–112
Muchhal US, Liu C, Raghothama K (1997) Ca^{2+}-ATPase is expressed differentially in phosphate starved roots of tomato. Physiol Plant 101:540–544
Mukherjee A, Ané JM (2011) Germinating spore exudates from arbuscular mycorrhizal fungi: molecular and developmental responses in plants and their regulation by ethylene. Mol Plant-Microbe Interact 24:260–270
Nagarajan VK, Smith AP (2012) Ethylene's role in phosphate starvation signaling: more than just a root growth regulator. Plant Cell Physiol 53:277–286
Nagy R, Karandashov V, Chague V et al (2005) The characterization of novel mycorrhiza-specific phosphate transporters from *Lycopersicon esculentum* and *Solanum tuberosum* uncovers func-

tional redundancy in symbiotic phosphate transport in solanaceous species. Plant J 42:236–250
Nagy R, Vasconcelos MJV, Zhao S et al (2006) Differential regulation of five Pht1 phosphate transporters from maize (*Zea mays* L.) Plant Biol 8:186–197
Nagy R, Drissner D, Amrhein N et al (2009) Mycorrhizal phosphate uptake pathway in tomato is phosphorus-repressible and transcriptionally regulated. New Phytol 181:950–959
Nakano-Hylander A, Olsson PA (2007) Carbon allocation in mycelia of arbuscular mycorrhizal fungi during colonisation of plant seedlings. Soil Biol Biochem 39:1450–1458
Narise T, Kobayashi K, Baba S et al (2010) Involvement of auxin signaling mediated by IAA14 and ARF7/19 in membrane lipid remodeling during phosphate starvation. Plant Mol Biol 72:533–544
Nielsen KL, Eshel A, Lynch JP (2001) The effect of phosphorus availability on the carbon economy of contrasting common bean (*Phaseolus vulgaris* L.) genotypes. J Exp Bot 52:329–339
Nukui N, Ezura H, Minamisawa K (2004) Transgenic Lotus japonicus with an ethylene receptor gene Cm-ERS1/H70A enhances formation of infection threads and nodule primordia. Plant Cell Physiol 45:427–435
O'Rourke JA, Yang SS, Miller SS et al (2013) An RNA-Seq transcriptome analysis of orthophosphate deficient white lupin reveals novel insights into phosphorus acclimation in plants. Plant Physiol 161:705–724
Oldroyd GE, Engstrom EM, Long SR (2001) Ethylene inhibits the nod factor signal transduction pathway of *Medicago truncatula*. Plant Cell 13:1835–1849
Pant BD, Buhtz A, Kehr J et al (2008) MicroRNA399 is a long-distance signal for the regulation of plant phosphate homeostasis. Plant J 53:731–738
Paszkowski U, Kroken S, Roux C et al (2002) Rice phosphate transporters include an evolutionarily divergent gene specifically activated in arbuscular mycorrhizal symbiosis. Proc Natl Acad Sci U S A 99:13324–13329
Pérez-Torres CA, López-Bucio J, Cruz-Ramírez A et al (2008) Phosphate availability alters lateral root development in Arabidopsis by modulating auxin sensitivity via a mechanism involving the TIR1 auxin receptor. Plant Cell 20:3258–3272
Popova Y, Thayumanavan P, Lonati E et al (2010) Transport and signaling through the phosphate – binding site of the yeast Pho84 phosphate transceptor. Proc Natl Acad Sci U S A 107:2890–2895
Pozo MJ, Loon LCV, Pieterse CMJ (2004) Jasmonates signals in plant microbe interactions. J Plant Growth Regul 23:211–222
Pratt J, Boisson AM, Gout E et al (2009) Phosphate (Pi) starvation effect on the cytosolic Pi concentration and Pi exchanges across the tonoplast in plant cells: an in vivo ^{31}P-nuclear magnetic resonance study using methylphosphonate as a Pi analog. Plant Physiol 151:1646–1657
Pumplin N, Harrison MJ (2009) Live-cell imaging reveals periarbuscular membrane domains and organelle location in *Medicago truncatula* roots during arbuscular mycorrhizal symbiosis. Plant Physiol 151:809–819
Raghothama KG (1999) Molecular regulation of phosphate acquisition in plants. In: Gissel-Nielsen G, Jensen A (eds) Plant nutrition—molecular biology and genetics. Springer, Netherlands, pp 95–103
Raghothama KG, Karthikeyan AS (2005) Phosphate acquisition. Plant Soil 274:37–49
Ramaekers L, Remans R, Rao IM et al (2010) Strategies for improving phosphorus acquisition efficiency of crop plants. Field Crop Res 117:169–176
Rausch C, Daram P, Brunner S et al (2001) A phosphate transporter expressed in arbuscule-containing cells in potato. Nature 414:462–466
Rochange S (2010) Strigolactones and their role in arbuscular mycorrhizal symbiosis. In: Kapulnik Y, Douds D (eds) Arbuscular mycorrhizas: physiology and function. Kluwer Academic Publishers, Amsterdam, pp 73–92
Rouached H, Arpat AB, Poirier Y (2010) Regulation of phosphate starvation responses in plants: signaling players and cross-talks. Mol Plant 3:288–299

Ruan HH, Shen WB, Xu LL (2004) Nitric oxide involved in the abscisic acid induced proline accumulation in wheat seedling leaves under salt stress. Acta Bot Sin 46:1307–1315
Rubio V, Linhares F, Solano R et al (2001) A conserved MYB transcription factor involved in phosphate starvation signaling both in vascular plants and in unicellular algae. Genes Dev 15:2122–2133
Rubio V, Bustos R, Irigoyen ML et al (2009) Plant hormones and nutrient signaling. Plant Mol Biol 69:361–373
Ruyter-Spira C, Al-Babili S, van der Krol S et al (2013) The biology of strigolactones. Trends Plant Sci 18:72–83
Ryan PR, Delhaize E (2001) Function and mechanism of organic anion exudation from plant roots. Annu Rev Plant Biol 52:527–560
Saito M (1995) Enzyme activities of the internal hyphae and germinated spores of an arbuscular mycorrhizal fungus, *Gigaspora margarita* Becker & Hall. New Phytol 129:425–431
Sakakibara H (2006) Cytokinins: activity, biosynthesis, and translocation. Annu Rev Plant Biol 57:431–449
Schachtman DP, Shin R (2007) Nutrient sensing and signaling: NPKS. Annu Rev Plant Biol 58:47–69
Schmidt W, Schikora A (2001) Control of root epidermal patterning: Effects of Fe- and P-deficiency. In: Horst WJ, Schenk MK, Bürkert A et al (Eds.), Plant Nutrition. Springer Netherlands, pp. 538–539
Schünmann PHD, Richardson AE, Smith FW et al (2004) Characterization of promoter expression patterns derived from the Pht1 phosphate transporter genes of barley (*Hordeum vulgare* L.) J Exp Bot 55:855–865
Shekoofeh E, Sepideh H, Roya R (2012) Role of mycorrhizal fungi and salicylic acid in salinity tolerance of *Ocimum basilicum* resistance to salinity. Afr J Biotechnol 11:2223–2235
Shin R (2011) Transcriptional regulatory components responding to macronutrient limitation. J Plant Biol 54:286–293
Shin R, Schachtman DP (2004) Hydrogen peroxide mediates plant root cell response to nutrient deprivation. Proc Natl Acad Sci U S A 101:8827–8832
Shin R, Berg RH, Schachtman DP (2005) Reactive oxygen species and root hairs in *Arabidopsis* root response to nitrogen, phosphorus and potassium deficiency. Plant Cell Physiol 46:1350–1357
Shin H, Shin HS, Chen R et al (2006) Loss of *At4* function impacts phosphate distribution between the roots and the shoots during phosphate starvation. Plant J 45:712–726
Shu B, Xia RX, Wang P (2012) Differential regulation of Pht1 phosphate transporters from trifoliate orange (*Poncirus trifoliata* L. Raf.) seedlings. Sci Hortic-Amsterdam 146:115–123
Shu B, Wang P, Xia RX (2014) Effects of mycorrhizal fungi on phytate-phosphorus utilization in trifoliate orange (*Poncirus trifoliata* L. Raf.) seedlings. Acta Physiol Plant 36:1023–1032
Shu B, Wang P, Xia RX (2015) Characterisation of the phytase gene in trifoliate orange (*Poncirus trifoliata* (L.) Raf.) seedlings. Sci Hortic-Amsterdam 194:222–229
Shu B, Li WC, Liu LQ et al (2016a) Transcriptomes of arbuscular mycorrhizal fungi and litchi host interaction after tree girdling. Front Microbiol 7:408
Shu B, Li WC, Liu LQ et al (2016b) Effects of girdling on arbuscular mycorrhizal colonization and root hair development of litchi seedlings. Sci Hortic-Amsterdam 210:25–33
Singh S, Katzer K, Lambert J et al (2014) CYCLOPS, a DNA-binding transcriptional activator, orchestrates symbiotic root nodule development. Cell Host Microbe 15:139–152
Sobkowiak L, Bielewicz D, Malecka EM et al (2012) The role of the P1BS element containing promoter-driven genes in Pi transport and homeostasis in plants. Front Plant Sci 3:58
Song F, Song G, Dong A et al (2011) Regulatory mechanisms of host plant defense responses to arbuscular mycorrhiza. Acta Ecol Sin 31:322–327
Steinkellner S, Lendzemo V, Langer I et al (2007) Flavonoids and strigolactones in root exudates as signals in symbiotic and pathogenic plant-fungus interactions. Molecules 12:1290–1306

Stenzel I, Hause B, Maucher H et al (2003) Allene oxide cyclase dependence of the wound response and vascular bundle-specific generation of jasmonates in tomato amplification in wound signalling. Plant J 33:577–589

Stumpe M, Carsjens JG, Stenzel I et al (2005) Lipid metabolism in arbuscular mycorrhizal roots of *Medicago truncatula*. Phytochemistry 66:781–791

Svistoonoff S, Creff A, Reymond M et al (2007) Root tip contact with low-phosphate media reprograms plant root architecture. Nat Genet 39:792–796

Teale WD, Ditengou FA, Dovzhenko AD et al (2008) Auxin as a model for the integration of hormonal signal processing and transduction. Mol Plant 1:229–237

Ticconi CA, Delatorre CA, Abel S (2001) Attenuation of phosphate starvation responses by phosphate in *Arabidopsis*. Plant Physiol 127:963–972

Trull MC, Guiltinan MJ, Lynch JP et al (1997) The responses of wild-type and ABA mutant *Arabidopsis thaliana* plants to phosphorus starvation. Plant Cell Environ 20:85–92

Turner BL, Papházy MJ, Haygarth PM et al (2002) Inositol phosphates in the environment. Philos Trans R Soc Lond B 357:449–469

Uhde-Stone C, Gilbert G, Johnson JMF et al (2003) Acclimation of white lupin to phosphorus deficiency involves enhanced expression of genes related to organic acid metabolism. Plant Soil 2048:99–116

Umehara M, Hanada A, Yoshida S et al (2008) Inhibition of shoot branching by new terpenoid plant hormones. Nature 455:195–200

Umehara M, Hanada A, Magome H et al (2010) Contribution of strigolactones to the inhibition of tiller bud outgrowth under phosphate deficiency in rice. Plant Cell Physiol 51:1118–1126

Valdés-LÓpez O, Arenas-Huertero C, Ramírez M et al (2008) Essential role of MYB transcription factor: PvPHR1 and microRNA: PvmiR399 in phosphorus-deficiency signalling in common bean roots. Plant Cell Environ 31:1834–1843

Van der Ent S, Van Wees SC, Pieterse CM (2009) Jasmonate signaling in plant interactions with resistance inducing beneficial microbes. Phytochemistry 70:1581–1588

Van Wees SC, Van der Ent S, Pieterse CM (2008) Plant immune responses triggered by beneficial microbes. Curr Opin Plant Biol 11:443–448

Vance CP, Uhde-Stone C, Allan DL (2003) Phosphorus acquisition and use: critical adaptations by plants for securing a nonrenewable resource. New Phytol 157:423–447

Varadarajan DK, Karthikeyan AS, Matilda PD et al (2002) Phosphite, an analog of phosphate, suppresses the coordinated expression of genes under phosphate starvation. Plant Physiol 129:1232–1240

Vierheilig H, Piché Y (2002) Signalling in arbuscular mycorrhiza: facts and hypotheses. Adv Exp Med Biol 505:23–39

Vieweg MF, Hohnjec N, Kuster H (2005) Two genes encoding different truncated hemoglobins are regulated during root nodule and arbuscular mycorrhiza symbioses of *Medicago truncatula*. Planta 220:757–766

Wang X, Yi K, Tao Y et al (2006) Cytokinin represses phosphate-starvation response through increasing of intracellular phosphate level. Plant Cell Environ 29:1924–1935

Wang X, Wang Y, Tian J et al (2009) Overexpressing AtPAP15 enhances phosphorus efficiency in soybean. Plant Physiol 151:233–240

Wasaki J, Shinano T, Onishi K et al (2006) Transcriptomic analysis indicates putative metabolic changes caused by manipulation of phosphorus availability in rice leaves. J Exp Bot 57:2049–2059

Wasternack C, Hause B (2013) Jasmonates: biosynthesis, perception, signal transduction and action in plant stress response, growth and development. An update to the 2007 review in annals of botany. Ann Bot 111:1021–1058

Wegmüller S, Svistoonoff S, Reinhardt D et al (2008) A transgenic *dTph1* insertional mutagenesis system for forward genetics in mycorrhizal phosphate transport of *Petunia*. Plant J 54:1115–1127

Xiao K, Liu J, Dewbre G et al (2006) Isolation and characterization of root-specific phosphate transporter promoters from *Medicago truncatula*. Plant Biol 8:439–449

Yang SY, Paszkowski U (2011) Phosphate import at the arbuscule: just a nutrient? Mol Plant-Microbe Interact 24:1296–1299

Yang SY, Grønlund M, Jakobsen I et al (2012) Nonredundant regulation of rice arbuscular mycorrhizal symbiosis by two members of the phosphate transporter1 gene family. Plant Cell 24:4236–4251

Yoneyama K, Xie X, Kusumoto D et al (2007) Nitrogen deficiency as well as phosphorus deficiency in sorghum promotes the production and exudation of 5-deoxystrigol, the host recognition signal for arbuscular mycorrhizal fungi and root parasites. Planta 227:125–132

Yoneyama K, Xie X, Sekimoto H et al (2008) Strigolactones, host recognition signals for root parasitic plants and arbuscular mycorrhizal fungi, from Fabaceae plants. New Phytol 179:484–494

Yoneyama K, Xie X, Kisugi T et al (2011) Characterization of strigolactones exuded by Asteraceae plants. Plant Growth Regul 65:495–504

Yoshida S, Kameoka H, Tempo M et al (2012) The D3 F-box protein is a key component in host strigolactone responses essential for arbuscular mycorrhizal symbiosis. New Phytol 196:1208–1216

Yu N, Luo D, Zhang X et al (2014) A DELLA protein complex controls the arbuscular mycorrhizal symbiosis in plants. Cell Res 24:130–133

Zhang C, Meng S, Li M et al (2016) Genomic identification and expression analysis of the phosphate transporter gene family in poplar. Front Plant Sci 7:1398

Zhang YJ, Lynch JP, Brown KM (2003) Ethylene and phosphorus availability have interacting yet distinct effects on root hair development. J Exp Bot 54:2351–2361

Zhang Z, Liao H, Lucas WJ (2014) Molecular mechanisms underlying phosphate sensing, signaling, and adaptation in plants. J Integr Plant Biol 56:192–220

Zhou K, Yamagishi M, Osaki M et al (2008a) Sugar signalling mediates cluster root formation and phosphorus starvation-induced gene expression in white lupin. J Exp Bot 59:2749–2756

Zhou J, Jiao F, Wu Z et al (2008b) *OsPHR2* is involved in phosphate-starvation signaling and excessive phosphate accumulation in shoots of plants. Plant Physiol 146:1673–1686

Zsögön A, Lambais MR, Benedito VA et al (2008) Reduced arbuscular mycorrhizal colonization in tomato ethylene mutants. Sci Agric 65:259–267

Chapter 6
Arbuscular Mycorrhizal Fungi and Tolerance of Fe Stress in Plants

Yong-Ming Huang and Qiang-Sheng Wu

Abstract Iron (Fe), an essential micronutrient for plant growth, has many important physiological and biochemical functions in plants, whereas Fe deficiency frequently occurred especially in calcareous and/or alkaline soils with high HCO_3^- and/or pH levels which easily generate slightly soluble Fe(III) compounds. Fe deficiency is usually characterized by chlorosis, which is an important indicator of the decreased chlorophyll level in leaf and mainly attributed to lower Fe uptake confined by Fe solubility in soils. Arbuscular mycorrhizal fungi (AMF), one of the soil beneficial microorganisms, represented positive contributions to soil nutrient uptake, especially in P-deficient soils, as well as alleviate the harmful effects of Fe deficiency. This review outlined the effects of AMF on Fe uptake, chlorophyll contents, photosynthesis, and Fe storage in glomalin under Fe stress. Plants have evolved two strategies for Fe uptake under Fe deficiency conditions, viz., Fe(III)-chelate reductase (FCR) and phytosiderophore (PS). AMF-inoculated plants have shown greater tolerance to Fe stress (Fe deficiency), which may be due to (1) the improved Fe uptake through direct absorption by extra-radical hyphae and (2) the Fe mobilization by increasing root FCR activities and decreasing soil pH value of mycorrhizal plants, thus resulting in higher Fe and chlorophyll contents, better photosynthetic performances, and higher plant biomass, but lower leaf chlorosis. Besides, glomalin released by mycorrhizas potentially participates in Fe availability and transferability in the mycorrhizosphere, but the underlying mechanisms need to be further analyzed.

Keywords Chlorophyll • Glomalin • Iron • Iron deficiency • Mycorrhiza • Fe(III)-chelate reductase

---

Y.-M. Huang
Institute of Fruit and Tea, Hubei Academy of Agricultural Sciences, Wuhan, Hubei 430064, China

Q.-S. Wu (✉)
College of Horticulture and Gardening, Yangtze University, Jingzhou, Hubei 434025, China

Institute of Root Biology, Yangtze University, Jingzhou, Hubei 434025, China

Department of Chemistry, Faculty of Science, University of Hradec Kralove, Hradec Kralove 50003, Czech Republic
e-mail: wuqiangsh@163.com

Q.-S. Wu (ed.), *Arbuscular Mycorrhizas and Stress Tolerance of Plants*,
DOI 10.1007/978-981-10-4115-0_6

6.1 Introduction

Iron (Fe), one of the necessary trace elements, has various important physiological and biochemical functions in plants growth. Generally, most of the leaf Fe (about 80%) is localized in the chloroplast (Terry and Abadia 1986) and in the molecular complexes participated in photosynthetic process reactions in plants, including the photosynthetic electron transport chain and chlorophyll synthesis (Bozorgi 2012; Vigani et al. 2013). Fe as a prosthetic group of ferriheme or chlorhematin in enzymatic systems can activate several enzymes (cytochrome oxidase and cytochrome), thus, improving the performance of photosystems (Hewit 1983; Malakouti and Tehrani 2005). Moreover, Fe is involved in the biosynthesis of chlorophyll, although it is not the component of chlorophyll. Chlorophyll biosynthesis is catalyzed by a series of enzyme complexes mainly comprising aconitase, which can be activated by Fe (Miller and Pushnik 1983). In addition to photosynthetic enzymes, antioxidant enzymes also contain Fe, either in heme (peroxidase, catalase) or nonheme (iron superoxide dismutases) form (Becana et al. 1998). Therefore, Fe plays an important role in the electron transport and enzymatic reaction, which further influence plant growth and resistance.

The vital role of Fe in the life cycle of plants is unable to replace or compensate by other nutrient elements. However, Fe uptake for plants is insufficient in some soil conditions (e.g., calcareous soils, alkaline soils), which is widespread in nature (Nenova 2006; Masalha et al. 2000). Thus, Fe deficiency characterized by Fe stress is a widespread agricultural problem, especially in calcareous soils with high HCO_3^- and pH levels (Abadía et al. 2011; Álvarez-Fernández et al. 2011; Nadi et al. 2013; Vigani et al. 2015). Earlier study of Lu et al. (2002) found that 34% of orchard soils in Hubei province were deficient in Fe. Under Fe stress condition, plants initiate various metabolic responses, mainly reflecting in the decrease of leaf chlorophyll content and photosynthetic performance. An easily observable symptom of Fe deficiency is leaf chlorosis, which is a major yield-limiting concern in crop production (Chen and Lu 2006). As a consequence, plant growth will be compromised by iron deficiency, finally, resulting in the reduction of crop yield and quality parameters. Generally, Fe is very abound in most soils, while its availability for plants is relatively low (Lucena et al. 2015). The solubility and availability of Fe in soils are affected by soil conditions, such as pH (Lindsay and Schwab 1982), moisture content, oxygen content (Guerinot and Yi 1994), microorganisms (Caris et al. 1998), etc.

Arbuscular mycorrhizal fungi (AMF), a kind of beneficial soil microorganisms, can build the symbiotic association with plant roots, usually known as arbuscular mycorrhizas (AMs), which play crucial roles in regulation of plant growth (Smith and Smith 2011; Wagg et al. 2011; Philippot et al. 2013) and the enhancement of stressful environmental resistance, including Fe stress (Caris et al. 1998). The alleviation of Fe stress in mycorrhizal soils may be due to the mobilization of Fe in the rhizosphere and direct absorption of Fe through the extra-radical hyphae of AMs, resulting in improving Fe uptake. In fact, it was demonstrated that AM fungi exhibited a greater Fe uptake under Fe-deficient condition (Haselwandter 1995). Wang et al. (2007) found that an AM fungus, *Glomus versiforme*, increased the Fe contents in trifoliate orange (*Poncirus trifoliata* L. Raf) under Fe deficiency induced by calcium bicarbonate stress. The level of extractable Fe in soils was significantly increased after 80 days growth of soybean (*Glycine max* L.) inoculated with either *Clariodeoglomus etunicatum* or *G. macrocarpum*, when compared to non-inoculated treatment (Nogueira et al. 2007). In addition, Li et al. (2015) reported that AMF-induced overproduction of secondary metabolites was involved in the response mechanism of plants to Fe stress. Inoculation with *G. versiforme* significantly promoted phenolic synthesis in *Poncirus trifoliata* seedlings under Fe deficiency, which might play a vital role in AMF-mediated alleviation of Fe stress. Therefore, AM plants develop positive responses to cope with Fe stress.

The purpose of this chapter is to review the potential roles involved in ameliorating effects of AMF toward Fe stress. Here, we focus on outlining the changes induced by AMF in Fe nutrition, chlorophyll content, photosynthetic performances, and glomalin-bound Fe ability under Fe deficiency. The different strategies of AM plants to enhance Fe absorption are discussed in the review.

6.2 Functionings of Mycorrhizas on Iron Uptake Under Iron Stress

6.2.1 Mechanisms of Plants on Iron Uptake

The available Fe content in soil is normally difficult to meet the needs of plant growth and development in the vast majority of natural farming soils. It is clear that Fe in rhizosphere soils can be activated by plants and subsequently absorbed by roots under Fe deficiency condition (Römheld and Marschner 1986). Plants in morphological and physiological variation, such as the increase of root hair number, the lateral root branching, and the formation of transfer cells in root tip (Römheld and Marschner 1981; Jin et al. 2008), the positive secretion of H^+ to the outside soil from root tip (Han et al. 1998), the increase in the production of phenols and organic acids (Landsberg 1984), and the enhancement of reductase activity in epidermal cell plasma membrane of root tip (Buckhout et al. 1989; Brüggemann et al. 1990), can enhance the ability of Fe uptake in root systems. To cope with Fe deficiency, plants have developed positive regulatory mechanisms that can make plants to facilitate mobilization and uptake of Fe, classified into strategy I species and strategy II species (Lucena et al. 2015).

Strategy I: For all higher plants except the Poaceae, the specific H^+-ATPase activated by Fe deficiency stress promotes soil acidification to increase iron solubility (Guerinot and Yi 1994), thus through the specific plasma membrane Fe(III)-chelate reductase (FCR) to reduce Fe^{3+} to Fe^{2+}, and then the absorbable Fe^{2+} will be taken up by Fe^{2+} transporter into root cells (Guerinot and Yi 1994; Santi and Schmidt 2009;

Kobayashi and Nishizawa 2012; Ivanov et al. 2012). The FCR and Fe transporter genes have been cloned from the same plant species. Eide et al. (1996) cloned the first Fe^{2+} transporter gene (*Irt* 1) in *Arabidopsis*. Meanwhile, Robinson et al. (1997) firstly obtained a FCR gene (*Fro* 2) through genetic screening from cDNA library of *Arabidopsis thaliana* roots, whose expression was significantly increased under Fe deficiency stress. Afterward, these genes were cloned from pea (Cohen et al. 2004), tomato (Li et al. 2004), and cucumber (Waters et al. 2007). In conclusion, these plants through three ways comprising protonation, chelation, and reduction dissolve Fe^{3+} oxides to form F^{2+} for Fe absorption.

Strategy II: For the Poaceae, these plants can secrete a kind of nonprotein amino acids that confirmed as mugineic acids (MAs) under Fe deficiency stress (Takagi 1993; Nozoye et al. 2011). MAs, a phytosiderophore which possesses a forceful affinity for Fe^{3+}, can form the ferric-chelate [Fe^{3+}-PS] (Takagi et al. 1984). Then, the [Fe^{3+}-MAs] complexes are absorbed by roots through the specific epidermal root cell plasma membrane transporters (Murata et al. 2006; Curie et al. 2009). Involvement of functional genomics also intensify the understanding of strategy II. The gene expression of nicotianamine synthase, a key enzyme in the biosynthetic pathway for phytosiderophores, was induced by Fe deficiency, which was firstly reported by Higuchi et al. (1999). Subsequently, Takahashi et al. (2001) found that the expression of nicotianamine aminotransferase genes, involved in the biosynthesis phytosiderophores, significantly enhanced the tolerance of rice to Fe deficiency in alkaline soils.

In addition, phytosiderophore in gramineous plants is less affected by soil pH and calcium carbonate content, which is why iron chlorosis less happens in gramineous plants compared to strategy I plants (Cao et al. 2002). Consequently, the Fe uptake ability of plant roots depends on plant species and soil physicochemical properties.

6.2.2 *Mycorrhiza and Iron Absorption*

It is well known that AM can enhance plant's ability to absorb mineral nutrients from soils. AMF-inoculated plants gain access to an alternative nutrient assimilation pathway, such as intra- and extra-radical hyphae (Parniske 2008; Smith and Read 2008). In response to low Fe availability environment, AM fungi can excrete Fe(III)-specific chelators (siderophores) to mobilize Fe nutrient, thus utilizing the reductive Fe assimilatory system (Haselwandter 1995; Haas 2003). The extra-radical hyphae of AM have a high cation exchange capacity (Joner et al. 2000) and also can extend to a broader area beyond depletion zones of plant rhizosphere for nutrient uptake, resulting in the increased concentration of nutrients in mycorrhizal plants than in non-mycorrhizal plants (Smith and Smith 2011; Abbaspour et al. 2012; Hart and Forsythe 2012). In addition to the direct absorption of extra-radical hyphae for Fe nutrient, metabolic enzymes associated with Fe^{3+} reducing in roots are stimulated by AMF under Fe deficiency stress. It was evaluated that root FCR activity would be increased under Fe deficiency condition, and FCR activity as a diagnostic index of Fe deficiency extensively applied in gramineous plants and xylophyta (Yi and Guerinot 1996; Chouliaras et al. 2004). Wang et al. (2008a) reported that *G. versiforme*-inoculated trifoliate orange showed a higher FCR activity in roots and lower ratios of P/Fe and 50(10P + K)/Fe in leaves, which subsequently increased the content of total Fe and active Fe in leaves. In addition, mycorrhizas can excrete H^+ which can increase Fe solubility in the rhizosphere soil, thus facilitating Fe acquisition (Li et al. 1991; Jin et al. 2014). Furthermore, AM improved plant root development and P uptake, which might be involved in enhanced Fe absorption (Clark and Zeto 1996). Thus, AM may possess multiple approaches to regulate Fe absorption of plants under Fe deficiency stress.

Recently, massive efforts have demonstrated a vital role of AM in promoting Fe uptake of host plant. AMF inoculation presented a significantly positive impact on Fe nutrition of crop plant (Yaseen et al. 2012; Baslam et al. 2013; Lehman and Rillig 2015). For example, Wu and Zou (2009) reported that *G. versiforme* colonization significantly increased Fe concentration in roots of trifoliate orange seedlings under drought stress conditions. In maize (*Zea mays* L.) plants, colonization with three AMF isolates (*C. etunicatum*, *Rhizoglomus diaphanum*, and *R. intraradices*) significantly increased the concentration of Fe (Clark and Zeto 1996). In addition, the root Fe concentration of chickpea with mycorrhizal inoculation was higher than those with non-mycorrhizal inoculation (Farzaneh et al. 2011). In contrast, Pacovsky and Fuller (1988) reported that soybean plants showed a decrease in shoot Fe concentrations after inoculating with *R. fasciculatus*. Inoculation with *Funneliformis mosseae* had no significant influence on Fe concentration in shoots of peanut plants (Caris et al. 1998). Moreover, the root FCR activity and leaf Fe content in trifoliate orange and red tangerine decreased with the increasing of soil pH, regardless of inoculation with or without AM fungus (*G. versiforme*) (Wang et al. 2008a). In conclusion, these results implied that the mycorrhizal effects on Fe uptake are possibly significantly dependent on AMF and plant species and several environmental factors, such as soil pH and soil P supply.

Although the level of available Fe was increased in the soil with mycorrhizal plants, the shoot Fe concentration in mycorrhizal plants was decreased by 45% than in non-mycorrhizal plants (Nogueira et al. 2007). Thereby, higher mobilization of Fe in the rhizosphere of mycorrhizal plants may not represent higher uptake ability by roots. This suggests that mycorrhizal colonization protects the host plant against excessive Fe toxicity. As a consequence, AM fungus processes the dual functions in Fe uptake, which depends on the nutritional status of Fe in plants.

6.3 Mycorrhiza-Induced Iron and Chlorophyll

Chlorophyll, a kind of important photosynthetic pigments, participates in the absorption and conversion of light energy and mainly comprises of chlorophyll a and chlorophyll b in higher plants (Stenbaek and Jensen 2010). Chlorophyll synthesis is affected by many factors, such as water status, light intensity, mineral element, etc. Although Fe is not the component of chlorophyll, Fe plays a vital role in the process of chlorophyll synthesis. The precursor of chlorophyll synthesis is blocked under Fe-deficient condition, which can lead to the decrease of chlorophyll concentration (Chatterjee et al. 2006). Fe is an important cofactor of many enzymes involved in the biosynthetic pathway of chlorophylls (Marschner 1995). Under Fe deficiency condition, the reduction of Fe concentration in leaf is often accompanied by a marked reduction of chlorophyll levels (Ranieri et al. 2001; Gogorcena et al. 2004). Moreover, most of the leaf Fe (about 80%) is localized in the chloroplast (Terry and Abadia 1986), and Fe deficiency primarily affects structure and functioning of the chloroplast (Molassiotis et al. 2006). The visible symptom of Fe deficiency mainly appeared as leaf chlorosis, which is due to the reduction of the leaf chlorophyll content (Ranieri et al. 2001; Gogorcena et al. 2004). Fe deficiency inducing the decreased content of photosynthetic pigments results in the relative enrichment of carotenoids over chlorophyll, which will lead to the yellow color (Nadi et al. 2013). Therefore, chlorophyll content is referred to as a conventional diagnostic indicator of Fe deficiency (Pestana et al. 2005).

AM plants usually present higher chlorophyll contents than non-AM plants, which may be due to the enhancement of Fe absorption (or other mineral nutrients, e.g., Mg) by extra-radical hyphae of AM. Wang et al. (2008b) observed an increase of chlorophyll concentration in trifoliate orange seedlings inoculated with *G. versiforme* under Fe deficiency condition. The result showed that AM possesses the ability to alleviate the nutritional deficiency symptom (leaf chlorosis) caused by Fe deficiency. Nevertheless, there is no information about the mechanism of AM-mediated changes of chlorophyll content in Fe deficiency plants. Whether AM-induced higher chlorophyll content is related to the increasing Fe acquisition, which is necessary to confirm this with more direct evidences.

6.4 Mycorrhiza-Induced Iron and Photosynthesis

Fe, as a form of molecular complexes, participated in the photosynthetic electron transport chain influencing photosynthesis (Vigani et al. 2013). Fe deficiency often causes the decline in the photosynthesis parameters, including photosynthetic rate and stomatal conductance, without intercellular CO_2 concentration (Molassiotis et al. 2006). According to Terry (1983), the decrease in photosynthesis per leaf area might be due to the lower photochemical activity in Fe-deficient plants. Nenova (2009) also found that leaf photosynthesis was significantly decreased under Fe deficiency. Iturbe-Ormaetxe et al. (1995) explained that the inhibition of net photosynthesis rate in the upper leaves of pea plants was caused by the decline of light harvesting and electron transport apparatus under Fe-deficient condition. On the other hand, the chlorophyll fluorescence analysis was used to obtain information about the state of photosynthetic apparatus and photosystem II (PS II) under Fe-deficient condition (Nenova 2006). A reduction of chlorophyll fluorescence parameters including the maximum quantum yield of PS II (F_v/F_m) and the ratio between variable and initial fluorescence was recorded in peach leaves under Fe deficiency tolerance (Molassiotis et al. 2006). Moreover, the decrease in F_v/F_m may indicate a sustained PS II downregulation or damage (Abadía et al. 1999). Therefore, Fe deficiency will destroy photosynthetic apparatus and induce disqualified chlorophyll fluorescence parameters, thus leading to the decline of photosynthesis.

AMF colonization was widely evidenced to increase plant photosynthesis. Under abiotic stress, AM inoculation was reported to facilitate the stomatal conductance (Sheng et al. 2008), improve the transpiration rate (Hameed et al. 2014), and boost the performance of PS I and PS II (Talaat and Shawky 2014). However, the information about the effect of AMF on the variation of photosynthesis under Fe deficiency condition is scarce. AM promoted plant photosynthesis performances under Fe stress possibly due to AM-induced greater water and iron uptake and higher chlorophyll content like under other abiotic stresses. Overall, it is speculated that the potential role of AM in inducing better photosynthesis is closely related to the amendatory preliminary basis of photosynthesis, which extremely likely owing to higher iron and chlorophyll contents when plants subjected to Fe deficiency.

6.5 Mycorrhiza and Plant Growth Under Iron Deficiency

Fe deficiency is a limiting factor for plant growth and adversely affected crop yield (Kobayashi and Nishizawa 2012). Fe deficiency significantly decreased the growth reflecting in root and shoot lengths and fresh and dry biomass of pea plants (Nenova 2009). Lack of Fe consequently causes young leaves to yellow photosynthesis activity, which result in reduced biomass (Briat et al. 2007). Sometimes, severe chlorosis may develop without producing any effect on growth under Fe deficiency (Gogorcena et al. 2001). On the other hand, growth depression can occur before the start of the leaf yellowing or even without yellowing (Kosegarten et al. 1998; Gruber and Kosegarten 2002). In addition, application of Fe through ferrous sulfate or chelated Fe significantly increased the growth characters and yield components in black gram (Balachander et al. 2003), pea (Thapu et al. 2003), chickpea (Sahu et al. 2008; Kumar et al. 2009), pigeon pea crop (Sharma et al. 2010), etc. Thus, Fe plays a crucial role in promoting plants growth.

AMF usually maintains a positive role in regulating plants growth under Fe deficiency. Li et al. (2015) reported that *D. versiforme* colonization markedly promoted *Poncirus trifoliata* growth under Fe deficiency, showing higher biomass in shoot and root, compared to non-mycorrhizal plants, regardless of Fe supply levels. Wang et al. (2007, 2008b) also reported that inoculation of *Poncirus trifoliata* with *G. versiforme* resulted in higher plant height, stem diameter, and dry weights of both shoots and roots compared to non-mycorrhizal seedlings under Fe deficiency. AMF-induced greater growth plants may be due to the improvement of primary metabolism, such as photosynthesis which provides the material energy for plant growth and reproduction (Birhane et al. 2012; Gamalero et al. 2009). Consequently, the increased plant growth parameters induced by AMF under Fe deficiency may be due to the improvement of Fe nutrient and photosynthesis.

6.6 Glomalin Contribution to Iron

Glomalin, a metal-sorbing glycoprotein produced by AM hyphae, widely coats on the surface of mycorrhizal hyphae (Wright and Upadhyaya 1996; Wright and Upadhyaya 1998; Rillig 2004), quantified operationally in rhizosphere soil as glomalin-related soil protein (GRSP) (Driver et al. 2005). Glomalin is not only a hyphal wall component (Driver et al. 2005) but also mainly contributes to the soil organic matter (Rillig et al. 2003), enhances the soil aggregate stability (Wright et al. 1999), and sequesters the metal cations (e.g., Fe) in soils (Nichols 2003). For example, Wu et al. (2016) found that some particles on hyphal surface according to elemental analysis contained certain amounts of Fe (with wt% values of 0.9–4.2%), indicating that these particles probably include GRSP. In addition, Nichols (2003) reported that GRSP was a compound tightly bound Fe (0.04–8.8%). Fe is a major component of glomalin molecule described by González-Chávez et al. (2004), who reported that Fe concentration in glomalin produced by *Gigaspora rosea* from polluted soil was 44 mg Fe/g glomalin.

Although previous studies proved that glomalin can sequester or bind Fe, there is less information about the effects of GRSP on the effective Fe content and the alleviation of Fe deficiency in soil. Generally, metallic elements (e.g., Pb, Cd, Cu) like Fe can be sequestered by glomalin, but their effectiveness in soil especially in an excessive level will be greatly reduced by glomalin to alleviate their toxic effects for mycorrhizas and plants (Göhre and Paszkowski 2006). Approximately, glomalin-binding Fe on the surface of hyphae may be used as an iron pool which can supply Fe for direct absorption or transportation of hyphae to plants. Therefore, glomalin potentially as a catcher of the metallic elements (including Fe) plays an important role in nutrient elements uptake by hyphae. Wang et al. (2010) found that easily extractable glomalin (EEG) and total glomalin (TG) were significantly negatively correlated with chelating Fe content in glomalin, which might be due to the competitive relationship between Fe and Cu or Zn. Even so, the relationship between mycorrhiza-released glomalin and the binding ability of Fe by glomalin will be changed with the variation of Fe content in soils.

6.7 Conclusions

AM symbiosis can alleviate the adverse effects of Fe stress (Fe deficiency) on plant growth and development and is referred to as a vital contributor in protecting plants against Fe stress. Mycorrhizal symbiosis through the direct impact of extra-radical hyphae on Fe uptake in combination with mobilization of sparingly available Fe by

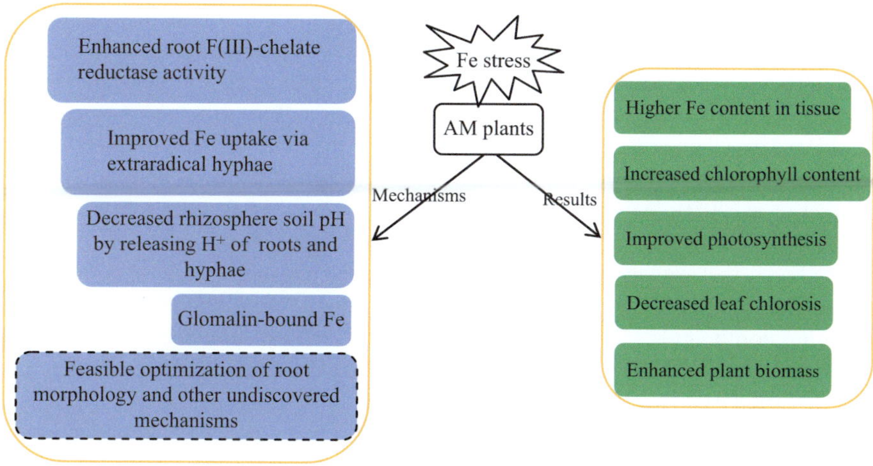

Fig. 6.1 The major mechanisms of AM-mediated plant tolerance to Fe stress

increasing the production of Fe^{3+}-chelate reductase and H^+ would result in higher total Fe, active Fe and chlorophyll contents in plants, greater photosynthetic performances, and higher plant biomass, but lower leaf chlorosis. Glomalin as a metal-sorbing glycoprotein originated from AM hyphae is closely related with Fe in soils, but the effect of glomalin on Fe availability and transferability especially in Fe deficiency soil is poorly known (Fig. 6.1).

6.8 Further Perspectives

Although arbuscular mycorrhiza played a crucial role in plants against Fe stress, the underlying mechanisms mainly focused on Fe acquisition stimulated by the direct absorption of AM hyphae and the enhancement of root FCR production and H^+ excretion. There are still many potential response mechanisms induced by AM against Fe stress, which still remain largely unknown and need to further study.

AM plant-exhibited greater root system architecture (RSA) dramatically increased the absorptive area and efficiency of water and nutrients from soil and have been widely demonstrated in many plant species growing in both biotic and abiotic stress conditions (Orfanoudakis et al. 2010; Wu et al. 2012a, b). Fe deficiency changed RSA by altering the roots number and length (Lucena et al. 2015) and other responses (see above). Nevertheless, the effect of AMF on RSA under Fe stress has not been reported. The variation of RSA induced by AMF, which is a response mechanism to Fe deficiency, also contributes to Fe absorption, whereas it is needed to dissect in the further studies.

Generally, the plants response to Fe deficiency can be predicted to increase Fe acquisition via an intricate signaling process (Jin et al. 2014). It has been demonstrated that several hormones and signaling compounds play an important role in triggering Fe deficiency response. These compounds include ethylene (García et al. 2011), auxin (Chen et al. 2010), and nitric oxide (Graziano M and Lamattina 2007), which activate Fe deficiency responses through the upregulation of Fe-related genes (García et al. 2010; Wu et al. 2012a, b; Meng et al. 2012). In addition, auxin and nitric oxide, along with other hormonal compounds, have been implicated in signaling pathway that was used by AM to increase plant resistance against unfavorable environmental conditions (Zhang et al. 2013; López-Ráez 2016). Whether these signaling compounds are involved in the AM-induced Fe deficiency tolerance has not been explored. Are there AM-induced signaling compounds to activate Fe deficiency responses?

Glomalin released from AM hyphae into soil is characterized by long decomposition time in soils approximately up to 6 ~ 42 years (Rillig et al. 2001), which absolutely impacts the revolution of metal elements binding in the glomalin in soil, although it has been reported to perform a positive role in soil carbon and nitrogen metabolisms. In addition, the functional verification of glomalin on Fe availability and transferability especially in Fe deficiency soil will provide a new evidence in AM enhancing plants tolerance against Fe stress.

The molecular information of AM-induced plant tolerance against Fe stress is rarely reported. Ruzicka et al. (2013) discovered two putative Fe transporters (contig26941 and contig02193) in *R. intraradices* using a deep sequencing metatranscriptomics approach, which might play roles in mineral nutrition in mycorrhizal plants. In addition, Li et al. (2015) used the real-time quantitative PCR to analyze the *pal*1 gene expression related to phenolic compounds. The gene expression level in *Poncirus trifoliata* roots significantly promoted by *R. versiformis* may be a response mechanism to Fe stress. In the further studies, the involvement of functional genomics in controlling Fe stress responses, in combination with physiological approach, will expand our perspectives in AM contribution to Fe tolerance in plants.

Acknowledgment This study was supported by the Plan in Scientific and Technological Innovation Team of Outstanding Young, Hubei Provincial Department of Education (T201604).

References

Abadía J, Morales F, Abadía A (1999) Photosystem II efficiency in low chlorophyll, iron-deficient leaves. Plant Soil 215:183–192

Abadía J, Vázquez S, Rellán-Álvarez R et al (2011) Towards a knowledge-based correction of iron chlorosis. Plant Physiol Biochem 49:471–482

Abbaspour H, Saeidi-Sar S, Afshari H et al (2012) Tolerance of mycorrhiza infected pistachio (*Pistacia vera* L.) seedling to drought stress under glasshouse conditions. J Plant Physiol 169:704–709

Álvarez-Fernández A, Melgar JC, Abadía J et al (2011) Effects of moderate and severe iron deficiency chlorosis on fruit yield, appearance and composition in pear (*Pyrus communis* L.) and peach (*Prunus persica* (L.) Batsch). Environ Exp Bot 71:280–286

Balachandar D, Nagarajan P, Gunasekaran S (2003) Effect of organic amendments and micronutrients on nodulation and yield of black gram in acid soil. Legume Res 26:192–195

Baslam M, Garmendia I, Goicoechea N (2013) Enhanced accumulation of vitamins, nutraceuticals and minerals in lettuces associated with arbuscular mycorrhizal fungi (AMF): a question of interest for both vegetables and humans. Agriculture 3:188–209

Becana M, Moran J, Iturbe-Oimaexte I (1998) Iron-dependent oxygen free radical generation in plants subjected to environmental stress: toxicity and antioxidant protection. Plant Soil 201:137–147

Birhane E, Sterck F, Fetene M et al (2012) Arbuscular mycorrhizal fungi enhance photosynthesis, water use efficiency, and growth of frankincense seedlings under pulsed water availability conditions. Oecologia 169:895–904

Bozorgi HR (2012) Effects of foliar spraying with marine plant *Ascophyllum nodosum* extract and nano iron chelate fertilizer on fruit yield and several attributes of eggplant (*Solanum melongena* L.). J Agric Biol Sci 7(5)

Briat JF, Curie C, Gaymard F (2007) Iron utilization and metabolism in plants. Curr Opin Plant Biol 10:276–282

Brüggemann W, Moog PR, Nakagawa H et al (1990) Plasma membrane-bound NADH: Fe^{3+}-EDTA reductase and iron deficiency in tomato (*Lycopersicon esculentum*). Is there a turbo reductase? Physiol Plant 79:339–346

Buckhout TJ, Luster DG, Chaney RL (1989) Iron-stress induced redox activity in tomato (*Lycopersicon esculentum* M.) is localized on the plasma membrane. Plant Physiol 90:151–156

Cao H, Han ZH, Xu XF et al (2002) Iron nutrition in higher plants. Plant Physiol Commun 38(2):180–186. (in Chinese with English abstract)

Caris C, Hördt W, Hawkins HJ et al (1998) Studies of iron transport by arbuscular mycorrhizal hyphae from soil to peanut and sorghum plants. Mycorrhiza 8:35–39

Chatterjee C, Gopal R, Dube BK (2006) Impact of iron stress on biomass, yield, metabolism and quality of potato (*Solanum tuberosum* L.) Sci Hortic 108:1–6

Chen F, Lu JW (2006) Effect of iron chelate application on citrus in the three gorges area. Southeast China 1:33–35

Chen WW, Yang JL, Qin C et al (2010) Nitric oxide acts downstream of auxin to trigger root ferric-chelate reductase activity in response to iron deficiency in *Arabidopsis*. Plant Physiol 154:810–819

Chouliaras V, Therios I, Molassiotis A et al (2004) Iron chlorosis in grafted sweet orange (*Citrus sinensis* L.) plants: physiological and biochemical responses. Biol Plantarum 48(1):141–144

Clark RB, Zeto SK (1996) Mineral acquisition by mycorrhizal maize grown on acid and alkaline soil. Soil Biol Biochem 28:1495–1503

Cohen CK, Garvin DF, Kochian LV (2004) Kinetic properties of a micronutrient transporter from *Pisum sativum* indicate a primary function in Fe uptake from the soil. Planta 218:784–792

Curie C, Cassin G, Couch D et al (2009) Metal movement within the plant: contribution of nicotianamine and yellow stripe 1-like transporters. Ann Bot 103:1–11

Driver JD, Holben WE, Rillig MC (2005) Characterization of glomalin as a hyphal wall component of arbuscular mycorrhizal fungi. Soil Biol Biochem 37:101–106

Eide D, Broderius M, Fett J et al (1996) A novel iron-regulated metal transporter from plants identified by functional expression in yeast. Proc Natl Acad Sci U S A 93:5624–5628

Farzaneh M, Vierheilig H, Lössll A et al (2011) Arbuscular mycorrhiza enhances nutrient uptake in chickpea. Plant Soil Environ 57(10):465–470

Gamalero E, Lingua G, Berta G et al (2009) Beneficial role of plant growth promoting bacteria and arbuscular mycorrhizal fungi on plant responses to heavy metal stress. Can J Microbiol 55(5):501–514

García MJ, Lucena C, Romera FJ et al (2010) Ethylene and nitric oxide involvement in the up-regulation of key genes related to iron acquisition and homeostasis in *Arabidopsis*. J Exp Bot 61:3885–3899

García MJ, Suárez V, Romera FJ et al (2011) A new model involving ethylene, nitric oxide and Fe to explain the regulation of Fe-acquisition genes in strategy I plants. Plant Physiol Bioch 49:537–544

Gogorcena Y, Molias N, Larbi A et al (2001) Characterization of the responses of cork oak (*Quercus suber*) to iron deficiency. Tree Physiol 21:1335–1340

Gogorcena Y, Abadía A, Abadía A (2004) A new technique for screening iron-efficient genotypes in peach rootstocks: elicitation of root ferric chelate reductase by manipulation of external iron concentrations. J Plant Nutr 27:1701–1715

Göhre V, Paszkowski U (2006) Contribution of the arbuscular mycorrhizal symbiosis to heavy metal phytoremediation. Planta 223(6):1115–1122

González-Chávez MC, Carrillo-González R, Wright SF et al (2004) The role of glomalin, a protein produced by arbuscular mycorrhizal fungi, in sequestering potentially toxic elements. Environ Pollut 130:317–323

Graziano M, Lamattina L (2007) Nitric oxide accumulation is required for molecular and physiological responses to iron deficiency in tomato roots. Plant J 52:949–960

Gruber B, Kosegarten H (2002) Depressed growth of non-chlorotic vine grown in calcareous soil is an iron deficiency symptom prior to leaf chlorosis. J Plant Nutr Soil Sci 165:11–117

Guerinot ML, Yi Y (1994) Iron: nutritious, noxious, and not readily available. Plant Physiol 104:815–820

Haas H (2003) Molecular genetics of fungal siderophore biosynthesis and uptake: the role of siderophores in iron uptake and storage. Appl Microbiol Biotechnol 62:316–330

Hameed A, Dilfuza E, Abd-Allah EF et al (2014) Salinity stress and arbuscular mycorrhizal symbiosis in plants. In: Miransari M (ed) Use of microbes for the alleviation of soil stresses, vol 1. Springer Science+Business Media, New York, pp 139–159

Han ZH, Shen T, Korcak RF et al (1998) Iron absorption by iron-efficient and inefficient species of apples. J Plant Nutr 21:181–190

Hart MM, Forsythe JA (2012) Using arbuscular mycorrhizal fungi to improve the nutrient quality of crops; nutritional benefits in addition to phosphorus. Sci Hortic 148:206–214

Haselwandter K (1995) Mycorrhizal fungi: siderophore production. Crit Rev Biotechnol 15:287–291

Hewit EJ (1983) Essential and functional metals in plants. In: Robb DA, Pierpoint WS (eds) Metals and micronutrients uptake and utilization by plants. Academic, New York, pp 277–323

Higuchi K, Suzuki K, Nakanishi H et al (1999) Cloning of nicotianamine synthase genes: novel genes involved in the biosynthesis of phytosiderophores. Plant Physiol 119(2):471–480

Iturbe-Ormaexte I, Moran JF, Arrese-Igor C et al (1995) Activated oxygen and antioxidant defenses in iron deficient pea plants. Plant Cell Environ 18:421–429

Ivanov R, Brumbarova T, Bauer P (2012) Fitting into the harsh reality: regulation of iron-deficiency responses in dicotyledonous plants. Mol Plant 5:27–42

Jin CW, Chen WW, Meng ZB et al (2008) Iron deficiency-induced increase of root branching contributes to the enhanced root ferric chelate reductase activity. J Integr Plant Biol 50:1557–1562

Jin CW, Ye YQ, Zheng SJ (2014) An underground tale: contribution of microbial activity to plant iron acquisition via ecological processes. Ann Bot 113:7–18

Joner EJ, Briones R, Leyval C (2000) Metal-binding capacity of arbuscular mycorrhizal mycelium. Plant Soil 226:227–234

Kobayashi T, Nishizawa NK (2012) Iron uptake, translocation, and regulation in higher plants. Ann Rev Plant Biol 63:131–152

Kosegarten H, Wilsona GH, Escha A (1998) The effect of nitrate nutrition on iron chlorosis and leaf growth in sunflower (*Helianthus annuus* L.) Eur J Agron 8:283–292

Kumar V, Dwivedi VN, Tiwari DD (2009) Effect of phosphorus and iron on yield and mineral nutrition in chickpea. Ann Plant Soil Res 11:16–18

Landsberg EC (1984) Regulation of iron stress response by whole-plant capacity. J Plant Nutr 7:609–621

Lehman A, Rillig MC (2015) Arbuscular mycorrhizal contribution to copper, manganese and iron nutrient concentrations in crops–a meta-analysis. Soil Biol Biochem 81:147–158

Li XL, Geoege E, Marschner H (1991) Phosphorus depletion and pH decrease at the root–soil and hyphae–soil inter-faces of VAM white clover fertilized with ammonium. New Phytol 119:397–404

Li L, Cheng X, Ling HQ (2004) Isolation and characterization of Fe(III)-chelate reductase gene LeFRO1 in tomato. Plant Mol Biol 54:125–136

Li JF, He XH, Li H et al (2015) Arbuscular mycorrhizal fungi increase growth and phenolics synthesis in *Poncirus trifoliata* under iron deficiency. Sci Hortic 183:87–92

Lindsay WL, Schwab AP (1982) The chemistry of iron in soils and its availability to plants. J Plant Nutr 5:821–840

López-Ráez JA (2016) How drought and salinity affect arbuscular mycorrhizal symbiosis and strigolactone biosynthesis? Planta 243:1375–1385

Lu JW, Chen F, Wang FH (2002) Study of classification of the soil nutrient status of citrus orchard in Hubei Province. Plant Nutr Fert Sci 8(4):390–394. (in Chinese with English abstract)

Lucena C, Romera FJ, García MJ et al (2015) Ethylene participates in the regulation of Fe deficiency responses in strategy I plants and in rice. Front Plant Sci 6:1056. doi:10.3389/fpls.2015.01056

Malkaouti M, Tehrani M (2005) Micronutrient role in increasing yield and important the quality of agricultural products. Tarbiat Modarres, Tehran

Marschner H (1995) Mineral nutrition of higher plants, 2nd edn. Academic, London

Masalha J, Kosegarten H, Elmaci Ö et al (2000) The central role of microbial activity for iron acquisition in maize and sunflower. Biol Fert Soils 30:433–439

Meng ZB, Chen LQ, Suo D et al (2012) Nitric oxide is the shared signalling molecule in phosphorus- and iron deficiency-induced formation of cluster roots in white lupin (*Lupinus albus*). Ann Bot 109:1055–1064

Miller GW, Pushnik JC (1983) Iron chlorosis: the role of iron in chlorophyll formation. Utah Sci 44:99–103

Molassiotis A, Tanou G, Diamantidis G et al (2006) Effects of 4-month Fe deficiency exposure on Fe reduction mechanism, photosynthetic gas exchange, chlorophyll fluorescence and antioxidant defense in two peach rootstocks differing in Fe deficiency tolerance. J Plant Physiol 163:176–185

Murata Y, Ma JF, Yamaji N et al (2006) A specific transporter for iron(III)-phytosiderophore in barley roots. Plant J 46:563–572

Nadi E, Aynehband A, Mojaddam M (2013) Effect of nano-iron chelate fertilizer on grain yield, protein percent and chlorophyll content of faba bean (*Vicia faba* L.) Int J Biosci 3(9):267–272

Nenova V (2006) Effect of iron supply on growth and photosystem II efficiency of pea plants. Gen Appl Plant Physiol 32:81–90

Nenova VR (2009) Growth and photosynthesis of pea plants under different iron supply. Acta Physiol Plant 31:385–391

Nichols KA (2003) Characterization of glomalin, a glycoprotein produced by arbuscular mycorrhizal fungi. Dissertation, University of Maryland, College Park

Nogueira MA, Nehls U, Hampp R et al (2007) Mycorrhiza and soil bacteria influence extractable iron and manganese in soil and uptake by soybean. Plant Soil 298:273–284

Nozoye T, Nagasaka S, Kobayashi T et al (2011) Phytosiderophore efflux transporters are crucial for iron acquisition in graminaceous plants. J Biol Chem 286:5446–5454

Orfanoudakis M, Wheeler CT, Hooker JE (2010) Both the arbuscular mycorrhizal fungus *Gigaspora rosea* and *Frankia* increase root system branching and reduce root hair frequency in *Alnus glutinosa*. Mycorrhiza 20:117–126

Pacovsky RS, Fuller G (1988) Mineral and lipid composition of *Glycine-Glomus-Bradyrhizobium* symbioses. Physiol Plant 72:733–746

Parniske M (2008) Arbuscular mycorrhiza: the mother of plant root endosymbioses. Nat Rev Microbiol 6:763–775

Pestana M, Varennes AD, Abad AJ et al (2005) Differential tolerance to iron deficiency of citrus rootstocks grown in nutrient solution. Sci Hortic 104:25–36

Philippot L, Raaijmakers JM, Lemanceau P et al (2013) Going back to the roots: the microbial ecology of the rhizosphere. Nat Rev Microbiol 11:789–799

Ranieri A, Castagna A, Baldan B et al (2001) Iron deficiency differently affects peroxidase isoforms in sunflower. J Exp Bot 52:25–35

Rillig MC (2004) Arbuscular mycorrhizae, glomalin and soil aggregation. Can J Soil Sci 84:355–363

Rillig MC, Wright SF, Nichols KA et al (2001) Large contribution of arbuscular mycorrhizal fungi to soil carbon pools in tropical forest soils. Plant Soil 233:167–177

Rillig MC, Ramsey PW, Morris S et al (2003) Glomalin, an arbuscular-mycorrhizal fungal soil protein, responds to land-use change. Plant Soil 253:293–299

Robinson NJ, Groom Sadjuga J, Groom QJ (1997) The *froh* gene family from *Arabidopsis thaliana*: putative iron-chelate reductases. Plant Soil 196:245–248

Römheld V, Marschner H (1981) Iron deficiency induced morphological and physiological changes in root tip of sunflower. Physiol Plant 53:353–360

Römheld V, Marschner H (1986) Mobilization of iron in the rhizosphere of different plant species. Adv Plant Nutri 2:155–204

Ruzicka D, Chamala S, Felipe H et al (2013) Inside arbuscular mycorrhizal roots-molecular probes to understand the symbiosis. Plant Genome 6:1–13

Sahu S, Lidder RS, Singh PK (2008) Effect of micronutrients and biofertilizers on growth, yield and nutrient uptake by chickpea (*Cicer arietinum* L.) in Vertisols of Madhya Pradesh. Adv Plant Sci 21:501–503

Santi S, Schmidt W (2009) Dissecting iron deficiency-induced proton extrusion in *Arabidopsis* roots. New Phytol 183:1072–1084

Sharma A, Nakul HT, Jelgeri BR et al (2010) Effect of micronutrients on growth, yield and yield components in pigeon pea (*Cajanus cajan* L.) Res J Agric Sci 1:142–144

Sheng M, Tang M, Chan H et al (2008) Influence of arbuscular mycorrhizae on photosynthesis and water status of maize plants under salt stress. Mycorrhiza 18:287–296

Smith SE, Read DJ (2008) Mycorrhizal symbiosis. Academic, San Diego/London

Smith SE, Smith FA (2011) Roles of arbuscular mycorrhizas in plant nutrient and growth: new paradigms from cellular to ecosystem scales. Annu Rev Plant Biol 62:227–250

Stenbaek A, Jensen PE (2010) Redox regulation of chlorophyll biosynthesis. Phytochemistry 71(8–9):853–859

Takagi SC (1993) Production of phytosiderophores. In: Barton LL, Hemming BC (eds) Iron chelation in plants and soil microorganisms. Academic, Clarendon, pp 111–130

Takagi SC, Nomoto K, Takemoto T (1984) Physiological aspect of mugineic acid, a possible phytosiderophore of graminaceous plants. J Plant Nutr 7:469–477

Takahashi M, NakanishiI H, Kawasaki S et al (2001) Enhanced tolerance of rice to low iron availability in alkaline soils using barley nicotianamine aminotransferase genes. Nat Biotechnol 19(5):466–469

Talaat NB, Shawky BT (2014) Protective effects of arbuscular mycorrhizal fungi on wheat (*Triticum aestivum* L.) plants exposed to salinity. Environ Exp Bot 98:20–31

Terry N (1983) Limiting factors in photosynthesis: IV. Iron stress-mediated changes in light-harvesting and electron transport capacity and its effects on photosynthesis *in vivo*. Plant Physiol 71:855–860

Terry N, Abadia J (1986) Function of iron in chloroplasts. J Plant Nutr 9:609–649

Thapu U, Rai P, Suresh CP et al (2003) Effect of micronutrients on growth and yield of pea in gangetic alluvial of West Bengal. Environ Ecol 21:179–182

Vigani G, Zocchi G, Bashir K et al (2013) Signal for chloroplast and mitochondria for homeostasis regulation. Trends Plant Sci 18:305–311

Vigani G, Faoro F, Ferretti AM et al (2015) Three-dimensional reconstruction, by TEM tomography, of the ultrastructural modifications occurring in *Cucumis sativus* L. mitochondria under Fe deficiency. PLoS One 10(6):e0129141. doi:10.1371/journal.pone.0129141

Wagg C, Jansa J, Schmid B et al (2011) Belowground biodiversity effects of plant symbionts support aboveground productivity. Ecol Lett 14:1001–1009

Wang MY, Xia RX, Hu LM et al (2007) Arbuscular mycorrhizal fungi alleviate iron deficient chlorosis in *Poncirus trifoliata* L. Raf under calcium bicarbonate stress. J Hortic Sci Biotechnol 82(5):776–780

Wang MY, Christie P, Xiao ZY et al (2008a) Arbuscular mycorrhizal enhancement of iron concentration by *Poncirus trifoliata* L. Raf and *Citrus reticulata* Blanco grown on sand medium under different pH. Biol Fertil Soils 45:65–72

Wang MY, Xia RX, Wang YS et al (2008b) Effects of arbuscular mycorrhizal fungi on growth of *Poncirus trifoliata* seedlings under iron deficiency and heavy bicarbonate stresses. Acta Hortic Sinica 35(4):469–474. (in Chinese with English abstract)

Wang MY, Xia RX, Wang P (2010) Effects of arbuscular mycorrhizal fungi on available iron and metals sequestered by glomalin in different rhizospheric soil of *Poncirus trifoliata*. J Fujian Agric For Univ 39(1):42–46. (in Chinese with English abstract)

Waters BM, Lucena C, Romera FJ et al (2007) Ethylene involvement in the regulation of the H^+-ATPase *CsHA1* gene and of the new isolated ferric reductase *CsFRO1* and iron transporter *CsIRT1* genes in cucumber plants. Plant Physiol Biochem 45:293–301

Wright SF, Upadhyaya A (1996) Extraction of an abundant and unusual protein from soil and comparison with hyphal protein of arbuscular mycorrhizal fungi. Soil Sci 161:575–586

Wright SF, Upadhyaya A (1998) A survey of soils for aggregate stability and glomalin, a glycoprotein produced by hyphae of arbuscular mycorrhizal fungi. Plant Soil 198:97–107

Wright SF, Starr JL, Paltineanu IC (1999) Changes in aggregate stability and concentration of glomain during tillage management transition. Soil Sci Soc Amer J 63:1825–1829

Wu QS, Zou YN (2009) Mycorrhizal influence on nutrient uptake of citrus exposed to drought stress. Philipp Agric Sci 92:33–38

Wu T, Zhang HT, Wang Y et al (2012a) Induction of root Fe(III) reductase activity and proton extrusion by iron deficiency is mediated by auxin-based systemic signalling in *Malus xiaojinensis*. J Exp Bot 63:859–870

Wu QS, He XH, Zou YN et al (2012b) Arbuscular mycorrhizas alter root system architecture of *Citrus tangerine* through regulating metabolism of endogenous polyamines. Plant Growth Regul 68:27–35

Wu SL, Zhang X, Sun YQ et al (2016) Chromium immobilization by extra- and intraradical fungal structures of arbuscular mycorrhizal symbioses. J Hazard Mater 316:34–42

Yaseen T, Burni T, Hussain F (2012) Effect of arbuscular mycorrhizal inoculation on nutrient uptake, growth and productivity of chickpea (*Cicer arietinum*) varieties. Int J Agron Plant Prod 3:334–345

Yi Y, Guerinot ML (1996) Genetic evidence that induction of root Fe^{3+} chelate reductase activity is necessary for iron deficiency. Plant J 10:835–844

Zhang RQ, Zhu HH, Zhao HQ et al (2013) Arbuscular mycorrhizal fungal inoculation increases phenolic synthesis in clover roots via hydrogen peroxide, salicylic acid and nitric oxide signaling pathways. J Plant Physiol 170(1):74–79

Chapter 7
Arbuscular Mycorrhizal Fungi and Heavy Metal Tolerance in Plants

Mohammad Miransari

Abstract Arbuscular mycorrhizal (AM) fungi are soil fungi developing symbiotic association with most terrestrial plants. In such a symbiosis, the fungi provide the host plant with water and nutrients in exchange for carbon. The beneficial effects of AM fungi on the growth of the host plant under stresses such as salinity, drought, heavy metals, etc. have been indicated by research work. However, more has yet to be elucidated on the mechanisms, which may increase the host plant as well as the fungal tolerance under stress. The alleviating effects of the fungi on plant growth and the environment under stress are mostly due to the superb abilities of the fungi in developing physiological and morphological mechanisms. The stress of heavy metals is among the most important stresses adversely affecting plant growth and the environment. The use of biological methods including the use of soil microbes such as AM fungi, plant growth-promoting rhizobacteria (PGPR), and endophytic bacteria has been proved to be among the most effective ones alleviating the adverse effects of stress on plant growth and the environment. The details of bioremediation mechanisms used by the fungi in association with the host plant including the expression of stress genes, the production of glomalin, the fungal phylogeny, and the allocation of heavy metals to different parts of mycorrhizal plant under the stress of heavy metals and some of the most recent advancement in this respect are presented in this chapter.

Keywords Arbuscular mycorrhizal (AM) fungi • Bioremediation • Heavy metals • Stress mechanisms

---

M. Miransari (✉)
Department of Book & Article, AbtinBerkeh Scientific Ltd. Company, Isfahan, Iran
e-mail: miransari1@gmail.com; Info@AbtinBerkeh.com

Q.-S. Wu (ed.), *Arbuscular Mycorrhizas and Stress Tolerance of Plants*,
DOI 10.1007/978-981-10-4115-0_7

7.1 Introduction

Plants and environments are subjected to different types of stresses such as salinity, drought, heavy metals, etc. adversely affecting plant growth and the environment. The presence of different microbes in the soil can be used as a great tool to increase the efficiency of plants and the environment. Different soil microbes including arbuscular mycorrhizal (AM) fungi, plant growth-promoting rhizobacteria (PGPR), and endophytic bacteria can positively enhance plant growth under different conditions including stress as well as the properties of the environment (Glick 2003; Gill and Tuteja 2010).

The stress of heavy metals is among the most important stresses decreasing the efficiency of plant and the environment. Heavy metals (53 elements) are determined according to their density (>5 g/cm^3) (Holleman and Wiberg 1985). Some of the heavy metals including iron (Fe), zinc (Zn), manganese (Mn), copper (Cu), and nickel (Ni) are essential for plant growth and yield production. Among the most important functions of such heavy metals in plant are (1) the catabolization of enzymatic and redox activities, (2) the transfer of electron, and (3) as the main part of DNA and RNA metabolism (Zenk 1996; Jamal et al. 2002). The other heavy metals such as lead (Pb), nickel (Ni), arsenic (As), etc., which are also the sources of contamination, do not have any functions in living organisms. The high concentrations of heavy metals can adversely affect the metabolism of plant and soil microbes and hence the food chain and the ecosystem (Öpik et al. 2006; Friedlova 2010).

The sources of heavy metals in the soil are industrialization and urbanization as well as the unsuitable disposal of wastes, which has increased the concentrations of heavy metals in the environment. Among such activities the anthropogenic activities including fertilization, the use of herbicides, wastes, and sewage sludge are also among the most important sources of environmental contamination. However, mining is the most important source of trace element in the environment (Whitmore 2006).

If present at extra amount in the environment, heavy metals can adversely affect plant and microbial activities as well as the environment (Miransari 2011). The high concentration of heavy metals negatively affects the enzymatic structure and functionality by altering the protein structure or substituting a necessary element. Accordingly, the structure of plasma membrane proteins including H^+-ATPases is altered affecting the plasma membrane permeability and functioning. The production of reactive oxygen species under heavy metal stress can also cause oxidative stress with negative effects on cellular structure and functioning and hence plant growth (Schützendübel and Polle 2002; Sajedi et al. 2010).

Biological methods, which are mainly the combined use of plants and microbes, can be suitably used for the bioremediation of contaminated areas with heavy metals. The two most important and applicable techniques, which are used for the removal of heavy metals from the environment, are bioremediation and phytoremediation, which are incorporated with the use of soil microbes (Lebeau et al. 2008). The phytoremediation techniques which are conducted in situ include:

1. Immobilization of metals in the soil by plants and microbes
2. Phytovolatilization, which is the volatilization of pollutants into the atmosphere
3. Rhizofiltration, which is the absorption of heavy metals from water
4. Phytodegradation, which is the degradation of pollutants by plants and soil microbes
5. Phytoextraction, which is the absorption of heavy metals (Cabral et al. 2015; Miransari 2016)

The most researched processes of phytoremediation are phytoextraction and phytostabilization. The second process does not eliminate the heavy metals from the environment, but immobilizes them in plant roots, decreasing their allocation to the plant aerial part; with time the heavy metals spread into the environment. In the process of phytoextraction, plants concentrate the heavy metals in their aerial parts by removing them from the soil. Some specific species of plants with suitable size and fast growth rate have the ability of greatly absorbing contaminants (phytoextraction) (Glick 2010).

Plants use different mechanisms to regulate ion homeostasis resulting in cellular detoxification under heavy metal stress (Clemens 2001). For example, the chelation of heavy metals by plant roots and their binding by cellular walls can regulate the concentration of heavy metals in plant cells. Plant cells are also able to produce metallothioneins and phytochelatins, which have a high affinity for the absorption of heavy metals. Subsequently their cellular concentration is controlled by their movement across the tonoplast and their sequestration in the vacuole (Hall 2002).

Different morphological and physiological activities of plant and microbes are adversely affected by the stress of heavy metals including the enzymatic activities, the uptake of nutrients, hormonal functioning, etc. Some heavy metals such as Zn and Cu are essential for plant growth; however, at higher concentration they negatively affect plant growth and the environment. Mycorrhizal fungi are able to increase the uptake of different micronutrients including Fe, Zn, Mn, and Cu under different conditions including stress (Miransari et al. 2009a, b; Lehmann et al. 2014; Lehmann and Rillig 2015).

Different techniques have been used for the alleviation of heavy metal stress; however, it has been indicated that the use of biological methods such as soil microbes including arbuscular mycorrhizal fungi, plant growth-promoting rhizobacteria (PGPR), and endophytic microbes are among the most effective ones. AM fungi are soil fungi developing symbiotic association with most of the terrestrial plants, and in such a symbiosis, the fungi provide the host plant with water and nutrients in the exchange for carbon. The fungal spores grow in the soil and develop the extensive network of hyphae with arbuscule and vesicle. Arbuscule is the structure, which is the interface for the exchange of water and nutrients; vesicle contains a high number of vacuoles and hence has a high absorbing potential for elements under different conditions including stress (Johnson et al. 2016).

In the following some of the most important details affecting the mycorrhizal role in alleviating heavy metal stress have been presented and analyzed.

7.2 Methods of Heavy Metal Alleviation by AM Fungi

Research work has indicated the alleviating effects of mycorrhizal fungi on plant growth and the environment under heavy metal stress. The fungi can help the host plant survive the stress of heavy metals by enhancing nutrient uptake and phytostabilization of heavy metals in the soil as well as by the absorption and detoxification of heavy metals. Such fungal activities can result in the bioremediation of the environment (Miransari 2011).

Research works have specified the alleviating effects of AM fungi on different stresses including salinity, drought, heavy metals, etc. The following mechanisms indicate how the fungi are able to alleviate the adverse effects of stress on plant growth and the environment: (1) the increased uptake of water and nutrients, (2) the production of plant hormones, (3) the interactions with the other soil microbes, (4) the ability for the uptake of heavy metals, and (5) the effects of the fungi on the activities of plant roots (Miransari 2010a, b; Garg and Pandey 2015).

Different mechanisms are used by mycorrhizal fungi to detoxify the adverse effects of heavy metals on the environment and on the growth of the host plant including (1) the production of chelating products, (2) the interaction with the plasma membrane, and (3) the effects of the cell wall components of the fungi and the plant. Such mechanisms can be conducted by the following:

1. Plant growth and the subsequent dilution of heavy metals in the tissues
2. The production of organic products such as organic acids by plant roots, which prevents the uptake of heavy metals from the rhizosphere by chelating, precipitating, and binding
3. The selective activity of plasma membrane in the absorption or desorption of heavy metals
4. The retention and immobilization of heavy metals by the fungal hyphae and plant roots
5. Chelating heavy metals by metallothioneins in the cytosol of both fungi and the host plant
6. The activity of specific or non-specific carriers and plasma membrane pores (both fungi and host plant)
7. Heavy metal sequestration by plant and fungi cellular vacuoles
8. Transfer of heavy metals by the fungal hyphae
9. The exchange of heavy metals between the fungi and the host plant
10. Active transfer of heavy metals by the specific and non-specific pathways in both fungi and the host plant (Kaldorf et al. 1999; Pawlowska and Charvat 2004; Miransari 2011)

The fungi are able to decrease the transfer of heavy metals to the host plant or from the roots to the aerial part. The higher uptake of nutrients including P can also enhance plant tolerance under heavy metal stress. The colonized roots of the host plant are able to retain heavy metals in the roots and the fungal hyphae and prevent their transfer to the plant aerial part. Among the most important mechanism by the fungi, which can protect the host plant from the adverse effects of heavy metals, is the immobilization of heavy metals in fungal hyphae. The fungal arbuscules, vesicles, and vacuoles can be the places for the accumulation of heavy metals and hence can prevent their transfer to the host plant and keep them away from the host plant (Kaldorf et al. 1999; Audet and Charest 2007).

The increased concentration of Cu in the spore vacuoles and the accumulation of Cd in the hyphal vacuoles of *Glomus intraradices* under heavy metal stress have been indicated by research work. Such structures of storage can make the fungi and the host plant tolerate the stress. The fungal cell wall is a suitable place for binding heavy metals because it contains hydroxyl, carboxyl, and free amino acids for the absorption of heavy metals. Such an absorbing potential is determined by the cell wall appearance and thickness. For example, the fungal species belonging to the family Glomeraceae do not have thick hyphae as their diameter ranges from 0.8 to 4.5 μm, and the maximum thickness of the hyphal cell wall is equal to 1.2 μm. However, the thickness of fungal hyphae from the Gigasporaceae family is higher with an average diameter of 20 μm. Interestingly, research work has indicated that the thinner hyphae can maintain higher amounts of heavy metals compared with the hyphae with a higher thickness, which is due to the higher surface area in the thinner hyphae. Accordingly, the rate of heavy metal (Pb, Cd, and Cu) absorption by *Rhizophagus clarus* with the average hyphal diameter of 4 μm was higher than *G. gigantean* with the average hyphal diameter of 13 μm. Research work has indicated that the root cortex in mycorrhizal plants contains much higher concentrations of heavy metals including Ni, Fe, and Zn; even different species of mycorrhizal fungi also have different potential of absorbing heavy metals (Kaldorf et al. 1999; Cabral et al. 2015).

The other interesting mechanism, which can make the mycorrhizal plants absorbing heavy metals, is the increased production of phytochelatins, glutathione-derived peptides, which are just produced by higher plants. Accordingly, plants will be able to chelate heavy metals and metalloids (Grill et al. 1987; Garg and Aggarwal 2011; Pollard et al. 2014). Among the parameters, which determine the alleviating effects of mycorrhizal fungi on the stress of heavy metals, are the species of the fungi and the types of heavy metals. The efficiency of AM fungi in the removal of heavy metals from the polluted soils is different, and some mycorrhizal fungi may be more efficient compared with the other types of fungal species. The allocation of the heavy metals by AM fungi to different parts of the host plant is also different as some heavy metals may concentrate in the roots and some may concentrate in plant aerial part (Wang et al. 2005, 2007).

The other important group of proteins (polypeptides with 70–80 amino acids and not with a high molecular weight), which can retain the heavy metals, are metallothioneins and the amino acids, which contain a high number of cysteine residues. Such proteins are produced in a wide variety of organisms, when subjected to heavy metal stress (Kumar et al. 2005). Metallothioneins not only regulate heavy metal homeostasis in plant, they can also alleviate the stress by controlling the oxidative stress caused by heavy metals (Maret 2000, 2003).

Although most fungal species are not tolerant under stress, and their abundance and number of species decrease, some fungal species, especially the ones which are present in the stressed environments (Glomeraceae is the dominant one), are tolerant and can be used for the inoculation of the host plant under stress. Because the fungi are able to grow fast and produce a high number of spores, they must have some morphological and physiological potential to be able to grow under stress (Lenoir et al. 2016). The fungi may prevent the host plant from contacting the heavy metals or may absorb the heavy metals from the soil. The interesting point is that the fungi are present in even the most unusable soils and can establish a symbiotic association with most terrestrial plants in highly polluted soils and hence can be used for the bioremediation of such soils (Cabral et al. 2015).

Although mycorrhizal fungi are present under different conditions including stress, their growth stages including spore germination and colonization, hyphal growth, and sporulation may be adversely affected by stress. Such negative effects can directly affect the fungi or the root growth of the host plant (Jacobson 1997; Liu et al. 2004; Wu and Xia 2006; Debiane et al. 2011; Zhu et al. 2012). Stress results in the induction of antioxidant activities and hence production of reactive oxygen species such as hydrogen peroxide (H_2O_2), superoxide anion radical ($O2^-$), and hydroxyl radicals (OH), which their extra amounts adversely affect cellular structure and activities (Apel and Hirt 2004; Fobert and Despres 2005).

Different research work has indicated that *Glomus* species are able to tolerate the stresses such as salinity and drought (Juniper and Abbott 2006; Sánchez-Castro et al. 2012). *F. mosseae* is distributed worldwide and can tolerate different types of stresses. These fungal species can quickly colonize the host plant roots and are resistant against different disturbances of soil such as salinity, heavy metals, drought, cold, etc. (Mohammad et al. 2003).

It has been indicated that mycorrhizal fungi have the Mn-dependent superoxide dismutase (SOD) gene, which is able to catalyze the superoxide anion to hydrogen peroxide and detoxify the adverse effect of reactive oxygen species under stress. It has also been indicated that mycorrhizal fungi produce other scavenging products such as thioredoxins and glutaredoxins, which are able to result in cellular redox homeostasis by regulating the redox potential of protein thiols and subsequently protecting the fungi from oxidative damage. There are about 30 genes, which are able to activate the antioxidant activity of mycorrhizal fungi (Tisserant et al. 2013).

The other molecules, which are also able to act as antioxidant in AM fungi, are vitamins B6, C and E, and glutathione. Under heavy metal stress, the induction of glutaredoxins has been indicated, and because glutathione can act as a redox donor for antioxidant enzymes, the scavenging role of glutathione has been suggested. The modulating effect of B6 on reactive oxygen species has also been indicated (González-Guerrero et al. 2007).

The adverse effects of pollutants on the membrane lipids of mycorrhizal fungi have been shown; membrane lipids are essential for the establishment of mycorrhizal fungal symbiosis (Debiane et al. 2011; Calonne et al. 2012). Different genes have been found, which are able to transfer heavy metals including Zn (*GintZnT1*), As (*GiArsA*), Cu, and Cd (*GintABC1*) (González-Guerrero et al. 2005, 2010).

The alleviating effects of mycorrhizal fungi under As stress indicated that after 1 h arsenic was detectable in mycorrhizal hyphae; however, after 3 days there was no As indicating that it was removed from the mycorrhizal hyphae. It was also revealed that the fungi were able to reduce As and make it less toxic. The fungi can overexpress the production of glomalin under Cu stress and hence protect the fungi from the adverse effects of Cu by fixing the fungal proteins (Ferrol et al. 2009).

A suitable method for cropping mycorrhizal plants under stress is the isolation of indigenous or stress-adapted fungi from stressful environments such as salty or polluted areas. The fungi not only are able to improve the quality of the environment and play as a bioindicator but can also act as an interface between the soil and the plant and hence transfer pollutants from the soil to the plant (Lenoir et al. 2016).

In brief the following mechanisms are used by mycorrhizal fungi to avoid stress: (1) morphological alteration; (2) production of different molecules such as antioxidants, trehalose, and chaperone proteins; (3) expression of different genes; (4) transport of pollutants; and (5) compartmentalization (Millar and Bennett 2016). However, more molecular details on the response of mycorrhizal fungi under stress must be elucidated. One step forward in this respect is the recent genetic identification of mycorrhizal fungi by researchers (Tisserant et al. 2013).

7.2.1 Expression of Mycorrhizal Fungal Genes

Mycorrhizal fungi are able to establish a favorable environment for the growth and activities of other soil microbes by increasing the rate of soil organic matter and enhancing the root growth and activity (Khan et al. 2000; Silva et al. 2006). Although the use of fungi is a suitable method for the bioremediation of the soil, their use is not the only method, and if combined with the other methods of bioremediation, the method may act more effectively.

The important point about the use of mycorrhizal fungi in contaminated areas is their single use in the absence of the host plant. Although the fungal spores can germinate in the absence of the host plant, they can only grow and produce the extensive network of hyphae, if the host plant is present. However, if the fungi are produced under monoxenic or axenic conditions, their single use may be possible for the bioremediation of contaminated areas (Mugnier and Mosse 1987; Declerck et al. 2005; Cabral et al. 2010).

Under heavy metal stress, certain mycorrhizal genes are activated, some are upregulated, and some are downregulated. The activation of the genes such as metallothionein, Zn transporter, glutathione S-transferase, and the stress protein genes in *G. intraradices* under the stress of Cd, Zn, and Cu has been indicated by research work. For example, the strong upregulation of metallothionein in the hyphae of *G. intraradices* subjected to Cu (20 μM) stress was the fungal response. However, this was not the case when the fungi were subjected to Cd (2.5 μM) stress as the metallothionein gene was not expressed (Lenoir et al. 2016).

The two most important mechanisms by which the mycorrhizal fungi are able to reduce the contamination of polluted areas are phytostabilization and phytoextraction. During the uptake of heavy metals by mycorrhizal fungal hyphae, a couple of physiological mechanisms including (1) the expression of proteins such as metallothionein and glomalin, (2) the retention of heavy metals by fungal spores, (3) chelation/complexation of heavy metals in the cellular membrane, and (4) the molecular expression of the genes results in the activation of the phytostabilization process (Behera 2014; Yang et al. 2014).

However, during the symbiosis of the hyperaccumulating plants with mycorrhizal fungi, the absorption and translocation of heavy metals increase and result in the process of phytoextraction. The processes of phytostabilization or phytoextraction have been found in different plants including herbaceous plants and crop plants including corn (Hetrick et al. 1994; Weissenhorn et al. 1995; Soares and Siqueira 2008). The efficiency of such mechanisms is determined by parameters such as plant properties including biomass production and growth rate and the tolerating ability of plant under high concentration of heavy metals as well as the bioavailability of heavy metals.

According to Cabral et al. (2010) mycorrhizal fungi had a high retention potential for Cu and Zn in their tissues; however, it was less for Cd and Pb. It is also important to evaluate the fungal genes, which are activated under heavy metal stress, and make effort to improve their activity under the stress. Accordingly the new genes, which are more efficient under heavy metal stress can be produced and used for biotechnological purposes (Cabral et al. 2015).

7.2.2 Production of Glomalin

The glycoprotein, glomalin, is a protein produced by AM fungi and is able to bind the soil particles and improve the properties of soil. However, it is also able to bind heavy metals in the soil. The glomalins are produced in the fungal hyphae, which is the place for the accumulation of a high concentration of heavy metals. Wu et al. (2014) investigated the effects of glomalin produced by AM fungi on the absorption of Pb and Ni in the soil and compared their results with the absorption of such heavy metals by organic matter. They found that after 140 days glomalin was able to absorb Pb at 0.21–1.78% of the Pb in the soil and 0.38–0.98% of Ni. The authors found that just 4% of the total Pb and Ni in the soil was absorbed with glomalin related to the 40–54% of the Pb and Cd, absorbed by the humin and fulvic acids in the soil organic matter. Their results indicated that although glomalin is able to absorb a part of Pb and Ni in the soil, its rate of absorption is negligible compared with the part of Pb and Ni absorbed by the soil organic matter.

Because chromium (Cr) is highly toxic to human health, Gil-Cardeza et al. (2014) investigated the distribution of Cr in the rhizosphere of *Ricinus communis* and *Conium maculatum* as well as in control soil from an industrial area in Argentina. The concentrations of total Cr, Cr (VI), and Cr (III) were evaluated using three different soil fractions: total, extractable, and bound to total glomalin-related soil protein (T-GRSP). The highest rate of total Cr and total CR (VI) was found in control soil. However, the concentrations of total Cr (VI) were not higher than the standard rate and not toxic for human health (8 μg Cr(VI) g^{-1} soil), but the respective rate in the control soil was 1.8 higher. There were concentrations of total Cr higher than the standard value (250 μg Cr g^{-1} soil) in all treatments. The Cr extractable and bound to T-GRSP was not detectable, and the highest rate of Cr (III) bound to T-GRSP was resulted by the control soil. Both plant species developed a symbiosis with AM fungi, and the rate of Cr was higher in the roots than the aerial part. The authors indicated that this is the first time that Cr is found in the T-GRSP of the soil organic matter. Accordingly, this mechanism is used by mycorrhizal *R. communis* and *C. maculatum* to decrease the concentration of Cr in their rhizosphere. However, the role of AM on the phytoremediation of Cr in the rhizosphere of these plant species has yet to be more investigated.

7.2.3 Mycorrhizal Phylogeny

How the phylogeny of mycorrhizal fungi may affect the bioremediation of heavy metals is an important research question. He et al. (2014) investigated the phylogenetic effects of Glomeraceae and non-Glomeraceae on the growth of the host plant under the stress of heavy metal. Their meta-analysis indicated that the effects of both phylogenetic mycorrhizal fungi were positive on the growth of the host plant under the heavy metal stress although variable under different levels of heavy metal. Glomeraceae was more advantageous to the host plant under the heavy metal stress, and non-Glomeraceae was more beneficial to the host plant under the non-stress conditions. The response of different plant species was different to the Glomeraceae under heavy metal stress; legumes grew better than nonlegumes under heavy metal stress. However, the response of non-Glomeraceae was more pronounced in legumes than nonlegumes under non-stress conditions. The authors accordingly indicated that the combined use of legumes with Glomeraceae can be a suitable way for the bioremediation of heavy metal stress.

Wu et al. (2016a) indicated that because the alleviating effects of mycorrhizal fungi on the stress of heavy metals are by improving the host plant nutrition, especially P, it is possible to improve plant growth under environmental contamination using exogenous P. However, their results showed although the fungi were able to enhance plant growth under heavy metal stress by increasing the uptake of nutrients including P, the exogenous use of P was not able to alleviate the stress like the fungi. The fungi were able to immobilize Cr in the host plant roots. The experiment also revealed the distribution and stabilization of Cr in the host plant roots. The research also indicated that the mycorrhizal roots are able to absorb Cr; however the fungi are able to inhibit the transfer of Cr to the aerial parts and hence controlled the phototoxic effect of Cr.

Hristozkova et al. (2015) investigated the effects of mycorrhization on the removal of Cd and Pb from a contaminated soil. The authors examined the effects of mycorrhizal fungi on the oil composition of *Origanum majorana* L., as well as the antioxidant activity and the uptake and accumulation of heavy metals by plants. Two strains of *Claroideoglomus claroideum* (S1 and S2) were isolated from a highly contaminated soil with heavy metals. The third strain (S3) (*Funneliformis mosseae*) was isolated from an industrial area. S1 resulted in the highest rate of root colonization, and there were not any significant differences between S2 and S3. The highest rate of total essential oil was resulted by S3 with a high rate of Pb in the roots. The content of essential oil was changed by mycorrhization. The use of S1 and S2 significantly decreased the concentration of Cd and Pb in the roots and the aerial parts of the host plant compared with the non-mycorrhizal plants. The antioxidant activity of the host plant increased due to inoculation with S2 and S3, which was mainly because of the increased levels of phenolic compounds.

7.2.4 Research Experiments

In the following, some research examples, which indicate the alleviating effects of mycorrhizal fungi on the stress of heavy metals, are presented.

Shabani et al. (2016) investigated the effects of mycorrhization (*Funneliformis mosseae* = *Glomus mosseae*) on the stress of nickel (Ni) using tall fescue (*Festuca arundinacea* = *Schedonorus arundinaceus*). The growth of tall fescue and the photosynthetic pigments as well as mycorrhizal colonization were adversely affected by Ni. Mycorrhization significantly increased plant phosphorus content but not the root carbohydrate. Although Ni addition to the soil enhanced Ni concentration in different parts of the plant, mycorrhization significantly decreased the Ni content of plant aerial part compared with plant roots. The transporter of ABC and the transcripts of metallothionein were significantly higher in mycorrhizal plants than non-mycorrhizal plants. The authors accordingly indicated that mycorrhizal fungi are able to alleviate the stress of Ni on the growth of tall fescue by decreasing the transport of Ni from the roots to the plant aerial part.

Research work has indicated that plants such as wheat are able to accumulate heavy metals in their tissues at extra amounts. Accordingly, Kanwal et al. (2016) investigated the effects of mycorrhization on the growth of wheat plants in the presence of different Zn concentrations (0, 100, 300, and 900 $mgkg^{-1}$). Different plant parameters including plant growth, root colonization, the content of heavy metals, and other biochemical parameters were investigated. The most optimum response of mycorrhizal plants was resulted at the moderate level of Zn (300 $mgkg^{-1}$). Mycorrhization significantly decreased the concentration of Zn in plant aerial part compared with plant roots. There was a high concentration of P in mycorrhizal plants related to non-mycorrhizal plants. The highest level of Zn (900 $mgkg^{-1}$) decreased plant nutrient content, plant growth, and the activity of antioxidant enzymes. The authors accordingly indicated that (1) mycorrhizal fungi use different strategies to alleviate the stress of heavy metals and (2) they can effectively alleviate the stress of Zn in contaminated soils by the process of Zn phytostabilization, which is important for the increased production of food as well as for food safety.

Wu et al. (2016b) investigated the effects of mycorrhizal fungi on the removal of chromium from a polluted soil. The authors investigated the speciation and distribution of Cr (VI) in the mycorrhizal roots and in the extraradical mycelium. They found that Cr stress resulted in the production of some extracellular polymeric substances on the fungal surface resulting in the significant immobilization of Cr and its subsequent reduction. There was a significant correlation between Cr and P indicating that phosphate ions may act as a counter for Cr (III). The analysis also indicated the reduction of Cr (VI) to Cr (III) in the fungal structure, including intraradical mycelium, arbuscules, etc., and the cell walls of mycorrhizal roots. The research work revealed how Cr is immobilized at a cellular level by mycorrhizal hyphae and mycorrhizal roots, and it also indicated the mechanisms, which may contribute to such Cr immobilization.

In a research work by Schneider et al. (2016), the role of mycorrhizal fungi on plant diversity and bioremediation of areas contaminated with Pb was investigated. A total of 39 mycorrhizal fungal species were found among which *Glomus* genera were the most abundant ones. All the herbs were colonized with mycorrhizal fungi. The highest rate of Pb concentration in the plant roots ranged from 1106 to 15,433 ($mgkg^{-1}$), and in the plant aerial part, it was in the range of 625 to 934 ($mgkg^{-1}$). The area heterogeneity determined the diversity of mycorrhizal fungi, which was also a function of Pb concentration. The diversity of the fungi also affected the diversity of plant species in areas with a high concentration of Pb. Investigating the diversity of mycorrhizal fungi in the areas contaminated with Pb may more clearly reveal the interactions between the fungi and heavy metals and hence the subsequent use of the fungi for the use of bioremediation techniques such as biostabilization.

Abd Allah et al. (2015) investigated the effects of mycorrhizal fungi on the growth and the biochemical properties of sunflower (*Helianthus annuus* L.) under cadmium stress. Sunflower is an important crop plant with high oil content and suitable for phytoremediation. Cadmium stress adversely affected plant growth, the chlorophyll content, and the stability of the cellular membrane. However, the fungi were able to alleviate the stress by positively affecting the abovementioned parameters.

Under cadmium stress, the rate of lipid peroxidation and hydrogen peroxide as well as the production of proline, total phenols, and fatty acid content increased. The stress also increased the production of antioxidant enzymes, which also increased in the presence of the fungi. The mycorrhizal plants had a higher rate of acid and alkaline phosphatase activities, which decreased under the stress. Stress negatively affected the production of oleic ($C_{18:1}$), palmitoleic acid ($C_{16:1}$), linoleic ($C_{18:2}$), and linolenic acid ($C_{18:3}$); however, the fungi were able to mitigate such adverse effects (Abd Allah et al. 2015).

In another study, Zhang et al. (2015) evaluated the effects of mycorrhization on the growth of *Medicago truncatula* under As stress by two pot experiments. A wild type and a non-mycorrhizal mutant of the plant were used for the experiments to investigate the effects of the fungi on the accumulation and speciation of As. The fungi were able to increase plant growth and the P concentration significantly and decrease As concentration in plant. The fungi also increased the rate of arsenite in total As compared with the other types of As in different parts of the plant; however, dimethylarsinic acid (DMA) just increased in the aerial part of the plant. The authors accordingly indicated that the fungi are able to alleviate the stress of As by the methylation of inorganic As into less toxic organic products (DMA) and by converting arsenate to arsenite. The results indicated the detoxification processes used by the fungi to alleviate As stress in plants. The use of mutant *M. truncatula* was also a useful tool revealing the positive effects of mycorrhizal fungi on process of As detoxification in the plants.

7.3 Conclusions

Heavy metal stress, which is the result of industrialization and excess use of chemicals, is a major concern affecting large areas of the world. Although different methods have been used to remediate the contaminated environments with heavy metals, the use of biological methods, which is the single or the combined use of plants and microbes, has been among the most effective ones. In this chapter the alleviating effects of mycorrhizal fungi, which are the most abundant type of soil fungi, developing a symbiotic association with their host plant, on plant growth and on the bioremediation of contaminated areas have been presented and analyzed. The fungi have some superb abilities, which make them survive the stress of heavy metals and alleviate the adverse effects of the stress on plant growth and on the environment. The fungi can affect the chemical properties of heavy metals in the soil, their absorption by the host plant, and their allocation to different plant parts, affecting plant growth and the bioremediation process. Some of the latest progress and finding in the field of mycorrhization and the related mechanisms affecting the process of heavy metal bioremediation have been analyzed and presented in this chapter.

References

Abd Allah EF, Abeer H, Alqarawi AA et al (2015) Alleviation of adverse impact of cadmium stress in sunflower (*Helianthus annuus* L.) by arbuscular mycorrhizal fungi. Pak J Bot 47:785–795

Apel K, Hirt H (2004) Reactive oxygen species: metabolism, oxidative stress, and signal transduction. Ann Rev Plant Biol 55:373–399

Audet P, Charest C (2007) Dynamics of arbuscular mycorrhizal symbiosis in heavy metal phytoremediation: meta-analytical and conceptual perspectives. Environ Pollut 147:609–614

Behera K (2014) Phytoremediation, transgenic plants and microbes. In: Sustainable Agriculture Reviews. Springer, Cham, pp 65–85

Cabral L, Siqueira J, Soares C et al (2010) Retention of heavy metals by arbuscular mycorrhizal fungi mycelium. Química Nova 33:25–29

Cabral L, Soares C, Giachini A et al (2015) Arbuscular mycorrhizal fungi in phytoremediation of contaminated areas by trace elements: mechanisms and major benefits of their applications. World J Microbiol Biotechnol 31:1655–1664

Calonne M, Sahraoui AL-H, Campagnac E et al (2012) Propiconazole inhibits the sterol 14a-demethylase in *Glomus irregulare* like in phytopathogenic fungi. Chemosphere 87:376–383

Clemens S (2001) Molecular mechanisms of plant metal tolerance and homeostasis. Planta 212:475–486

Debiane D, Calonne M, Fontaine J et al (2011) Lipid content disturbance in the arbuscular mycorrhizal, Glomus Irregulare grown in monoxenic conditions under PAHs pollution. Fungal Biol 115:782–792

Declerck S, Strullu D, Fortin J (2005) In vitro culture of mycorrhizas. Springer, New York

Ferrol N, González-Guerrero M, Valderas A et al (2009) Survival strategies of arbuscular mycorrhizal fungi in Cu-polluted environments. Phytochem Rev 8:551–559

Fobert PR, Despres C (2005) Redox control of systemic acquired resistance. Curr Opin Plant Biol 8:378–382

Friedlova M (2010) The influence of heavy metals on soil biological and chemical properties. Soil Water Res 5:21–27

Garg N, Aggarwal N (2011) Effects of interactions between cadmium and lead on growth, nitrogen fixation, phytochelatin, and glutathione production in mycorrhizal *Cajanus cajan* (L.) Millsp. J Plant Growth Regul 30:286–300

Garg N, Pandey R (2015) Effectiveness of native and exotic arbuscular mycorrhizal fungi on nutrient uptake and ion homeostasis in salt-stressed *Cajanus cajan* L. (Millsp.) genotypes. Mycorrhiza 25:165–180

Gil-Cardeza ML, Ferri A, Cornejo P et al (2014) Distribution of chromium species in a Cr-polluted soil: presence of Cr (III) in glomalin related protein fraction. Sci Total Environ 493:828–833

Gill SS, Tuteja N (2010) Reactive oxygen species and antioxidant machinery in abiotic stress tolerance in crop plants. Plant Physiol Biochem 48:909–930

Glick BR (2003) Phytoremediation: synergistic use of plants and bacteria to clean up the environment. Biotechnol Adv 21:383–393

Glick BR (2010) Using soil bacteria to facilitate phytoremediation. Biotechnol Adv 28:367–374

González-Guerrero M, Azcón-Aguilar C, Mooney M et al (2005) Characterization of a Glomus Irregularis gene encoding a putative Zn transporter of the cation diffusion facilitator family. Fungal Gen Biol 42:130–140

González-Guerrero M, Cano C, Azcón-Aguilar C et al (2007) GintMT1 encodes a functional metallothionein in *Glomus irregularis* that responds to oxidative stress. Mycorrhiza 17:327–335

González-Guerrero M, Benabdellah K, Valderas A et al (2010) GintABC1 encodes a putative ABC transporter of the MRP subfamily induced by Cu, Cd, and oxidative stress in *Glomus irregularis*. Mycorrhiza 20:137–146

Grill E, Winnacker E, Zenk M (1987) Phytochelatins, a class of heavy metal binding peptides from plants are functionally analogous to metallothioneins. Proc Nat Acad Sci USA 84:439–443

Hall JL (2002) Cellular mechanisms for heavy metal detoxification and tolerance. J Exp Bot 53:1–11

He L, Yang H, Yu Z et al (2014) Arbuscular mycorrhizal fungal phylogenetic groups differ in affecting host plants along heavy metal levels. J Environ Sci 26:2034–2040

Hetrick B, Wilson G, Figge D (1994) The influence of mycorrhizal symbiosis and fertilizer amendments on establishment of vegetation in heavy metal mine spoil. Environ Pollut 86:171–179

Holleman A, Wiberg E (1985) Lehrbuch der Anorganischen Chemie. Nabu Press, Berlin

Hristozkova M, Geneva M, Stancheva I et al (2015) Aspects of mycorrhizal colonization in adaptation of sweet marjoram (*Origanum majorana* L.) grown on industrially polluted soil. Turk J Biol 39:461–468

Jacobson K (1997) Moisture and substrate stability determine VA-mycorrhizal fungal community distribution and structure in an arid grassland. J Arid Environ 35:59–75

Johnson N, Gehring C, Jansa J (2016) Mycorrhizal mediation of soil: fertility, structure, and carbon storage. Elsevier, Cambridge, MA
Jamal A, Ayuba N, Usmana M (2002) Arbuscular mycorrhizal fungi enhance zinc and nickel uptake from contaminated soil by soyabean and lentil. Int J Phytoremed 4:205–221
Juniper S, Abbott L (2006) Soil salinity delays germination and limits growth of hyphae from propagules of arbuscular mycorrhizal fungi. Mycorrhiza 16:371–379
Kaldorf M, Kuhn AJ, Schroder WH et al (1999) Selective element deposits in maize colonized by a heavy metal tolerance conferring arbuscular mycorrhizal fungus. J Plant Physiol 154:718–728
Kanwal S, Bano A, Malik RN (2016) Role of arbuscular mycorrhizal fungi in phytoremediation of heavy metals and effects on growth and biochemical activities of wheat (*Triticum aestivum* L.) plants in Zn contaminated soils. Afr J Biotechnol 15:872–883
Khan A, Kuek C, Chaudhry T et al (2000) Role of plants, mycorrhizae and phytochelators en heavy metal contaminated land remediation. Chemosphere 41:197–207
Kumar K, Dayananda S, Subramanyam C (2005) Copper alone, but not oxidative stress, induces copper–metallothionein gene in *Neurospora crassa*. FEMS Microbiol Lett 242:45–50
Lebeau T, Braud A, Jezequel K (2008) Performance of bioaugmentation-assisted phytoextraction applied to metal contaminated soils: a review. Environ Poll 153:497–522
Lehmann A, Rillig MC (2015) Arbuscular mycorrhizal contribution to copper, manganese and iron nutrient concentrations in crops–a meta-analysis. Soil Biol Biochem 81:147–158
Lehmann A, Veresoglou SD, Leifheit EF et al (2014) Arbuscular mycorrhizal influence on zinc nutrition in crop plants–a meta-analysis. Soil Biol Biochem 69:123–131
Lenoir I, Fontaine J, Sahraoui A (2016) Arbuscular mycorrhizal fungal responses to abiotic stresses: a review. Phytochemistry 123:4–15
Liu A, Wang B, Hamel C (2004) Arbuscular mycorrhiza colonization and development at suboptimal root zone temperature. Mycorrhiza 14:93–101
Maret W (2000) The function of zinc metallothionein: a link between cellular zinc and redox state. J Nutrit 130:1455S–1458S
Maret W (2003) Cellular zinc and redox states converge in the metallothionein/thionein pair. J Nutrit 133:1460S–1462S
Millar NS, Bennett AE (2016) Stressed out symbiotes: hypotheses for the influence of abiotic stress on arbuscular mycorrhizal fungi. Oecologia 182:625–641
Miransari M, Bahrami HA, Rejali F et al (2009a) Effects of arbuscular mycorrhiza, soil sterilization, and soil compaction on wheat (*Triticum aestivum* L.) nutrients uptake. Soil Till Res 104:48–55
Miransari M, Bahrami HA, Rejali F et al (2009b) Effects of soil compaction and arbuscular mycorrhiza on corn (*Zea mays* L.) nutrient uptake. Soil Till Res 103:282–290
Miransari M (2010a) Contribution of arbuscular mycorrhizal symbiosis to plant growth under different types of soil stresses. Plant Biol 12:563–569
Miransari M (2010b) Arbuscular mycorrhizal fungi and heavy metals. In: Gupta VK, Tuohy M, Gaur RK (eds) Fungal biochemistry and biotechnology. LAMBERT Academic Publishing AG & Co. KG, Saarbrücken
Miransari M (2011) Hyperaccumulators, arbuscular mycorrhizal fungi and stress of heavy metals. Biotechnol Adv 29:645–653
Miransari M (2016) Soybean production and heavy metal stress. In: Miransari M (ed) Abiotic and biotic stresses in soybean production. Soybean production. Elsevier, Amsterdam, pp 197–216
Mohammad M, Hamad S, Malkawi H (2003) Population of arbuscular mycorrhizal fungi in semiarid environment of Jordan as influenced by biotic and abiotic factors. J Arid Environ 53:409–417
Mugnier J, Mosse B (1987) Vesicular-arbuscular mycorrhizal infection in transformed root-inducing T-DNA roots grown axenically. Phytopathol 77:1045–1050
Öpik M, Moora M, Liira J et al (2006) Composition of root-colonizing arbuscular mycorrhizal fungal communities in different ecosystems around the globe. J Ecol 94:778–790

Pawlowska T, Charvat I (2004) Heavy-metal stress and developmental patterns of arbuscular mycorrhizal fungi. Appl Environ Microbiol 70:6643–6649
Pollard AJ, Reeves RD, Baker AJ (2014) Facultative hyperaccumulation of heavy metals and metalloids. Plant Sci 217:8–17
Sajedi NA, Ardakani MR, Rejali F et al (2010) Yield and yield components of hybrid corn (*Zea mays* L.) as affected by mycorrhizal symbiosis and zinc sulfate under drought stress. Physiol Mol Biol Plants 16:343–351
Sánchez-Castro I, Ferrol N, Barea J (2012) Analyzing the community composition of arbuscular mycorrhizal fungi colonizing the roots of representative shrubland species in a Mediterranean ecosystem. J Arid Environ 80:1–9
Schneider J, Bundschuh J, do Nascimento C (2016) Arbuscular mycorrhizal fungi-assisted phytoremediation of a lead-contaminated site. Sci Total Environ 572:86–97
Schützendübel A, Polle A (2002) Plant responses to abiotic stresses: heavy metal-induced oxidative stress and protection by mycorrhization. J Exp Bot 53:1351–1365
Shabani L, Sabzalian MR, Mostafavi Pour S (2016) Arbuscular mycorrhiza affects nickel translocation and expression of ABC transporter and metallothionein genes in *Festuca arundinacea*. Mycorrhiza 26:67–76
Silva S, Siqueira J, Soares C (2006) Mycorrhizal fungi influence on brachiariagrass growth and heavy metal extraction in a contaminated soil. Pesq Agr Bras 41:1749–1757
Soares C, Siqueira J (2008) Mycorrhiza and phosphate protection of tropical grass species against heavy metal toxicity in multi-contamined soil. Biol Fert Soils 44:833–841
Tisserant E, Malbreil M, Kuo A et al (2013) Genome of an arbuscular mycorrhizal fungus provides insight into the oldest plant symbiosis. Proc Nat Acad Sci USA 110:20117–20122
Wang F, Lin X, Yin R (2005) Heavy metal uptake by arbuscular mycorrhizas of *Elsholtzia splendens* and the potential for phytoremediation of contaminated soil. Plant Soil 269:225–232
Wang F, Lin X, Yin R (2007) Role of microbial inoculation and chitosan in phytoextraction of Cu, Zn, Pb and Cd by *Elsholtzia splendens*—a field case. Environ Pollut 147:248–255
Weissenhorn I, Leyval C, Berthelin J (1995) Bioavailability of heavy metals and abundance of arbuscular in soil polluted by atmospheric deposition from a smelter. Biol Fert Soils 19:22–28
Whitmore A (2006) The emperors new clothes: sustainable mining? J Clean Prod 14:309–314
Wu QS, Xia RX (2006) Arbuscular mycorrhizal fungi influence growth, osmotic adjustment and photosynthesis of citrus under well-watered and water stress conditions. J Plant Physiol 163:417–425
Wu Z, McGrouther K, Huang J et al (2014) Decomposition and the contribution of glomalin-related soil protein (GRSP) in heavy metal sequestration: field experiment. Soil Biol Biochem 68:283–290
Wu S, Zhang X, Chen B et al (2016a) Chromium immobilization by extraradical mycelium of arbuscular mycorrhiza contributes to plant chromium tolerance. Environ Exp Bot 122:10–18
Wu S, Zhang X, Sun Y et al (2016b) Chromium immobilization by extra-and intraradical fungal structures of arbuscular mycorrhizal symbioses. J Haz Mat 316:34–42
Yang W, Zhang T, Li S et al (2014) Metal removal from and microbial property improvement of a multiple heavy metals contaminated soil by phytoextraction with a cadmium hyperaccumulator *Sedum alfredii* H. J Soils Sedim 14:1385–1396
Zenk MH (1996) Heavy metal detoxification in higher plants: a review. Gene 179:21–30
Zhang X, Ren BH, Wu SL et al (2015) Arbuscular mycorrhizal symbiosis influences arsenic accumulation and speciation in *Medicago truncatula* L. in arsenic-contaminated soil. Chemosphere 119:224–230
Zhu X, Song F, Liu S et al (2012) Arbuscular mycorrhizae improves photosynthesis and water status of *Zea mays* L. under drought stress. Plant Soil Environ 58:186–191

Chapter 8
Arbuscular Mycorrhizal Fungi and Tolerance of Temperature Stress in Plants

Xiancan Zhu, Fengbin Song, and Fulai Liu

Abstract Temperature is one of the most important environmental factors that determine the growth and productivity of plants across the globe. Many physiological and biochemical processes and functions are affected by low and high temperature stresses. Arbuscular mycorrhizal (AM) symbiosis has been shown to improve tolerance to temperature stress in plants. This chapter addresses the effect of AM symbiosis on plant growth and biomass production, water relations (water potential, stomatal conductance, and aquaporins), photosynthesis (photosynthetic rate, chlorophyll, and chlorophyll fluorescence), plasma membrane permeability (malondialdehyde and ATPase), reactive oxygen species (ROS) and antioxidants, osmotic adjustment, carbohydrate metabolism, nutrient acquisition, and secondary metabolism under low or high temperature stress. The possible mechanisms of AM symbiosis improving temperature stress tolerance of the host plants via enhancing water and nutrient uptake, improving photosynthetic capacity and efficiency, protecting plant against oxidative damage, and increasing accumulation of osmolytes are discussed. This chapter also provides some future perspectives for better understanding the mechanisms of AM plant tolerance against temperature stress.

Keywords Mycorrhiza • Plasma membrane permeability • Secondary metabolism • Temperature

X. Zhu (✉) • F. Song
Northeast Institute of Geography and Agroecology, Chinese Academy of Sciences, Changchun 130102, People's Republic of China
e-mail: zhuxiancan@neigae.ac.cn

F. Liu
Department of Plant and Environmental Sciences, Faculty of Science, University of Copenhagen, Taastrup DK-2630, Denmark

Q.-S. Wu (ed.), *Arbuscular Mycorrhizas and Stress Tolerance of Plants*,
DOI 10.1007/978-981-10-4115-0_8

8.1 Introduction

Plants often encounter a wide range of environmental perturbations including abiotic and biotic stresses (Kaplan et al. 2004). Temperature is one of the most important environmental factors that determine the growth and productivity of plants across the globe (Zhu et al. 2011). In recent years, extreme temperature events have become more frequent along with global climate change. In Northeast China, for example, crops often suffer from the damage of low temperature during early spring, causing significant losses in grain yield.

Temperature stress includes both low temperature and high temperature events occurring during the growth season of the crops. In general, low but not freezing temperatures (i.e., 0–15 °C) is considered as low temperature stress (Theocharis et al. 2012), and the elevation in temperature beyond a threshold level (i.e., 10–15 °C above ambient) is considered as high temperature stress (Wahid et al. 2007). When plant is exposed to low or high temperature stress, many physiological and biochemical processes and functions will be disturbed. The injuries include damage of cell membrane structure and lipid composition, cellular leakage of electrolytes and amino acids, peroxidation of membrane lipids, a diversion of electron flow to alternate pathways, denaturation and aggregation of proteins, redistribution of intracellular calcium ions, inactivation of enzymes in chloroplast and mitochondria, and production of toxic compounds and reactive oxygen species (ROS) (Wahid et al. 2007; Ruelland and Zachowski 2010; Matteucci et al. 2011; Theocharis et al. 2012).

To overcome the temperature stress, plants have evolved some adaptive strategies and triggered a cascade of events that cause changes in the expression of a large number of temperature-induced genes and the production of various protein types, and induce biochemical and physiological modifications (Theocharis et al. 2012; Janmohammadi et al. 2015; Sharma and Laxmi 2016). The tolerance mechanisms involved in modification of plant cell membrane, accumulation of cytosolic calcium ion, acclimation of photosynthesis, activation of ROS scavenger systems, accumulation of compatible solutes such as proline and sugars, and induction of cold-related gene expression (Wahid et al. 2007; Theocharis et al. 2012). For example, in Arabidopsis, ICE-DREB1/CBF regulon plays a key role in tolerance to low temperature stress. *ICE1* is a MYC-type transcription factor that positively regulates and activates *CBF/DREB1* genes, and *CBF/DREB1* induces and regulates *COR* genes leading to tolerance to low temperature stress (Chinnusamy et al. 2007; Thomashow 2010).

It is well accepted that arbuscular mycorrhizal (AM) symbiosis is an effective strategy to improve tolerance to temperature stress in plants (Duhamel and Vandenkoornhuyse 2013). AM symbioses are common in nature and have been demonstrated to be beneficial for a sustainable agricultural ecosystem due to their functions in improving plant growth, enhancing nutrient uptake, increasing soil stability, and alleviating abiotic and biotic stress (Gianinazzi et al. 2010; Berta et al. 2014; Zhu et al. 2015). AM symbiosis can alter plant physiology to allow it to cope with stress conditions (Miransari et al. 2008). Thus, in this chapter, the effects of

AM on plants under low and high temperature stress are discussed, and the aim is to identify the underlying mechanisms contributing to plant tolerance to temperature stress.

8.2 Effect of Temperature Stress on AM Fungi

Environmental conditions affect the community and development of AM fungi and the formation of mycorrhizae (Liu et al. 2004). Using traditional morphological and molecular methods, numerous studies have investigated the diversity of AM fungi at varied environmental conditions (Lumini et al. 2011; Camenzind et al. 2014; Botnen et al. 2015). However, as far as we are aware, little information is available about the effect of temperature on AM community structure. Heinemeyer et al. (2003) reported that an increase of temperature by 3 °C above ambient results in no significant changes in the AM fungal community in a native grassland community. Gutknecht et al. (2012) quantified the response of soil microbial biomass and community structure from 2001 to 2006 to simulated global change at the Jasper Ridge Global Change Experiment. These authors found that AM fungi biomarker biomass was lower under elevated temperature (by 1 °C) only in 2006, and there was no significant difference in another 5 years. In addition, several studies have reported the effect of seasonal and climate changes including temperature fluctuations on AM fungal community structure (Dumbrell et al. 2011; Torrecillas et al. 2013; Bainard et al. 2014); however, until now there is no consensus about how AM fungal community structure and function respond to temperature stress.

Temperature altered mycelium growth, growth pattern, and phenology of intraradical and extraradical colonization of AM fungi (Gavito et al. 2005; Compant et al. 2010). Gavito et al. (2005) suggested that temperature optima for AM fungi development were between 18 and 30 °C. But the temperature optima were depending on the species of AM fungus. Soil warming stimulated external hyphal production and extraradical hyphal network growth in temperate conditions (Staddon et al. 2004; Hawkes et al. 2008). Low and high temperatures reduced AM fungal growth, inhibited extraradical hyphal network structure formation and AM fungal activity, although AM fungi response to temperature stress exist difference. Liu et al. (2004) reported that sporulation of *Glomus intraradices* was reduced at 15 °C, while spore metabolic activity was not reduced until an even lower temperature (10 °C) was reached. Schenck et al. (1975) found the germination of spores of *Glomus coralloidea* and *Glomus heterogama* were reduced above 34 °C.

Root colonization was also changed by low and high temperatures. Some researchers have reported that AM colonization of crop plants was strongly reduced at 15 °C compared with ambient temperature, and almost completely inhibited at 10 or 5 °C (Zhang et al. 1995; Liu et al. 2004; Zhu et al. 2010a). In contrast, some scientists found that AM colonization was unaffected by low temperature (Hayman 1974; Karasawa et al. 2012). At high temperature conditions, AM colonization was also affected. Haugen and Smith (1992) showed that colonization of cashew

(*Anacardium occidentale*) roots by *Glomus intraradices* was reduced at 38 °C compared to 22 °C; whereas Martin and Stutz (2004) reported that colonization of pepper (*Capsicum annuum*) roots by *Glomus* AZ112 was increased at 32.1–38 °C compared to 20.7–25.4 °C. Zhu et al. (2011) found that there was no significant change in colonization of maize (*Zea mays*) roots by *Glomus etunicatum* at 35 and 40 °C. The discrepancy of the results could be due to different AM fungi and plant materials as well as experimental conditions among the different studies.

8.3 AM and Tolerance of Temperature Stress in Plants

In the past three decades, several studies have investigated the effect of AM fungi on plant growth and physiology under low and high temperature stresses. Table 8.1 lists the published works of AM effect on the host plants exposed to low and high temperature stresses from 1976. Below, the AM plants' response to temperature stress and possible role of AM in tolerance of temperature stress are discussed.

8.3.1 *Plant Growth and Biomass*

Under low or high temperature stress, plant morphological features were affected and plant growth and biomass production could be inhibited. AM symbiosis has been shown to increase the fitness of the host plants via enhancing its growth and biomass production. It has been reported that AM plants grown better than non-AM plants under low or high temperature stress, which partly attributed to the enhanced photosynthesis and nutrient uptake, especially for P nutrition. Several researches have reported that shoot and root dry weights of AM plants were higher than the non-AM plants at low or high temperature conditions (Haugen and Smith 1992; Martin and Stutz 2004; Zhu et al. 2010b; Abdel Latef and Chaoxing 2011; Liu et al. 2014a). Matsubara et al. (2004) found AM fungi increased leaf and root number, crown diameter, and leaf area of strawberry (*Fragaria ananassa*) plants under high temperature stress. Bunn et al. (2009) observed AM *Dichanthelium lanuginosum* plants had higher root length and root diameter compared with the non-AM plants when exposed to high temperature. Hu et al. (2015) reported increased seed number and plant biomass of *Medicago truncatula* plants colonized by *Rhizophagus irregularis* when ambient night temperature increased by 1.53 °C.

In contrast, several studies have shown that AM symbiosis had no or negative effect on plant growth and biomass production under low and high temperature stress. Zhu et al. (2011, 2015) reported AM-inoculated maize had a similar shoot and root dry weights with the non-AM plants under high and low temperature stress. Maya and Matsubara (2013) found root and tuber dry weights of AM cyclamen (*Cyclamen persicum*) were lower than those of the non-AM plants under high

Table 8.1 AM effects on host plant parameters under temperature stress in reported paper since 1986

AM fungi	Host plants	Temperature	Length of stress	Parameters	AM effect	References
Glomus intraradices	*Gossypium hirsutum*	30 and 36 °C	60	Plant growth; %P, Cu, Zn, and Mn contents	Positive	Smith and Roncadori (1986)
Gl. ambisporum						
Gigaspora margarita						
Gl. intraradices	*Gossypium hirsutum*	18 °C	60	Zn and Mn contents	Positive	Smith and Roncadori (1986)
Gl. ambisporum				Plant growth; %P and Cu contents	No	
Gl. margarita	*Gossypium hirsutum*	18 °C	60	Plant growth; %P, Cu, Zn, and Mn contents	No	Smith and Roncadori (1986)
Gl. etunicatum	*Fraxinus pennsylvanica*	12, 16, and 20 °C	30	Leaf area growth rate; relative leaf area growth rate; leaf area; plant weight; leaf weight ratio (16 °C); mean leaf size	Positive	Andersen et al. (1987)
				Stem, leaf, and root weight ratio; root-shoot ratio; mean leaf size (12 °C)	No	
				Stem weight ratio (20 °C); root weight ratio (16 °C); root-shoot ratio (16 °C)	Negative	
AMF	*Hordeum vulgare*	12, 16, and 20 °C		Shoot mass	Positive	Volkmar and Woodbury (1989)
				Root mass and length	No	
Gl. fasciculatum	*Sorghum bicolor*	20 °C	48	Dry weight; root length; content of P, N, S, K, Ca, Mg, Mn, Zn, Fe, and Cu	Positive	Raju et al. (1990)
Gl. intraradices						
Gl. macrocarpum						

(continued)

Table 8.1 (continued)

AM fungi	Host plants	Temperature	Length of stress	Parameters	AM effect	References
Gl. fasciculatum	*Sorghum bicolor*	30 °C	48	Dry weight; root length; content of P, N, S, K, Ca, Mg, Mn, Zn, Fe, and Cu	Positive	Raju et al. (1990)
Gl. macrocarpum						
Gl. intraradices	*Sorghum bicolor*	30 °C	48	Dry weight; root length; content of P, N, S, K, Ca, Mg, Mn, Zn, Fe, and Cu	Negative	Raju et al. (1990)
Gl. intraradices	*Vigna radiata*	30 and 38 °C	14 and 42	Plant growth	Positive	Haugen and Smith (1992)
Gl. mosseae	*Zea mays*	10 °C	7	Shoot mass; sugars; protein of Pride 5; starch of Pioneer 3902	Positive	Charest et al. (1993)
				Chlorophyll concentration; protein of Pioneer 3902; starch of Pride 5	Negative	
Gl. intraradices	*Hordeum vulgare*	15 and 10 °C	48	Specific P uptake	Positive	Baon et al. (1994)
				P concentration; PUE	No	
				Plant biomass; P content	Negative	
Gl. mosseae	*Triticum aestivum* Glenlea	5 °C	7	Chlorophyll content	Positive	Paradis et al. (1995)
				Plant growth; content of total sugars, reducing sugars, and proteins	No	
				Nonreducing sugar content	Negative	
Gl. mosseae	*Triticum aestivum* AC Ron	5 °C	7	Plant growth; content of chlorophyll, total sugars, nonreducing sugar, reducing sugars, and proteins	No	Paradis et al. (1995)

Gl. versiforme	*Glycine max*	15 °C		Nodule weight; N concentration and content; root weight	Positive	Zhang et al. (1995)
				Nodule number	Negative	
Gl. versiforme	*Glycine max*	18.2 °C		N concentration and content; root weight	Positive	Zhang et al. (1995)
				Nodule number and weight	Negative	
Gl. versiforme	*Glycine max*	21.6 °C		Nodule number and weight; N content; root weight	Positive	Zhang et al. (1995)
				N concentration	Negative	
Gl. intraradices	*Phaseolus vulgaris*	4 °C	6	Leaf water potential	Positive	El-Tohamy et al. (1999)
				Electrolyte leakage	No	
Gl. margarita	*Asparagus officinalis*	15 and 30 °C	28	Plant height; number of shoots and crowns; dry weight; P concentration	Positive	Matsubara et al. (2000)
Gl. intraradices	*Allium porrum*	15 and 0 °C	7 and 14	^{32}P activity (15 °C for 14 d)	Positive	Wang et al. (2002)
				^{32}P activity	No	
Gl. intraradices	*Sorghum bicolor*	15 and 10 °C	35, 70, and 105	Root length and fresh weight; shoot dry weight	No	Liu et al. (2004)
				Root fresh weight (15 °C for 35 d)	Negative	
Gl. intraradices	*Capsicum annuum*	32.1–38 °C	56	Plant growth; root respiration	Positive	Martin and Stutz (2004)
				Leaf P content	Negative	
Gl. margarita, Gl. fasciculatum, Gl. mosseae, Gl. aggregatum	*Fragaria ananassa*	~35 °C		Number of leaves and roots; diameter of crown; leaf area; dry weight	Positive	Matsubara et al. (2004)
				P content	Negative	

(continued)

Table 8.1 (continued)

AM fungi	Host plants	Temperature	Length of stress	Parameters	AM effect	References
Gl. claroideum	*Gnaphalium norvegicum*	15 °C	30	Plant growth; N content; Pn; PNUE;	Positive	Ruotsalainen and Kytöviita (2004)
				gemination; SLA	No	
				N concentration	Negative	
Gl. claroideum	*Gnaphalium norvegicum*	8 °C	37	Gemination; plant growth; N content; Pn; PNUE; SLA	No	Ruotsalainen and Kytöviita (2004)
				N concentration	Negative	
Gl. intraradices	*Phaseolus vulgaris*	4 °C	2	Expression of *PvPIP1;3*; abundance of PIP protein	Positive	Aroca et al. (2007)
				RWC; Tr; osmotic root hydraulic properties; expression of *PvPIP1;1*, *PvPIP2;1* and *PvPIP1;2*;	No	
Gl. claroideum	*Potentilla crantzii*	17 °C	56	PPUE; P concentration	Positive	Kytöviita and Ruotsalainen (2007)
	Ranunculus acris		49	Pn; plant growth; PPUE; N concentration; N and P content	No	
Gl. hoi	*Plantago lanceolata*	7 °C	10	Proportion of respiratory capacity of roots; cytochrome c oxidase activity	No	Atkin et al. (2009)
				Abundance of cytochrome c oxidase subunit II and alternative oxidase	Negative	

AMF mixture	*Agrostis scabra*, *Mimulus guttatus*	From 30 to 50 °C at the surface to base of the pot	70 and 60	Total biomass; root length and diameter	No	Bunn et al. (2009)
AMF mixture	*Dichanthelium lanuginosum*	From 30 to 50 °C at the surface to base of the pot	80	Total biomass; root length and diameter	Positive	Bunn et al. (2009)
Gl. mosseae	*Citrus tangerine*	15 °C	55	Root length; Ca content	Positive	Wu and Zou (2010)
				Plant growth; Pn, Tr, gs; root area, diameter, and volume; content of P, K, Mg, Fe, Cu, Mn, and Zn	No	
Gl. etunicatum	*Zea mays*	15 and 5 °C	7	Root dry weight; WUE; chlorophyll content; Fm; Fv/Fm; Fv/Fo; Pn; Tr; gs	Positive	Zhu et al. (2010a)
				Shoot dry weight; RWC; WSD; Fo	No	
				Ci	Negative	
Gl. etunicatum	*Zea mays*	5, 15, 35, and 40 °C	7	Plant growth; root proline content (5 and 15 °C); root soluble sugar content; activity of SOD and CAT and root POD	Positive	Zhu et al. (2010b)
				Leaf soluble sugar content; root proline content (35 and 40 °C); leaf POD activity	No	
				Membrane relative permeability; MDA; leaf proline content	Negative	

(continued)

Table 8.1 (continued)

AM fungi	Host plants	Temperature	Length of stress	Parameters	AM effect	References
Gl. mosseae	*Lycopersicon esculentum*	8 °C	7	Plant growth; content of chlorophyll and soluble sugar; activity of SOD, POD, and APX	Positive	Abdel Latef and Chaoxing (2011)
				Content of total sugar and soluble protein; CAT activity	No	
				Content of insoluble sugar, proline, and MDA	Negative	
Gl. hoi	*Plantago lanceolata*	12 °C		N and ^{15}N content	Positive	Barrett et al. (2011)
Gl. etunicatum	*Zea mays*	35 and 40 °C	7	Pn; Tr; gs; Fm; Fv/Fm; Fv/Fo; chlorophyll and carotenoid contents; RWC; WUE	Positive	Zhu et al. (2011)
				Plant growth; Fo; chlorophyll a/b	No	
				Ci	Negative	
Gl. margarita	*Poa pratensis*	Ambient +3 °C		Community plant biomass and P stock	Positive	Büscher et al. (2012)
Gl. intraradices	*Lolium perenne*			Community N stock	No	
	Medicago lupulina					
	Lotus corniculatus					
	Rumex acetosa					
	Plantago lanceolata					

Gl. mosseae	*Plantago lanceolata*	Ambient +2.7 °C	16	P content	Positive	Karasawa et al. (2012)
				Plant growth; content of ^{13}C, N, K, Ca, Mg, and Zn; plant CO_2 uptake	No	
Gl. versiforme	*Tectona grandis*	6, 3, and 0 °C	0.5	Content of chlorophyll, soluble protein; activity of SOD and POD	Positive	Zhou et al. (2012)
				MDA	Negative	
Funneliformis mosseae	*Cucumis sativus*	15 °C	25	Plant growth; content of phenols, flavonoids, lignin, and phenolic acids; activity of DPPH, G6PDH, SKDH, PAL, CAD, PPO, and POD; expression of PR-1, CCOMT, G6PDH, CAD, LPO, PAL, WRKY30, and C4H	Positive	Chen et al. (2013)
				H_2O_2 content	Negative	
Gl. mosseae	*Oryza sativa*	15 °C	10	Root length; fresh weight; N and P content; NUE; soluble sugar content; activity of NR, GS, SPS, and SS; JA and NO contents	Positive	Liu et al. (2013)
				Dry weight	No	

(continued)

Table 8.1 (continued)

AM fungi	Host plants	Temperature	Length of stress	Parameters	AM effect	References
Gl. fasciculatum	*Cyclamen persicum*	30 °C	28	Shoot dry weight; activity of leaf and tuber SOD, APX and DPPH and root SOD and DPPH; leaf, root, and tuber ascorbic acid and polyphenol contents	Positive	Maya and Matsubara (2013)
				Root and tuber dry weight; activity of root APX	No	
Acaulospora scrobiculata	*Zea mays*	15 °C	7	Leaf soluble sugar and proline content; CAT and POD activities	Positive	Chen et al. (2014)
Gl. tortuosum				Plant growth; MDA content; SOD activity	No	
Gl. etunicatum	*Zea mays*	15 °C	7	Leaf soluble sugar, proline, and MDA content; CAT and POD activities	Positive	Chen et al. (2014)
				Plant growth; SOD activity	No	
Gl. intraradices	*Zea mays*	15 °C	7	Leaf soluble sugar, proline, and MDA content; CAT activity	Positive	Chen et al. (2014)
				Plant growth; SOD activity	No	
				POD activity	Negative	

Funneliformis mosseae	*Cucumis sativus*	15 °C	39	Plant growth; root activity; ATPase activity and concentration; plasma membrane protein content; expression of proton pump and calcium-transporting ATPase-related genes	Positive	Liu et al. (2014a)
				Water content	No	
				H_2O_2 content; NADPH oxidase activity	Negative	
Gl. intraradices	*Oryza sativa*	15 °C	7	RWC; root length; expression of *PIP1;1*, *PIP1;3*, *PIP2;1*, *PIP2;5*, *TPS1*, *TPS2*, and *TPP1*	Positive	Liu et al. (2014b)
				Expression of *PIP1;1* and *PIP2;3*	No	
Rhizophagus irregularis	*Medicago truncatula*	Ambient night temperature + 1.53 °C	58	Seed number; plant weight; shoot Zn, root P, Ca, glucose, and fructose concentration; expression of *MtSucS2*	Positive	Hu et al. (2015)
				Flower number; shoot P and K, root K and Zn, leaf sucrose, glucose, and fructose, and stem sucrose, glucose and fructose concentration; expression of *MtSucS3* and *MtSucS5*	No	
				Shoot Ca, and root sucrose concentration; expression of *MtSucS1* and *MtSucS4*	Negative	

(continued)

Table 8.1 (continued)

AM fungi	Host plants	Temperature	Length of stress	Parameters	AM effect	References
Gl. tortuosum	*Zea mays*	15 °C	14	Root dry weight; content of N, leaf soluble sugar and reducing sugar, root reducing sugar, sucrose, and fructose; activity of GS, GOT, GPT, SPS, and AMS; Pn	Positive	Zhu et al. (2015)
				Plant height; shoot dry weight; SS activity	No	
				Content of leaf sucrose and fructose, root soluble sugar	Negative	
AMF mixture: *R. irregularis (isolate*	*Triticum aestivum*	35 °C	7	Number of grains; concentration of Tiller K and Ca and spike C; spike biomass	Positive	Cabral et al. (2016)
BEG140), R. irregularis, Funneliformis mosseae (isolate BEG95), F. geosporum, Claroideoglomus claroideum				Grain weight; concentration of Tiller C, N, Mg, B and Mn, spike N, grain C, P, and K; tiller biomass; light-use efficiency	No	
				Concentration of Tiller P, K/Ca, Cu, Zn, and Fe, grain N	Negative	

Gl. tortuosum	*Zea mays*	15 °C	14	Concentration of shoot N, NO_3^--N, NH_4^+-N, P, K, Cu, and root N, P, Ca, Zn; NR activity	Positive	Liu et al. 2016
				Plant growth; concentration of shoot Ca, S, Mg, Na, Al, Fe, Mn, Zn, and root K, S, Mg, Na, Al, Fe, Mn, Cu	No	
Gl. tortuosum	*Zea mays*	15 °C	14	Concentration of Thr, Gly, His, Lys, total amino acid, and leaf Ile, Leu, Phe, Agr, and root Val	Positive	Zhu et al. (2016)
				Plant growth; concentration of Asp, Ser, Glu, Ala, Met, Tyr, Pro, and leaf Val and root Ile, Leu, Phe, Arg	No	

AMS amylase, *APX* ascorbate peroxidase, *CAD* cinnamyl alcohol dehydrogenase, *CAT* catalase, *Ci* intercellular CO_2 concentration, *Fm* maximal fluorescence, *Fo* primary fluorescence, *Fv/Fm* maximum quantum efficiency of photosystem II primary photochemistry, *Fv/Fo* potential photochemical efficiency, *G6PDH* glucose-6-phosphate dehydrogenase, *Gi* Gigaspora, *Gl* Glomus, *GOT* glutamate oxaloacetate transaminase, *GPT* glutamate pyruvate transaminase, *gs* stomatal conductance, *GS* glutamine synthetase, H_2O_2 hydrogen peroxide, *JA* jasmonic acid, *MDA* malondialdehyde, *NR* nitrate reductase, *NUE* nitrogen use efficiency, *PAL* phenylalanine ammonia-lyase, *PIP* plasma membrane intrinsic protein, *Pn* net photosynthetic rate, *PNUE* photosynthetic nitrogen use efficiency, *POD* peroxidase, *PPO* polyphenol oxidase, *PPUE* photosynthetic phosphorus use efficiency, *RWC* relative water content, *SKDH* shikimate dehydrogenase, *SLA* specific leaf area, *SOD* superoxide dismutase, *SPS* sucrose phosphate synthase, *SS* sucrose synthase, *Tr* transpiration rate, *WSD* water saturation deficit, *WUE* water use efficiency temperature stress. These results could be attributed to the high carbon cost-benefit ratio of AM fungi with the host plants or AM fungi failed to deliver P and other nutrition to the host plants (Martin and Stutz 2004; Chen et al. 2014).

8.3.2 Water Relations

Water status plays a great role in growth and physiological processes of plants. Plant water status is an important variable under changing ambient temperatures (Mazorra et al. 2002). AM and non-AM plants also often display different water status (Augé 2001). Under temperature stress, AM maize plants had higher water conservation, water holding capacity, and relative water content (Zhu et al. 2010a, 2011; Liu et al. 2014b). EI-Tohamy et al. (1999) reported that AM bean (*Phaseolus vulgaris*) plants had higher leaf water potential during chilling stress. These studies suggested that AM symbiosis could improve plant water status at low or high temperature stress, although several authors found water content in AM plants was similar to the non-AM plants under low temperature stress (Aroca et al. 2007; Liu et al. 2014a).

Under low or high temperature stress, the ability of roots to take up water was reduced. Root water uptake depends on root hydraulic conductivity. AM plants are found to have better water status which could be due to the enhanced water extraction by the external hyphae (Faber et al. 1991) and higher activity and hydraulic conductivity of the roots (Augé and Stodola 1990). Moreover, AM symbiosis was beneficial for stomatal opening in leaves and water flow through the plants to the evaporating surfaces in the leaves (Nelsen and Safir 1982). Zhu et al. (2010a, 2011) found that stomatal conductance and transpiration rate of AM maize plants were higher than the corresponding non-AM plants under low and high temperature stress, indicating that AM colonization could improve the gas exchange capacity through maintaining opened stomata, decreased stomatal resistances, and increased transpiration rates.

Water uptake and hydraulic conductance of roots are governed by aquaporins (Luu and Maurel 2005). Aquaporins belong to membrane intrinsic protein family that facilitating the passive water flow through membranes (Kruse et al. 2006). Plasma membrane intrinsic proteins (PIPs) have been shown to regulate the whole water transport through plant tissues (Aroca et al. 2007). Liu et al. (2014b) demonstrated that AM fungi not only transport more water to the host plant by regulating their own aquaporin activities but also regulate the expression of plant aquaporin genes to improve water transport of the host. Under low temperature stress, *PvPIP1;3* gene expression and PIP protein abundance were increased in AM plants (Aroca et al. 2007). Liu et al. (2014b) also found *PIP1;1*, *PIP1;3*, *PIP2;1*, and *PIP2;5* gene expression were upregulated in AM rice (*Oryza sativa*) plants. All these improved plant water status facilitated by AM colonization enable plants to use water more efficiently (Evelin et al. 2009; Zhu et al. 2010b, 2011).

8.3.3 Photosynthesis

It is well documented that photosynthesis is one of the most sensitive processes to temperature stress and can be depressed before other symptoms of the stress are detected (Berry and Björkman 1980). Any constraint in photosynthesis can limit plant growth at low or high temperatures (Wahid et al. 2007). AM plants often show different photosynthetic performances from the non-AM plants (Augé 2001). It has been reported that AM plants had higher net photosynthetic rate (Pn) than the non-AM plants under temperature stress (Ruotsalainen and Kytöviita 2004; Zhu et al. 2010a, 2011, 2015). Higher Pn implied greater CO_2 assimilation capacity in plants. Thus, AM colonization usually stimulated plant growth, although AM fungi cost excess carbon for its own growth. However, Wu and Zou (2010) observed AM inoculation did not affect Pn of citrus (*Citrus tangerine*) seedlings under low temperature stress.

Adverse temperature stress causes a decrease in chlorophyll concentration indicating a suppression of chlorophyll biosynthesis or higher rate of chlorophyll degradation. Several authors reported AM plants had higher chlorophyll concentration compared with the non-AM plants at suboptimal temperatures (Paradis et al. 1995; Zhu et al. 2010a, 2011). The results suggest that temperature stress interferes less with chlorophyll synthesis and light harvesting in AM than in non-AM plants, and AM symbiosis alleviates the damage of the mesophyll chloroplasts, thereby improving photosynthetic efficiency (Evelin et al. 2009). In contrast, Charest et al. (1993) found that chlorophyll concentration of AM maize plants was lower than the non-AM plants. Furthermore, carotenoid acts as accessory light harvesting pigments and plays an essential role in the photoprotection of photosynthetic apparatus (Young 1991). Zhu et al. (2011) reported AM maize plants had higher carotenoid concentration than the non-AM plants implying AM colonization stabilizes the lipid phase of the thylakoid membranes and provide photoprotection to cellular structures and photosynthetic apparatus (Karim et al. 1999; Wahid et al. 2007).

Photosystems, mainly photosystem II (PSII) with its oxygen-evolving complex is one of three primary site of injury to the photosynthetic machinery at low and high temperature stresses (Allakhverdiev et al. 2008). PSII photochemical reactions in thylakoid lamellae would inevitably be affected by temperature stress. Chlorophyll fluorescence has been used in physiological and ecophysiological studies to probe and elucidate the changes in the function of PSII and to reflect the primary photosynthetic processes under abiotic stress (Maxwell and Johnson 2000; Baker 2008; Hajiboland et al. 2010). The ratio of Fv/Fm (maximum quantum efficiency of PSII primary photochemistry) and of Fv/Fo (potential photochemical efficiency of PSII) is useful relative measurements of the capacity of primary photochemistry of PSII, which are reliable diagnostic indicators of damage caused by environmental stresses (Krause and Weis 1991; Maxwell and Johnson 2000). It has been shown that Fv/Fm and Fv/Fo of AM maize plants was significantly higher than that of the non-AM plants under low and high temperature stresses (Zhu et al. 2010a, 2011). This implies that AM symbiosis mitigates the toxic influence of temperature stress on the

PSII reaction center and the structural and functional disruption of photosynthetic apparatus. An increase in Fv/Fm may be due to the decrease in primary fluorescence (Fo) or increase in maximal fluorescence (Fm). Zhu et al. (2010a, 2011) found that AM maize plants had higher Fm and lower Fo compared with the non-AM plants when subjected to temperature stress. Temperature stress could destroy the PSII reaction center, reduce the number of open PSII units, inactive the PSII photochemical reaction, and disrupt electron transport in photosynthetic apparatus (Camejo et al. 2005; Baker 2008). This detrimental effect of temperature stress on PSII reaction center could be alleviated by AM symbiosis, consequently improving the PSII photochemistry efficiency and photosynthetic performance of AM plants. However, a recent study by Cabral et al. (2016) showed that there was no influence of AM on the effective quantum yield, electron transfer rate, and non-photochemical quenching in wheat (*Triticum aestivum*) plants under high temperature stress.

8.3.4 Plasma Membrane Permeability

The plasma membrane is considered to be the primary site of injury when plant subjected to temperature stress. There are many alterations in the composition, structure, and function of plasma membrane responding to temperature stress (Uemura et al. 2006). Such alterations increase the membrane permeability, as evident from increased loss of electrolytes (Wahid et al. 2007). Studies have shown that AM plants maintain higher electrolyte concentration than non-AM plants by keeping improved integrity and stability of the membrane (Evelin et al. 2009). EI-Tohamy et al. (1999) found no significant effects on electrolyte leakage in AM and non-AM bean plants under low temperature stress. However, Zhu et al. (2010b) opined that the membrane relative permeability in the leaves and roots of AM maize plants was lower than non-AM plants under low and high temperature stresses, which suggests AM symbiosis can decrease membrane electrolyte permeability and alleviate the adverse effects of temperature stress on cell membrane.

The plasma membrane lipids seem to be responsible for the fate of plasma membrane at temperature stresses (Uemura et al. 2006). Temperature stress causes the peroxidation of membrane lipids. The level of malondialdehyde (MDA) reflects the degree of membrane lipid peroxidation (Ali et al. 2005). Several studies have demonstrated that MDA content in AM plants was lower than that in the non-AM plants (Zhu et al. 2010b; Abdel Latef and Chaoxing 2011; Zhou et al. 2012; Chen et al. 2014), which indicates that AM symbiosis could alleviate the peroxidation of membrane lipids and maintain the fluidity of membrane.

Under low temperature stress, AM colonization induced plasma membrane ATPase activities and ATP accumulation in cucumber (*Cucumis sativus*) plants (Liu et al. 2014a). ATPase can regulate intracellular pH and generate an electrochemical gradient for secondary active transport (Kim et al. 2013). H^+-ATPase is active in the plant membrane around arbuscules in arbuscular mycorrhizae and plays a crucial role in plant-fungal interactions at the symbiotic interface (Gianinazzi-Pearson et al. 1995; Liu et al. 2014a). Liu et al. (2014a) also observed that the expression of proton pump and calcium-transporting ATPase-related genes (*CsHA2*, *CsHA3*, *CsHA4*, *CsHA8*, *CsHA9*, *CsHA10*, *CA1*, and *CA9*) in cucumber roots was significantly upregulated by AM inoculation under low temperature stress.

8.3.5 ROS and Antioxidants

When plants are exposed to low or high temperature stress, a variety of ROS such as superoxide anion radical ($O_2^{\bullet -}$), hydroxyl radicals ($OH^{\bullet}$), and hydrogen peroxide (H_2O_2) are induced, causing unbalance between production and detoxification in the cell or organism (Apel and Hirt 2004; Lenoir et al. 2016). The oxidative stress caused by ROS is one of the main damaging factors in plants (Wahid et al. 2007). An excess of ROS can react with DNA, lipids, and proteins, resulting in damage of cell structure and function, such as DNA/RNA nicking, membrane lipid peroxidation, protein denaturation, and enzyme inhibition (Maya and Matsubara 2013). A large number of studies have investigated the effect of AM inoculation on ROS in unstressed and stressed conditions. However, little information is known about the relation between AM and ROS under temperature stress. Only Chen et al. (2013) and Liu et al. (2014a) found AM colonization significantly reduced H_2O_2 content in cucumber leaves under low temperature stress. It is assumed that reduction in H_2O_2 is one of the mechanisms by which AM fungi protect host plants against temperature stress (Zhang et al. 2013). Low level of H_2O_2 induced by AM symbiosis may be acts as a signaling molecule in defense and adaptive responses (Chen et al. 2013). Moreover, Liu et al. (2014a) reported NADPH oxidase activity in AM cucumber plants was lower than the non-AM plants. The $O_2^{\bullet -}$ produced by NADPH oxidase can be converted to H_2O_2 by superoxide dismutase (SOD) in the plant apoplast. AM symbiosis could lessen H_2O_2 accumulation and enhance temperature tolerance via suppression of NADPH oxidase activity (Liu et al. 2014a).

To alleviate or prevent temperature stress induced oxidative injury, plant has evolved a cellular defense mechanism to scavenge these ROS by antioxidant systems (Zhou et al. 2012). The antioxidant systems involving antioxidative enzymes, such as SOD, ascorbate peroxidase (APX), peroxidase (POD), catalase (CAT), dehydroascorbate reductase or glutathione reductase, and low molecular weight antioxidant, such as ascorbic acid, glutathione, tocopherols, or polyphenols (Evelin et al. 2009; Zhu et al. 2010b; Maya and Matsubara 2013). For example, SOD will help detoxify O_2^- to H_2O_2, and the H_2O_2 produced will be detoxified by CAT, POD, or APX. The increased glutathione reductase will serve plant to minimize the production of O_2^- (Evelin et al. 2009). Several studies have demonstrated that AM fungi could assist host plant to alleviate temperature stress by increasing the activities of antioxidant enzymes. Zhu et al. (2010b) found that AM maize plants had higher SOD and CAT activities than the non-AM plants under low and high temperature stresses. Abdel Latef and Chaoxing (2011) reported the activities of SOD, POD, and APX in AM tomato (*Lycopersicon esculentum*) plants were higher than the non-AM plants under low temperature stress. Zhou et al. (2012) showed that AM colonization increased SOD and POD activity of teak (*Tectona grandis*) seedlings under low temperature stress. Maya and Matsubara (2013) observed activities of SOD and APX in leaf and tuber of cyclamen plants which were greater compared with those of the non-AM plants under high temperature stress. However, zero or negative effects of AM symbiosis on some enzymatic activities also were observed under temperature stress (Zhu et al. 2010b; Abdel Latef and Chaoxing 2011; Maya and Matsubara 2013; Chen et al. 2014). The aforementioned results suggest that AM plants possess higher antioxidant enzyme activities, but the response of the individual enzymes varies with respect to the AM fungal and the host plant species (Evelin et al. 2009). This variation may also be due to some enzymes such as SOD, CAT, and APX which are metalloenzymes whose activities can be determined by the availability of the metals they utilize (Evelin et al. 2009).

In addition to these enzymatic systems, AM symbiosis can induce accumulation of nonenzymatic antioxidant components to scavenge ROS. Chen et al. (2013) showed that production of phenols and flavonoids was enhanced in AM cucumber plants compared with that in the non-AM plants under low temperature stress. Maya and Matsubara (2013) also reported AM cyclamen plants had higher ascorbic acid and polyphenol contents than the non-AM plants under high temperature stress.

8.3.6 Osmotic Adjustment

Osmotic adjustment is considered a key tolerance mechanism in higher plants grown under low and high temperature stresses. In response to temperature stress, plants accumulate a variety of certain organic compounds of low molecular solutes, generally referred to as compatible osmolytes, such as sugars, proline, polyamines, betaines, and acylatedsterols (Wahid et al. 2007; Theocharis et al. 2012). The accumulation of these solutes could lower the osmotic potential in the cytosol, hereby maintaining positive turgor pressure of the cells.

Proline is known to accumulate in many plants as a nontoxic and protective osmolyte to maintain osmotic balance under temperature and osmotic stresses. Proline also serves as a sink for energy to regulate redox potentials, as a hydroxyl radical scavenger, as a solute that protects macromolecules against denaturation, and as a means of reducing acidity in the cell (Kishor et al. 2005; Sharma and Dietz 2006; Theocharis et al. 2012). Proline accumulation has been shown to increase when plant is colonized by AM fungi. Under low temperature stress, AM maize plants were found to have higher root proline content than the non-AM plants (Zhu et al. 2010b). Chen et al. (2014) also reported proline content in the AM maize leaves was higher compared with that in the non-AM plants at low temperature condition. However, Zhu et al. (2010b) reported proline content was lower in the AM maize leaves than that in the non-AM plants under low and high temperature stresses, and no significant difference between the AM and the non-AM maize roots was found under high temperature stress. Abdel Latef and Chaoxing (2011) also found AM tomato plants accumulated less proline than the non-AM plants under low temperature stress. Zhu et al. (2010b) suggested that the change of leaf proline level reflects the degree of injury of mycorrhizal plants by the stress, and if the stress was moderate, there was no need to synthesize more proline for osmotic adjustment protection.

Soluble sugars have been reported to have multiple roles in temperature tolerance and contributed to the preservation of water within plant cells as typical compatible osmolytes (Theocharis et al. 2012). It is well documented that plant accumulates soluble sugars to adjust the osmotic potential during temperature stress (Evelin et al. 2009). Many studies have reported that plants inoculated with AM fungi had higher levels of soluble sugars than the non-AM plants under unstressed and stressed conditions (Paradis et al. 1995; Zhu et al. 2010b, Abdel Latef and Chaoxing 2011; Liu et al. 2013; Chen et al. 2014). Moreover, increased soluble protein is also often associated with temperature tolerance. An alternation of protein content in AM plants under low temperature stress has been observed by Charest et al. (1993), Abdel Latef and Chaoxing (2011), and Zhou et al. (2012).

8.3.7 Carbohydrate Metabolism

In addition to acting as osmolytes, sugars may have a role as cryoprotectants to protect plant cell membrane, replacing water molecules in establishing hydrogen bonds with lipid molecules. They also serve as scavengers of ROS and contribute to increased membrane stabilization. As signaling molecules, they contribute to the regulation of growth and development and stress responses in plants associated with hormone signaling (Uemura et al. 2006; Zeng et al. 2011; Theocharis et al. 2012). Both AM inoculation and temperature stress could influence the accumulation of sugars in plants. Charest et al. (1993) reported that AM symbiosis increased sugar contents of maize plants under low temperature stress. Zhu et al. (2010b) found the root soluble sugar content in AM maize plants was higher than that in the non-AM plants. However, some authors also reported zero or even negative effect of AM inoculation on sugar accumulation in host plants under temperature stress (Paradis et al. 1995; Zhu et al. 2010b, 2015). The different alternations of the contents of reducing sugars, nonreducing sugars, glucose, fructose, and sucrose of plants inoculated with AM fungi under low or high temperature stress were also observed (Paradis et al. 1995; Hu et al. 2015; Zhu et al. 2015). Depending on the AM fungal and the host plant species, various forms of sugars are found to be involved in physiological reactions to temperature stress. Koide and Schreiner (1992) stressed that control over the activity of AM fungi by the host could occur via a regulation of carbohydrate transfer, one way being by the regulation of arbuscule number. The changes of carbohydrate contents could refer to carbohydrate sink-source relationship of AM fungi and plants. AM fungi are able to facilitate sugar transport and metabolism between source and sink organs via an increased exchange of carbohydrates and nutrients (Dodd and Perez-Alfocea 2012). Heinemeyer et al. (2006)

found that AM fungi can consume up to 20% photosynthates of the host plant. Photosynthates are transported via the phloem to the root and, subsequently, into the intraradical and extraradical mycelium, as well as in the spores of AM fungi.

Temperature stress may activate and enhance expression of genes encoding specific enzymes involved in the sink-source transition and sucrose metabolism in plants, such as sucrose synthase (SS) and sucrose phosphate synthase (SPS) (Li et al. 2008). Generally, SS plays an important role in sucrose hydrolysis rather than synthesis in sink tissues (Verma et al. 2011). In AM plants, sucrose is usually hydrolyzed to monosaccharide by SS or invertase prior to utilizing by AM fungi (Hu et al. 2015). Liu et al. (2013) reported AM rice plants had higher SS activity than the non-AM plants under low temperature stress. Hu et al. (2015) also found AM colonization enhanced expression of gene encoding SS, *MtSucS2* of *M. truncatula* plants grown at night warming condition. However, Zhu et al. (2015) showed that the SS activity was unaffected by low temperature stress and AM colonization. Hu et al. (2015) also reported that no or even negative AM effect on four genes encoding SS expression of *M. truncatula* plants under night warming. Hawkes et al. (2008) speculated that AM fungi consumed more sucrose and the resulting decrease in sucrose ultimately led to lower SS and downregulated the expression of gene encoding SS. Moreover, SPS, a key enzyme for sucrose synthesis, controls the flux of photosynthetic carbon into sucrose (Verma et al. 2011). Under low temperature stress, SPS activity was enhanced by AM fungi in rice and maize plants as reported by Liu et al. (2013) and Zhu et al. (2015), indicating that AM symbiosis enhanced sucrose metabolism under low temperature stress.

It is worth mentioning that trehalose is the main storage carbohydrate in AM fungi and plays an important role in stress tolerance by stabilizing dehydrated enzymes and membranes and protecting biological structures from desiccation damage (Evelin et al. 2009). In response to temperature stress, fungal cells react by activating transcriptionally and/or posttranscriptionally the trehalose metabolism enzymes, which results in trehalose accumulation (Lenoir et al. 2016). Liu et al. (2014b) reported that rice plants colonized by AM fungi had higher *trehalose phosphate synthase* (TPS) and *trehalose phosphate phosphatase* (TPP) transcript levels compared with the non-AM plants under low temperature stress. The increase in the expression of *OsTPS1*, *OsTPS2*, and *OsTPP1* could result in an increased trehalose biosynthesis and higher trehalose concentration in the AM rice plants under low temperature stress (Liu et al. 2014b). The finding suggests that AM symbiosis could enhance plant tolerance against temperature stress. Increases in trehalose concentration may also be involved in starch accumulation (Fernandez et al. 2010).

Starch accumulation is affected by AM symbiosis and temperature stress. Charest et al. (1993) reported that AM colonization altered starch content in maize plants under low temperature stress. Zhu et al. (2015) found that the activity of amylase was higher in AM maize than that in the non-AM plants under low temperature stress. Amylase is able to catalyze the hydrolysis of starch into sugars. The result suggests that AM roots require more carbohydrates by AM fungi and/or synthesize more sugar to cope with low temperature stress via the breakdown of leaf starch.

8.3.8 Nutrient Uptake

Improved nutrient uptake is one of the most important functions in AM symbiosis. AM symbioses are generally recognized to be involved in bidirectional nutrient exchange: AM fungi receive organic carbon from the plants, and in return, the hosts obtain a nutritional benefit from AM fungi by taking up P, N, and other macro- and micronutrients from the soil (Smith and Read 2008). Nutrients are mobilized and transported to the plants via direct pathway from the rhizosphere by root epidermal cells and root hairs and AM uptake pathway by the huge intraradical and extraradical mycorrhizal network (Smith and Smith 2012). It is well documented that the acquisition of soil nutrients by plant is suppressed by temperature stress, and several studies have indicated that AM symbiosis could improve nutrient uptake of the host plant under low and high temperature stresses.

N is one of the major limiting macronutrients for plant growth and development. It serves as a constituent of many plant cell components and the synthesis of proteins, nucleic acids, coenzymes, and many products and by-products of secondary metabolism (Evelin et al. 2009). Several studies have demonstrated that AM fungi can take up and transfer significant amounts of N (accounting for 20 to 50% of the total root N) to the host plant (Govindarajulu et al. 2005; Ngwene et al. 2013). A large body of evidence indicates that AM fungi have a potential to improve plant N acquisition under low or high temperature stress (Raju et al. 1990; Zhang et al. 1995; Ruotsalainen and Kytöviita 2004; Barrett et al. 2011; Liu et al. 2013, 2016 Zhu et al. 2015). On the contrary, some studies have reported that the N uptake was not enhanced by AM symbiosis under temperature stress (Raju et al. 1990; Kytöviita and Ruotsalainen 2007; Büscher et al. 2012; Karasawa et al. 2012). The differential AM effect on N uptake has been reported to be specific to the combination of genotypes of AM fungi and host plants, which may explain the different contributions of the symbiosis (Cabral et al. 2016).

The extraradical mycelium of AM fungi has been shown to take up and assimilate organic and inorganic N resources from the soil and to transfer N to the host plants (Govindarajulu et al. 2005). For both AM fungi and plants, the readily available inorganic N are NO_3^- and NH_4^+, and NO_3^- is the predominant form of N available in most agricultural soils (He et al. 2003). Under low temperature stress, AM maize plants had higher NO_3^--N and NH_4^+-N contents compared with the non-AM plants (Liu et al. 2016). NO_3^- reduction, the most important stage of N assimilation in plants, is first catalyzed by nitrate reductase (NR). Low temperature stress activated NR in wheat leaves via protein dephosphorylation (Yaneva et al. 2002). NR activity was found to be higher in AM rice and maize plants than in the non-AM plants under low temperature stress (Liu et al. 2013, 2016), indicating that an special capability of AM symbiosis to take up N and to reduce and assimilate NO_3^-.

Arginine is synthesized in the extraradical mycelium of AM fungi and is transported to the intraradical mycelium, where it is broken down to release N for transfer to the host plant (Govindarajulu et al. 2005). In plant, N metabolism involved in different pathways, such as glutamine synthetase (GS) and glutamate synthase cycles (Javaid 2009). Genetic and molecular approaches have demonstrated that GS plays a key role in N metabolism, which is essential for plant N uptake and use efficiency (Martin et al. 2006). Liu et al. (2013) reported AM rice plants had higher GS activity compared with the non-AM plants grown at low temperature condition. Zhu et al. (2015) also found the GS activity in AM maize plants was greater than the non-AM plants under low temperature stress. Furthermore, Zhu et al. (2015) showed that AM symbiosis increased glutamate oxaloacetate transaminase and glutamate pyruvate transaminase activities of maize plants under low temperature stress. These transaminases are involved in amino acid metabolism and are linked to the ornithine cycle. Increased amino acid content in maize plants inoculated AM fungus under low temperature stress was observed by Zhu et al. (2016), which indicates that the AM fungi facilitated the uptake and synthesis of amino acids.

P is an essential macronutrient for plant growth and crop production: forms an integral component of the key structure element in nucleic acids, phospholipids, and enzymes; and is also involved in the metabolism of matter and energy and signal transduction cascades (Karandashov and Bucher 2005). Plant P uptake, translocation, and metabolism are inhibited by temperature stress. However, AM symbiosis is able to sustain plant P acquisition by the hyphal network of AM fungus and the increased surface of plant roots (Karasawa et al. 2012). AM symbiosis can increase the inorganic P gradient from root to leaves because AM plants enhance the conversion of P from inorganic to organic forms in the leaves, consequently increasing the P sink and then enhancing P uptake (Javaid 2009). AM hyphae also can store larger amounts of absorbed P than the roots, facilitating the continued movement of P into the hyphae. The kinetics of P uptake into hyphae differ from those of roots because of its higher affinity for P ions or lower threshold concentration for uptake (Evelin et al. 2009). To date, most of studies have reported that AM plants had higher P content compared with the non-AM plant under low or high temperature stress (Smith and Roncadori 1986; Matsubara et al. 2000; Wang et al. 2002; Büscher et al. 2012; Karasawa et al. 2012; Liu et al. 2013, 2016; Hu et al. 2015), suggesting that AM symbiosis enhanced the availability and uptake of P under adverse conditions. However, some studies have demonstrated that the contribution of AM symbiosis to P uptake maybe small because the P content in AM plants was lower than the non-AM plants (Baon et al. 1994; Matsubara et al. 2004; Kytöviita and Ruotsalainen 2007).

In addition to N and P, other nutrients also play a key role in plant growth and development and are involved in all physiological metabolic processes and cellular functions. Plants show different needs for these nutrients (Hänsch and Mendel 2009). Several studies have reported the effect of AM symbiosis on these nutrients under low or high temperature stress, though the effect varies among different nutrients. Wu and Zou (2010) found only Ca content in AM citrus plants was higher than the non-AM plants, whereas no effect of AM on the content of K, Mg, Fe, Cu, Mn, and Zn was noticed under low temperature stress. Karasawa et al. (2012) reported contents of K, Ca, Mg, and Zn in AM *Plantago lanceolata* plants that were similar with those of the non-AM plants under lowered soil temperature. AM inoculation increased shoot Zn and root Ca contents, decreased shoot Ca content, and has no effect on K and root Zn content in *M. truncatula* plants under increased night temperature (Hu et al. 2015). AM wheat plants had higher tiller K and Ca; lower tiller Cu, Zn, and Fe concentrations; and similar tiller Mg, B and Mn, and grain K compared with the non-AM plants under high temperature stress (Cabral et al. 2016). Liu et al. (2016) also observed AM colonization had a positive effect on the concentrations of shoot K and Cu and root Ca and Zn, but no effect on the concentrations of shoot Ca, S, Mg, Na, Al, Fe, Mn and Zn and root K, S, Mg, Na, Al, Fe, Mn, and Cu of maize plants under low temperature stress. Moreover, the nutrient uptake of the host plants has been shown to vary with the isolates of AM fungi. Smith and Roncadori (1986) reported that contents of Cu, Zn, and Mn were not significantly changed in cotton plants inoculated with *Gigaspora margarita*, contents of Zn and Mn were increased when plants were inoculated with *Glomus intraradices* and *Glomus ambisporum* mixtures under low temperature stress, and the contents of Cu, Zn, and Mn were both increased when plants were inoculated with these three AM fungal mixtures under high temperature stress. Raju et al. (1990) also found that sorghum plants colonized by *Glomus intraradices* had lower S, K, Ca, Mg, Mn, Zn, Fe, and Cu contents than the non-AM plants under high temperature stress, but when sorghum plants were colonized by *Glomus fasciculatum* and *Glomus macrocarpum* mixtures, the contents of S, K, Ca, Mg, Mn, Zn, Fe, and Cu were increased. The different effects of AM on nutrient uptake in these studies may be due to the variations of experimental design and biological materials.

8.3.9 Secondary Metabolism

Plant secondary metabolism produces a large number of specialized compounds that are required for the interaction of plant with its environment, for example, in defense reactions against herbivores and microbial pathogens as well as in mutualistic symbiotic associations, such as with AM (Strack and Fester 2006). There is an increasing evidence that plant secondary metabolites play an important role in the development and functional regulation of AM symbiosis, such as flavonoids and strigolactones involved in spore germination and hyphal branching in AM fungi (Akiyama 2007). AM fungi have also been shown to increase secondary metabolites contents in the host plants, such as phenolics, triterpenoids, and spermine (Evelin et al. 2009; Chen et al. 2013). However, the information about AM symbiosis influence on secondary metabolism under environmental stress is limited. Under low temperature stress, only Chen et al. (2013) reported that AM inoculation induced the accumulation of phenolics, flavonoids, and lignin in cucumber leaves. They also observed that activities of glucose-6-phosphate dehydrogenase (G6PDH), shikimate dehydrogenase, phenylalanine ammonia-lyase (PAL), cinnamyl alcohol dehydrogenase (CAD), polyphenol oxidase, guaiacol peroxidase, caffeic acid peroxidase, and chlorogenic acid peroxidase of AM cucumber plants were greater than the non-AM plants under low temperature stress. These enzymes are involved in secondary metabolism and are the key enzymes of the pentose phosphate pathway, the phenylpropanoid pathway, and the biosynthesis of secondary metabolites. In addition, Chen et al. (2013) also reported AM cucumber seedlings had higher expression levels of *WRKY30*, *PR-1*, *C4H*, *CCOMT*, *CAD*, *G6PDH*, *PAL*, *LPO*, and *POD* encoding genes compared with the non-AM plants grown in low temperature. These genes related to the regulation of the phenylpropanoid pathway and plant defense. The activation of these secondary metabolism genes is important for successful plant growth under low temperature stress (Chen et al. 2013). Thus, accumulation of secondary metabolites and activation of secondary metabolism related enzymes are linked to plant antioxidants system and defence system which inferred that AM symbiosis enhanced plant tolerance to temperature stress.

8.4 Conclusions and Future Perspectives

Based on the described above, the possible mechanism of AM symbiosis in improving plant tolerance to temperature stress includes enhancing water and nutrient uptake, improving photosynthetic capacity and efficiency, protecting plant against oxidative damage, and increasing accumulation of osmolytes (Fig. 8.1). To date, however, knowledge about AM plants' response to temperature stress is not sufficient. Most of the aforementioned studies have investigated the effect of AM fungi on plant growth under temperature stress; little attention is paid on the physiological, biochemical, and molecular effects. For example, it remains largely unknown about the effect of AM on plant endogenous hormones and AM plants' responses to application of exogenous hormones under temperature stress. Also, the ultrastructural changes in AM plants under temperature stress have not been addressed by far. Thus, the underlying physiological mechanisms of AM symbiosis that protect plant against temperature stress remain yet to be elucidated.

Our knowledge of the molecular mechanisms involved in AM plants' responses to temperature stress is very rare. The roles of temperature-tolerant related genes with respect to AM symbiosis and AM-induced signaling and symbiotic genes by temperature stress need to be explored. Applications of genomics, transcriptomics,

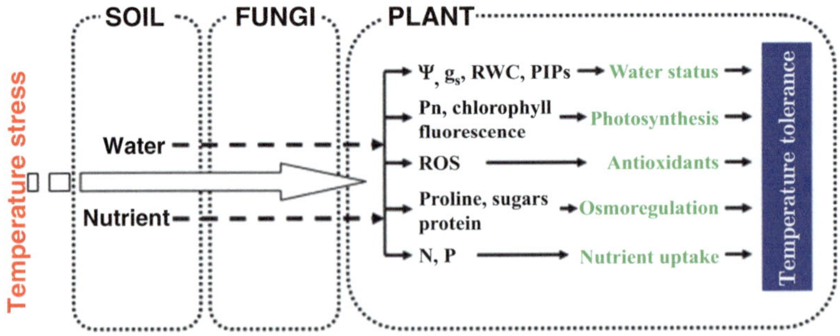

Fig. 8.1 Possible mechanisms of AM fungi improving plant tolerance to temperature stress

and proteomics approaches to a better understanding of the molecular basic of AM plant response to temperature stress and AM plant tolerance against temperature stress are imperative.

Although AM fungi are widespread in nature and are not specific to its host plants, the effectiveness of AM symbiosis is different in adverse temperature-stressed conditions. Thus, it is of great important to screen indigenous and presumably temperature-stressed-tolerant AM fungi isolates for the inoculation of plants adapted to temperature stress in future research.

Global climate change scenarios reveal temperature will increase and extremes of temperature will be occurring more frequently, so the chance of crops being exposed to adverse temperatures will be more likely. AM symbiosis has been shown to provide a superior road in the adaption of adverse environmental conditions in the future agriculture. However, numerous challenges of AM application in agriculture remain to be met, such as large-scale cultivation of AM fungi, and hypercompetition from other organisms in the complex soil environment.

Acknowledgements We are very grateful for the support by the "One-Three-Five" Strategic Planning Program of Chinese Academy of Sciences (IGA-135-04), the Science Foundation of Chinese Academy of Sciences (XDB15030103), and the National Natural Science Foundation of China (31370144; 41571255).

References

Abdel Latef AA, Chaoxing H (2011) Arbuscular mycorrhizal influence on growth, photosynthetic pigments, osmotic adjustment and oxidative stress in tomato plants subjected to low temperature stress. Acta Physiol Plant 33:1217–1225

Akiyama K (2007) Chemical identification and functional analysis of apocarotenoids involved in the development of arbuscular mycorrhizal symbiosis. Biosci Biotechnol Biochem 71(6):1405–1414

Ali MB, Hahn E, Paek K (2005) Effects of temperature on oxidative stress defense systems, lipid peroxidation and lipoxygenase activity in *Phalaenopsis*. Plant Physiol Biochem 43:213–223

Allakhverdiev SI, Kreslavski VD, Klimov VV et al (2008) Heat stress: an overview of molecular responses in photosynthesis. Photosynth Res 98:541–550

Andersen CP, Sucoff EI, Dixon RK (1987) The influence of low soil temperature on the growth of vesicular-arbuscular mycorrhizal *Fraxinus pennsylvanica*. Can J For Res 17:951–956

Apel K, Hirt H (2004) Reactive oxygen species: metabolism, oxidative stress, and signal transduction. Annu Rev Plant Biol 55:373–399

Aroca R, Porcel R, Ruiz-Lozano JM (2007) How does arbuscular mycorrhizal symbiosis regulate root hydraulic properties and plasma membrane aquaporins in *Phaseolus vulgaris* under drought, cold or salinity stresses? New Phytol 173:808–816

Atkin OK, Sherlock D, Fitter AH et al (2009) Temperature dependence of respiration in roots colonized by arbuscular mycorrhizal fungi. New Phytol 182:188–199

Augé RM (2001) Water relations, drought and vesicular-arbuscular mycorrhizal symbiosis. Mycorrhiza 11:3–42

Augé RM, Stodola JW (1990) An apparent increase in symplastic water contributes to greater turgor in mycorrhizal roots of droughted *Rosa* plants. New Phytol 115:285–295

Bainard LD, Bainard JD, Hamel C et al (2014) Spatial and temporal structuring of arbuscular mycorrhizal communities is differentially influenced by abiotic factors and host crop in a semi-arid prairie agroecosysten. FEMS Microbiol Ecol 88:333–344

Baker NR (2008) Chlorophyll fluorescence: a probe of photosynthesis in vivo. Annu Rev Plant Biol 59:89–113

Baon JB, Smith SE, Alston AM (1994) Phosphorus uptake and growth of barley as affected by soil temperature and mycorrhizal infection. J Plant Nutr 17:479–492

Barrett G, Campbell CD, Fitter AH et al (2011) The arbuscular mycorrhizal fungus *Glomus hoi* can capture and transfer nitrogen from organic patches to its associated host plant at low temperature. Appl Soil Ecol 48:102–105

Berry J, Björkman O (1980) Photosynthetic response and adaptation to temperature in high plants. Annu Rev Plant Physiol 31:491–543

Berta G, Copetta A, Gamalero E et al (2014) Maize development and grain quality are differentially affected by mycorrhizal fungi and a growth-promoting pseudomonad in the field. Mycorrhiza 24:161–170

Botnen S, Kauserud H, Carlsen T et al (2015) Mycorrhizal fungal communities in coastal sand dunes and heaths investigated by pyrosequencing analyses. Mycorrhiza 25(6):447–456

Bunn R, Lekberg Y, Zabinski C (2009) Arbuscular mycorrhizal fungi ameliorate temperature stress in thermophilic plants. Ecology 90(5):1378–1388

Büscher M, Zavalloni C, de Boulois HD et al (2012) Effects of arbuscular mycorrhizal fungi on grassland productivity are altered by future climate and below-ground resource availability. Environ Exp Bot 81:62–71

Cabral C, Ravnskov S, Tringovska I et al (2016) Arbuscular mycorrhizal fungi modify nutrient allocation and composition in wheat (*Triticum aestivum* L.) subjected to heat-stress. Plant Soil 408(1):385–399

Camejo D, Rodríguez P, Morales MA et al (2005) High temperature effects on photosynthetic activity of two tomato cultivars with different heat susceptibility. J Plant Physiol 162:281–289

Camenzind T, Hempel S, Homeier J et al (2014) Nitrogen and phosphorus additions impact arbuscular mycorrhizal abundance and molecular diversity in a tropical montane forest. Global Change Biol 20(12):3646–3659

Charest C, Dalpé Y, Brown A (1993) The effect of vesicular-arbuscular mycorrhizae and chilling on two hybrids of *Zea mays* L. Mycorrhiza 4:89–92

Chen S, Jin W, Liu A et al (2013) Arbuscular mycorrhizal fungi (AMF) increase growth and secondary metabolism in cucumber subjected to low temperature stress. Sci Hortic 160:222–229

Chen XY, Song FB, Liu FL et al (2014) Effect of different arbuscular mycorrhizal fungi on growth and physiology of maize at ambient and low temperature regimes. Sci World J 2014:956141

Chinnusamy V, Zhu J, Zhu JK (2007) Cold stress regulation of gene expression in plants. Trends Plant Sci 12:444–451

Compant S, van der Heijden MGA, Sessitsch A (2010) Climate change effects on beneficial plant-microorganism interactions. FEMS Microbial Ecol 73:197–214

Dodd IC, Perez-Alfocea F (2012) Microbial amelioration of crop salinity stress. J Exp Bot 63:3415–3428

Duhamel M, Vandenkoornhuyse P (2013) Sustainable agriculture: possible trajectories from mutualistic symbiosis and plant neodomestication. Trends Plant Sci 18:597–600

Dumbrell AJ, Ashton PD, Aziz N et al (2011) Distinct seasonal assemblages of arbuscular mycorrhizal fungi revealed by massively parallel pyrosequencing. New Phytol 190:794–804

El-Tohamy W, Schnitzler WH, El-Behairy U et al (1999) Effect of VA mycorrhiza on improving drought and chilling tolerance of bean plants (*Phaseolus vulgaris*). J Appl Bot 73:178–183

Evelin H, Kapoor R, Giri B (2009) Arbuscular mycorrhizal fungi in alleviation of salt stress: a review. Ann Bot 104:1263–1280

Faber BA, Zasoski RJ, Munns DN et al (1991) A method for measuring hyphal nutrient and water uptake in mycorrhizal plants. Can J Bot 69:87–94

Fernandez O, Bethencourt L, Quero A et al (2010) Trehalose and plant stress responses: friend or foe? Trends Plant Sci 15:409–417

Gavito ME, Olsson PA, Rouhier H et al (2005) Temperature constraints on the growth and functioning of root organ cultures with arbuscular mycorrhizal fungi. New Phytol 168:179–188
Gianinazzi-Pearson V, Gollotte A, Tisserant B et al (1995) Cellular and molecular approaches in the characterization of symbiotic events in functional arbuscular mycorrhizal associations. Can J Bot 73:526–532
Gianinazzi S, Gollotte A, Binet M et al (2010) Agroecology: the key role of arbuscular mycorrhizas in ecosystem services. Mycorrhiza 20:519–530
Govindarajulu M, Pfeffer PE, Jin HR et al (2005) Nitrogen transfer in the arbuscular mycorrhizal symbiosis. Nature 435:819–823
Gutknecht JLM, Field CB, Balser TC (2012) Microbial communities and their responses to simulated global change fluctuate greatly over multiple years. Global Change Biol 18:2256–2269
Hajiboland R, Aliasgharzadeh N, Laiegh SF et al (2010) Colonization with arbuscular mycorrhizal fungi improves salinity tolerance of tomato (*Solanum lycopersicum* L.) plants. Plant Soil 331:313–327
Hänsch R, Mendel RR (2009) Physiological functions of mineral micronutrients (Cu, Zn, Mn, Fe, Ni, Mo, B, Cl). Curr Opin Plant Biol 12:259–266
Haugen LM, Smith SE (1992) The effect of high temperature and fallow period on infection of mung bean and cashew roots by the vesicular-arbuscular mycorrhizal fungus *Glomus intraradices*. Plant Soil 145:71–80
Hawkes CV, Hartley IP, Ineson P et al (2008) Soil temperature affects carbon allocation within arbuscular mycorrhizal networks and carbon transport from plant to fungus. Global Change Biol 14:1181–1190
Hayman DS (1974) Plant growth responses to vesicular-arbuscular mycorrhiza. VI. Effect of light and temperature. New Phytol 73:71–80
He XH, Critchley C, Bledsoe C (2003) Nitrogen transfer within and between plants through common mycorrhizal networks (CMNs). Crit Rev Plant Sci 22:531–567
Heinemeyer A, Ridgway KP, Edwards EJ et al (2003) Impact of soil warming and shading on colonization and community structure of arbuscular mycorrhizal fungi in roots of a native grassland community. Global Change Biol 10:52–64
Heinemeyer A, Ineson P, Ostle N et al (2006) Respiration of the external mycelium in the arbuscular mycorrhizal symbiosis shows strong dependence on recent photosynthates and acclimation to temperature. New Phytol 171:159–170
Hu Y, Wu S, Sun Y et al (2015) Arbuscular mycorrhizal symbiosis can mitigate the negative effects of night warming on physiological traits of *Medicago truncatula* L. Mycorrhiza 25(2):131–142
Janmohammadi M, Zolla L, Rinalducci S (2015) Low temperature tolerance in plants: changes at the protein level. Phytochemistry 117:76–89
Javaid A (2009) Arbuscular mycorrhizal mediated nutrition in plant. J Plant Nutr 32:1595–1618
Kaplan F, Kopka J, Haskell DW et al (2004) Exploring the temperature-stress metabolome of Arabidopsis. Plant Physiol 136:4159–4168
Karandashov V, Bucher M (2005) Symbiotic phosphate transport in arbuscular mycorrhizas. Trends Plant Sci 10:22–29
Karasawa T, Hodge A, Fitter AH (2012) Growth, respiration and nutrient acquisition by the arbuscular mycorrhizal fungus *Glomus mosseae* and its host plant *Plantago lanceolata* in cooled soil. Plant Cell Environ 35:819–828
Karim MA, Fracheboud Y, Stamp P (1999) Photosynthetic activity of developing leaves of *Zea mays* is less affected by heat stress than that of developed leaves. Physiol Plant 105:685–693
Kim HS, Oh JM, Luan S et al (2013) Cold stress causes rapid but differential changes in properties of plasma membrane H^+-ATPase of camelina and rapeseed. J Plant Physiol 170:828–837
Kishor PBK, Sangama S, Amrutha RN et al (2005) Regulation of proline biosynthesis degradation, uptake and transport in higher plants: its implications in plant growth and abiotic stress tolerance. Curr Sci 88:424–438
Koide RT, Schreiner RP (1992) Regulation of the vesicular-arbuscular mycorrhizal symbiosis. Annu Rev Plant Physiol Plant Mol Biol 43:557–581

Krause G, Weis E (1991) Chlorophyll fluorescence and photosynthesis: the basics. Annu Rev Plant Physiol Plant Mol Biol 42:313–349

Kruse E, Uehlein N, Kaldenhoff R (2006) The aquaporins. Genome Biol 7:206

Kytöviita M, Ruotsalainen AL (2007) Mycorrhizal benefit in two low arctic herbs increases with increasing temperature. Am J Bot 94(8):1309–1315

Lenoir I, Fontaine J, Sahraoui AL (2016) Arbuscular mycorrhizal fungal responses to abiotic stresses: a review. Phytochemistry 123:4–15

Li JY, Liu XH, Cai QS et al (2008) Effects of elevated CO_2 on growth, carbon assimilation, photosynthate accumulation and related enzymes in rice leaves during sink-source transition. J Integr Plant Biol 50:723–732

Liu A, Wang B, Hamel C (2004) Arbuscular mycorrhiza colonization and development at suboptimal root zone temperature. Mycorrhiza 14:93–101

Liu ZL, Li YJ, Hou HY et al (2013) Differences in the arbuscular mycorrhizal fungi-improved rice resistance to low temperature at two N levels: aspects of N and C metabolism on the plant side. Plant Physiol Biochem 71:87–95

Liu A, Chen S, Chang R et al (2014a) Arbuscular mycorrhizae improve low temperature tolerance in cucumber via alterations in H_2O_2 accumulation and ATPase activity. J Plant Res 127:775–785

Liu ZL, Ma LN, He XY et al (2014b) Water strategy of mycorrhizal rice at low temperature through the regulation of PIP aquaporins with the involvement of trehalose. Appl Soil Ecol 84:185–191

Liu N, Chen X, Song F et al (2016) Effects of arbuscular mycorrhiza on growth and nutrition of maize plants under low temperature stress. Philipp Agric Sci 99(3):246–252

Lumini E, Vallino M, Alguacil MM et al (2011) Different farming and water regimes in Italian rice fields affect arbuscular mycorrhizal fungal soil communities. Ecol Appl 21(5):1696–1707

Luu DT, Maurel C (2005) Aquaporins in a challenging environment: molecular gears for adjusting plant water status. Plant Cell Environ 28:85–96

Martin CA, Stutz JC (2004) Interactive effects of temperature and arbuscular mycorrhizal fungi on growth, P uptake and root respiration of *Capsicum annuum* L. Mycorrhiza 14:241–244

Martin A, Lee J, Kichey T et al (2006) Two cytosolic glutamine synthetase isoforms of maize (*Zea mays* L.) are specifically involved in the control of grain production. Plant Cell 18:3252–3274

Matsubara Y, Kayukawa Y, Fukui H (2000) Temperature-stress tolerance of asparagus seedling through symbiosis with arbuscular mycorrhizal fungus. J Japan Soc Hort Sci 69(5):570–575

Matsubara Y, Hirano I, Sassa D et al (2004) Alleviation of high temperature stress in strawberry plants infected with arbuscular mycorrhizal fungi. Environ Control Biol 42(2):105–111

Matteucci M, D'Angeli S, Errico S et al (2011) Cold affects the transcription of fatty acid desaturases and oil quality in the fruit of *Olea europaea* L, genotypes with different cold hardiness. J Exp Bot 62:3403–3420

Maxwell K, Johnson GN (2000) Chlorophyll fluorescence-a practical guide. J Exp Bot 51:659–668

Maya MA, Matsubara Y (2013) Influence of arbuscular mycorrhiza on the growth and antioxidative activity in cyclamen under heat stress. Mycorrhiza 23:381–390

Mazorra LM, Nunez M, Echerarria E et al (2002) Influence of brassinosteriods and antioxidant enzymes activity in tomato under different temperatures. Plant Biol 45:593–596

Miransari M, Bahrami HA, Rejali F et al (2008) Using arbuscular mycorrhiza to alleviate the stress of soil compaction on wheat (*Triticum aestivum* L.) growth. Soil Biol Biochem 40:1197–1206

Nelsen CE, Safir GR (1982) The water relations of well-watered, mycorrhizal and non-mycorrhizal onion plants. J Am Soc Hortic Sci 107:271–274

Ngwene B, Gabriel E, George E (2013) Influence of different mineral nitrogen sources (NO_3^--N vs. NH_4^+-N) on arbuscular mycorrhiza development and N transfer in a *Glomus intraradices*-cowpea symbiosis. Mycorrhiza 23:107–117

Paradis R, Dalpé Y, Charest C (1995) The combined effect of arbuscular mycorrhizas and short-term cold exposure on wheat. New Phytol 129:637–642

Raju PS, Clark RB, Ellis JR et al (1990) Effects of species of VA-mycorrhizal fungi on growth and mineral uptake of sorghum at different temperatures. Plant Soil 121:165–170
Ruelland E, Zachowski A (2010) How plants sense temperature. Environ Exp Bot 69:225–232
Ruotsalainen AL, Kytöviita M (2004) Mycorrhiza does not alter low temperature impact on *Gnaphalium norvegicum*. Oecologia 140:226–233
Schenck NC, Graham SO, Green NE (1975) Temperature and light effects on contamination and spore germination of vesicular-arbuscular mycorrhizal fungi. Mycologia 67:1189–1192
Sharma SS, Dietz KJ (2006) The significance of amino acids and amino-acid derived molecules in plant responses and adaptation to heavy metal stress. J Exp Bot 57:711–726
Sharma M, Laxmi A (2016) Jasmonates: emerging players in controlling temperature stress tolerance. Front Plant Sci 6:1129
Smith SE, Read DJ (2008) Mycorrhizal symbiosis, 3rd edn. Academic, London
Smith GS, Roncadori RW (1986) Responses of three vesicular-arbuscular mycorrhizal fungi at four soil temperatures and their effects on cotton growth. New Phytol 104:89–95
Smith SE, Smith FA (2012) Fresh perspectives on the roles of arbuscular mycorrhizal fungi in plant nutrition and growth. Mycologia 104(1):1–13
Staddon PL, Gregersen R, Jakobsen I (2004) The response of two *Glomus* mycorrhizal fungi and a fine endophyte to elevated atmospheric CO2, soil warming and drought. Global Change Biol 10:1909–1921
Strack D, Fester T (2006) Isoprenoid metabolism and plastid reorganization in arbuscular mycorrhizal roots. New Phytol 172:22–34
Theocharis A, Clement C, Barka EA (2012) Physiological and molecular changes in plants grown at low temperatures. Planta 235:1091–1105
Thomashow MF (2010) Molecular basis of plant cold acclimation: insights gained from studying the CBF cold response pathway. Plant Physiol 154:571–577
Torrecillas E, Torres P, Alguacil MM et al (2013) Influence of habitat and climate variables on arbuscular mycorrhizal fungus community distribution, as revealed by a case study of facultative plant epiphytism under semiarid conditions. Appl Environ Microbiol 79(23):7203–7209
Uemura M, Tominaga Y, Nakagawara C et al (2006) Responses of the plasma membrane to low temperatures. Physiol Plant 126:81–89
Verma AK, Upadhyay SK, Verma PC et al (2011) Functional analysis of sucrose phosphate synthase (SPS) and sucrose synthase (SS) in sugarcane (*Saccharum*) cultivars. Plant Biol 13:325–332
Volkmar KM, Woodbury W (1989) Effects of soil temperature and depth on colonization and root and shoot growth of barley inoculated with vesicular-arbuscular mycorrhizae indigenous to Canadian prairie soil. Can J Bot 67:1702–1707
Wahid A, Gelani S, Ashraf M et al (2007) Heat tolerance in plants: an overview. Environ Exp Bot 61:199–223
Wang B, Funakoshi DM, Dalpé Y et al (2002) Phosphorus-32 absorption and translocation to host plants by arbuscular mycorrhizal fungi at low root-zone temperature. Mycorrhiza 12:93–96
Wu QS, Zou YN (2010) Beneficial roles of arbuscular mycorrhizas in citrus seedlings at temperature stress. Sci Hortic 125:289–293
Yaneva IA, Hoffmann GW, Tischner R (2002) Nitrate reductase from winter wheat leaves is activated at low temperature via protein dephosphorylation. Physiol Plant 114:65–72
Young A (1991) The photoprotective role of carotenoids in higher plants. Physiol Plant 83:702–708
Zeng Y, Yu J, Cang J et al (2011) Detection of sugar accumulation and expression levels of correlative key enzymes in winter wheat (*Triticum aestivum*) at low temperatures. Biosci Biotechnol Biochem 75:681–687
Zhang F, Hamel C, Kianmehr H et al (1995) Root-zone temperature and soybean [*Glycine max*. (L.) Merr.] vesicular-arbuscular mycorrhizae: development and interactions with the nitrogen fixing symbiosis. Environ Exp Bot 35:287–298

Zhang RQ, Zhu HH, Zhao HQ et al (2013) Arbuscular mycorrhizal fungal inoculation increases phenolic synthesis in clover roots via hydrogen peroxide, salicylic acid and nitric oxide signaling pathways. J Plant Physiol 170:74–79

Zhou Z, Ma H, Liang K et al (2012) Improved tolerance of teak (*Tectona grandis* L.f.) seedlings to low-temperature stress by the combined effect of arbuscular mycorrhiza and paclobutrazol. J Plant Growth Regul 31:427–435

Zhu XC, Song FB, Xu HW (2010a) Arbuscular mycorrhizae improves low temperature stress in maize via alterations in host water status and photosynthesis. Plant Soil 331:129–137

Zhu XC, Song FB, Xu HW (2010b) Influence of arbuscular mycorrhizae on lipid peroxidation and antioxidant enzyme activity of maize plants under temperature stress. Mycorrhiza 20:325–332

Zhu XC, Song FB, Liu SQ et al (2011) Effects of arbuscular mycorrhizal fungus on photosynthesis and water status of maize under high temperature stress. Plant Soil 346:189–199

Zhu XC, Song FB, Liu FL et al (2015) Carbon and nitrogen metabolism in arbuscular mycorrhizal maize plants under low-temperature stress. Crop Pasture Sci 66(1):62–70

Zhu XC, Song FB, Liu FL (2016) Altered amino acid profile of arbuscular mycorrhizal maize plants under low temperature stress. J Plant Nutr Soil Sci 179:186–189

Chapter 9
Arbuscular Mycorrhizal Fungi as Potential Bioprotectants Against Aerial Phytopathogens and Pests

M. Comby, G. Mustafa, M. Magnin-Robert, B. Randoux, J. Fontaine, Ph. Reignault, and A. Lounès-Hadj Sahraoui

Abstract In the context of an increasing worldwide food requirement, the control of crop diseases is crucial to guarantee high and stable yield, as well as sanitary quality. An environmentally friendly contribution to this could be biocontrol using beneficial microorganisms, such as arbuscular mycorrhizal fungi (AMF). AMF establish symbiosis with their host plants, thus influencing their growth, but they also induce tolerance to environmental stresses. Among stresses that can be alleviated through AMF inoculation, plant attacks by aerial pathogens and pests have so far been underestimated. Therefore, we present here an overview of studies focusing on AMF-mediated bioprotection against aerial pathogens and pests. Obtained protection is mainly due to changes in host nutrition and induction of defense following the establishment of arbuscular mycorrhizal symbiosis. This protection can vary greatly depending on different factors such as host genotype, AMF species involved, pest and pathogen lifestyles, interactions between AMF and other microorganisms, or even crop management practices. Finally, some future challenges for the use of AMF in biocontrol are discussed.

Keywords Arbuscular mycorrhizal fungi • Aerial phytopathogens • Aerial pests • Biocontrol • Mycorrhiza-induced resistance

---

M. Comby and G. Mustafa are co-first authors; A. Lounès-Hadj Sahraoui and Ph. Reignault are co-last authors

M. Comby • G. Mustafa • M. Magnin-Robert • B. Randoux • J. Fontaine • P. Reignault
A. Lounès-Hadj Sahraoui (✉)
UCEIV-EA 4492, SFR Condorcet FR CNRS 3417, University of the Littoral Côte d'Opale, CS 80699, F-62228, Calais cedex, France
e-mail: lounes@univ-littoral.fr

Q.-S. Wu (ed.), *Arbuscular Mycorrhizas and Stress Tolerance of Plants*,
DOI 10.1007/978-981-10-4115-0_9

9.1 Introduction

Using alternative control strategies compatible with a more sustainable agriculture, such as biocontrol, has become an urgent need to reduce the systematic and extensive use of fungicides which causes both human health problems (reprotoxicity, neurotoxicity, and carcinogenicity) and environmental pollutions (Oruc 2010). Biocontrol is based on the use of living microorganisms or derived molecules to control plant pathogens and pests and relies on natural mechanisms occurring during interactions between microorganisms and host plants (Alabouvette et al. 2006). Even though the presence of microorganisms in plants has long been regarded as responsible for diseases, it is now admitted that the vast majority of encountered microorganisms are not pathogenic (Mendes et al. 2013), most of them being neutral or beneficial for the health and growth of plants. Plants continuously interact with microorganisms; thus, rhizospheric microbiome has received substantial attention in recent years as it is a key for plant health and productivity (Berendsen et al. 2012; Turner et al. 2013). Plants may even modulate their own rhizospheric microbial communities by recruiting beneficial microorganisms through the release of primary and secondary metabolites from their roots (Berendsen et al. 2012; Cameron et al. 2013; Lakshmanan et al. 2014). Among associations of plants with beneficial microorganisms, symbiosis with arbuscular mycorrhizal fungi (AMF) is probably the oldest and most widespread in terms of geographical distribution and phylogenetic coverage within the plant kingdom (Gutjahr and Parniske 2013). AMF, obligate biotrophs from the phylum Glomeromycota, are able to establish symbiosis with over 80% of land plant species, conferring a set of physiological benefits to host plants (Smith et al. 2010). They mainly supply mineral nutrients such as phosphorous or nitrogen to the host plant which increases its growth and fitness, and in return the plant provides carbohydrates to the fungus, up to 20% of its fixed carbon (Parniske 2008; Smith and Read 2008). In addition to this nutrient supply improvement, AMF also impact the plant host ability to overcome abiotic and biotic stresses (Miransari 2010). More specifically, the establishment of arbuscular mycorrhizal symbiosis can improve host resistance to pathogens and pests (Whipps 2004; Pozo and Azcón-Aguilar 2007; Jung et al. 2012). AMF are interesting putative biocontrol agents, since they are natural soilborne inhabitants that establish stable and long-lasting mutualistic symbiosis with the roots of most plant species, including major crops such as wheat, maize, rice, soybean, potato, tomato, onion, and pulses (Bonfante and Perotto 1995; Garcia-Garrido et al. 2000; Smith and Read 2008).

The impact of AMF on root-infecting pathogens has been largely documented, and most studies on AMF in bioprotection have focused on the control of nematodes and soilborne fungi, oomycota, or bacteria (Whipps 2004; Wehner et al. 2010; Cameron et al. 2013; Baum et al. 2015). The AMF-mediated bioprotection against root-infecting pathogens is explained by various mechanisms, like direct competition for space and nutrients between AMF and soilborne pathogens, the qualitative and quantitative modifications of root exudates stimulating the production of some antimicrobial compounds by soil bacteria, the root system transformation slowing down the disease progression, the damage compensation, and the induction of plant defense mechanisms (Whipps 2004; Pozo et al. 2009; Wehner et al. 2010). In the present review, we will focus on the contribution of AMF in the interactions between host plants and aboveground pathogens and pests, hereafter referred to as "aerial" pathogens and pests, which have been far less investigated. Thus, this review aims to provide an overview of the plant protection by AMF against aerial phytopathogens and pests, mechanisms involved in such a protection, as well as factors influencing AMF protective efficacy.

9.2 AMF-Mediated Protection Against Aerial Pathogens and Pests

While most data about bioprotection conferred by AMF were obtained from root pathogens (Whipps 2004; Wehner et al. 2010; Cameron et al. 2013; Baum et al. 2015), Table 9.1 summarizes the main studies reporting the involvement of AMF in the control of aerial phytopathogens and pests. However, the impact of AMF on pathogen infection and pest infestation in aboveground parts of plants is variable, and the establishment of mycorrhizal symbioses does not always increase disease resistance, as reported below.

9.2.1 Plant Virus Diseases

Very little information is available about interactions between AMF and plant viral infections. In 1999, Shaul et al. reported that disease severity was enhanced in mycorrhizal plants of tobacco colonized by *Rhizophagus irregularis* and infected with *Tobacco mosaic virus* (TMV), compared to non-mycorrhizal plants. Accordingly, Sipahioglu et al. (2009) have shown that potato plants colonized by *R. irregularis* were more susceptible to the *Potato virus Y* (PVY), with an enhanced PVY multiplication. The higher physiological potential of arbuscular mycorrhizal plants may promote virus multiplication and should be responsible for a better spread within the whole plants (Dehne 1982). More recently, Miozzi et al. (2011) observed that colonization by *Funneliformis mosseae* increased the multiplication of *Tomato spotted wilt virus* (TSWV) in tomato plants, due to an attenuation of the regulation of plant genes responding to virus infection, leading to a lower accumulation of pathogenesis-related (PR) proteins. Elsharkawy et al. (2012) observed no significant differences on *Cucumber mosaic virus* infection between cucumber plants colonized by *F. mosseae* and control plants. Only Maffei et al. (2014) have reported a protective effect of arbuscular mycorrhizal on viruses' infection, with tomato plants colonized by *F. mosseae* less infected by the *Tomato yellow leaf curl*

Table 9.1 Table summarizing the main studies reported in the literature on the involvement of arbuscular mycorrhizal fungi (AMF) in the control of aerial pathogens and pests

Pathogens and pests		Diseases	Host plants	AMF	AMF effect	Defense mechanisms	References
Fungi	*Botrytis cinerea/N*	Gray mold	Tobacco – *Nicotiana tabacum* cv. Xanthi nc.	*Rhizophagus irregularis*	–		Shaul et al. (1999)
			Roses	*Funneliformis mosseae*	Ø		Møller et al. (2009)
			Tomato – *Solanum lycopersicum*	*F. mosseae*	+	Induction of systemic resistance	Pozo et al. (2010)
			Tomato – *Solanum lycopersicum* cv. Moneymaker	*F. mosseae*	+	Lower content of abscisic acid in mycorrhizal plant leaves	Fiorilli et al. (2011)
	Alternaria solani/N	Early blight of tomato	Tomato – *Solanum lycopersicum* cv. Frembgens Rheinlands Ruhm	*R. irregularis*	+	Not determined	Fritz et al. (2006)
			Tomato – *Solanum lycopersicum* cv. Amalia	*Glomus fasciculatum*	+	Induction of systemic resistance	De la Noval et al. (2007)
				Glomus clarum	–		
			Tomato – *Solanum lycopersicum* cv. Jin Bao	*F. mosseae*	+	Priming via jasmonic acid pathway	Song et al. (2015)
	Alternaria alternata/N	Stem canker disease	Tomato- *Solanum lycopersicum* cv. Pusa Ruby	*G. fasciculatum*	+	Induction of systemic resistance, with a possible role of methyl jasmonate	Nair et al. (2015b)

	Sclerotinia sclerotiorum /N	White mold disease	Common bean *Phaseolus vulgaris* cv. A-55, Az Reg87 and Az Hig	*R. irregularis*	+	Induction of systemic resistance	Mora-Romero et al. (2015a)
	Fusarium oxysporum f. sp. *lini*/H	Flax wilt	Flax – *Linum usitatissimum*	*R. irregularis*	+	Induction of systemic resistance	Dugassa et al. (1996)
		Fusarium *wilt disease*	Tomato – *Solanum lycopersicum* cv. Pusa Ruby	*G. fasciculatum*	+	Priming via jasmonic acid pathway	Nair et al. (2015a)
	Colletotrichum orbiculare/H	Anthracnose	Cucumber – *Cucumis sativus* cv. Eun Sung	*R. irregularis*	+	Callose accumulation at pathogen entry site	Lee et al. (2005)
			Cucumber – *Cucumis sativus* cv. Jibai	*F. mosseae*	Ø		Chandanie et al. (2006)
			Cucumber – *Cucumis sativus* cv. Tokiwa Jibai	*F. mosseae*	+	Induction of systemic resistance	Saldajeno and Hyakumachi (2011)
	Magnaporthe oryzae/H	Rice blast	Rice – *Oryza sativa* cv. Senia	*R. irregularis*	+	Induction of systemic resistance	Campos-Soriano et al. (2012)

(continued)

Table 9.1 (continued)

Pathogens and pests		Diseases	Host plants	AMF	AMF effect	Defense mechanisms	References
	Oidium lini/B	Powdery mildew	Flax – *Linum usitatissimum*	*R. irregularis*	–		Dugassa et al. (1996)
	Oidium begoniae/B		*Begonia hiemalis*	*Glomus etunicatum*	+	Not determined	Feldmann and Boyle (1998)
	Blumeria graminis f. sp. *tritici/B*		Wheat – *Triticum aestivum* cv. Orvantis and Lord	*F. mosseae*, *R. irregularis*, *Glomus sp.*	+	Accumulation of H_2O_2 and phenolic compounds. Overexpression of peroxidase, phenylalanine ammonia-lyase, and chitinase-encoding genes	Mustafa et al. (2016, 2017)
	Erysiphe graminis f. sp. *hordei/B*		Barley – *Hordeum vulgare* cv. Aura	*G. etunicatum*	–		Gernns et al. (2001)
	Podosphaera xanthii/B		Cucumber – *Cucumis sativus*	*R. irregularis*	Ø		Larsen and Yohalem (2004)
Oomycota	*Phytophthora cactorum/H*	Crown rot	Strawberry – *Fragaria x ananassa* cv. Jonsok	*F. mosseae*, *Glomus hoi*, *G. fistulosum*	–		Vestberg et al. (1994)
	Phytophthora infestans/H	Late blight	Tomato – *Solanum lycopersicum* cv. Amalia	*G. fasciculatum*, *G. clarum*	Ø		De la Noval et al. (2007)
			Potato – *Solanum tuberosum* cv. Bintje	*Glomus* sp.	+	Priming. Accumulation of salicylic acid and overexpression of PR protein genes	Gallou et al. (2011)

	Phytophthora parasitica/H	Late blight	Tomato – *Solanum lycopersicum* cv. Amalia	*G. fasciculatum*, *G. clarum*	Ø		De la Noval et al. (2007)
	Phytophthora capsici/H	Pepper blight	Ancho chili plants – *Capsicum annuum*	*G. fasciculatum*	+	Fast accumulation of H_2O_2 and increased activities of peroxidase and superoxide dismutase	Alejo-Iturvide et al. (2008)
	Phytophthora sojae/H	Stem rot	Soybean – *Glycine max*	*R. irregularis*	+	Accumulation of H_2O_2 and jasmonic acid. Changes in host nutrition	Li et al. (2013)
Phytoplasmas	*Phytoplasma* (16Sr XII-A subgroup)		Tomato – *Solanum lycopersicum* cv. Early Mech	*F. mosseae*	+	Possible involvement of phytohormones in the degeneration of phytoplasmas	Lingua et al. (2002)
	Phytoplasma (Stolbur group)	Chrysanthemum yellows disease	*Chrysanthemum carinatum*	*F. mosseae*	+	Not determined	D'Amelio et al. (2007)
				R. irregularis	–		

(continued)

Table 9.1 (continued)

Pathogens and pests		Diseases	Host plants	AMF	AMF effect	Defense mechanisms	References
Bacteria	*Xanthomonas campestris* pv. *alfalfae*	Bacterial spot	Alfalfa – *Medicago truncatula*	*R. irregularis* *Gigaspora gigantea*	+	Overexpression of defense genes encoding glycosyltransferase, kinase, calcium-binding protein, ubiquitin-protein ligase	Liu et al. (2007)
				G. versiforme			
	Xanthomonas campestris pv. *vesicatoria*		Tomato – *Solanum lycopersicum*	*R. irregularis*	+	Induction of systemic resistance	Mora-Romero et al. (2015a)
			cv. Micro-Tom and Missouri				
			Tomato – *Solanum lycopersicum* var. Missouri	*R. irregularis*	+	Induction of systemic resistance	Cervantes-Gamez et al. (2016)
Viruses	*Tobacco mosaic virus*	Tobacco mosaic disease	Tobacco – *Nicotiana tabacum* cv. Xanthinc	*R. irregularis*	–		Shaul et al. (1999)
	Potato virus Y (PVY)	Potato mosaic disease	Potato – *Solanum tuberosum* cv. Marfona	*R. irregularis*	–		Sipahioglu et al. (2009)
	Tomato spotted wilt virus (TSWV)	Tomato spotted wilt disease	Tomato – *Solanum lycopersicum*	*F. mosseae*	–		Miozzi et al. (2011)
	Cucumber mosaic virus (CMV)	Cucumber mosaic disease	Cucumber – *Cucumis sativus* cv. Tokiwa Jibai	*F. mosseae*	Ø		Elsharkawy et al. (2012)
	Tomato yellow leaf curl Sardinia virus (TYLCSV)	Tomato yellow leaf curl Sardinia disease	Tomato – *Solanum lycopersicum* cv. Moneymaker	*F. mosseae*	+	Not determined	Maffei et al. (2014)

Pest insects	*Arctia caja*		*Plantago lanceolata*	Indigenous AMF	+	Not determined	Gange and West (1994)
	Myzus persicae		*Plantago lanceolata*	Indigenous AMF	–		Gange and West (1994)
	Phlogophora meticulosa		Ryegrass – *Lolium perenne*	*F. mosseae*	+	Not determined	Vicari et al. (2002)
	Spodoptera littoralis		*Plantago lanceolata*	*R. irregularis*	+	Reduced emission of sesquiterpenes, involved in indirect defense	Fontana et al. (2009)

For true fungal pathogens, lifestyle is indicated following species name (*N* necrotroph, *H* hemibiotroph, *B* biotroph). The effect of AMF on pathogens and pests is either positive (+), negative (–), or neutral (Ø). In case of positive effects, resulting in protection of mycorrhizal plants, defense mechanisms involved in the AMF-mediated protection are indicated

Sardinia virus (TYLCSV). In their study, mycorrhizal tomato plants exhibited milder symptoms and presented a lower concentration of viral DNA.

9.2.2 Bacterial and Phytoplasma Diseases

Few examples of protection conferred by AMF against phytopathogenic bacteria exist in the literature, but several studies demonstrated a protective effect of AMF against *Xanthomonas campestris*, leading to fewer necrotic lesions in alfalfa colonized by various AMF (Liu et al. 2007) or tomato plants colonized by *R. irregularis* (Mora-Romero et al. 2015a; Cervantes-Gámez et al. 2016). Mycorrhization also led to a reduction of symptoms caused by phytoplasmas, which are specialized bacteria-like obligate parasites of plant phloem tissues, transmitted by insect vectors. A reduction of the disease caused by a phytoplasma of the Stolbur group was observed in tomato plants colonized by *F. mosseae* which exhibited less severe symptoms (Lingua et al. 2002). Moreover, D'Amelio et al. (2007) showed that the colonization of chrysanthemum by *F. mosseae* reduced the number of plants infected with the chrysanthemum yellows phytoplasma. In addition, the presence of the AMF slowed down the disease development. Taken together, these studies suggest that mycorrhization confers resistance against phytoplasmas.

9.2.3 Fungal Diseases

Most studies on the role of AMF in bioprotection against aerial pathogens focused on fungal agents, mainly from the genera *Alternaria* (de la Noval et al. 2007; Nair et al. 2015b; Song et al. 2015), *Fusarium* (Dugassa et al. 1996; Nair et al. 2015a), *Botrytis* (Shaul et al. 1999; Møller et al. 2009; Pozo et al. 2010; Fiorilli et al. 2011), and *Colletotrichum* (Lee et al. 2005; Chandanie et al. 2006; Saldajeno and Hyakumachi 2011) or on oomycota from the genus *Phytophthora* (Vestberg et al. 1994; de la Noval et al. 2007; Alejo-Iturvide et al. 2008; Gallou et al. 2011; Li et al. 2013). The level of protection conferred by AMF against such pathogens seems to strongly depend on the attacker's lifestyle, being either necrotroph, hemibiotroph, or biotroph (Pozo and Azcón-Aguilar 2007; Jung et al. 2012).

Concerning necrotrophic fungi, which kill host tissues and extract nutrients from dead host cells, the presence of AMF sometimes increased the susceptibility of plants to diseases (Shaul et al. 1999; de la Noval et al. 2007). Møller et al. (2009) showed that *F. mosseae* alone increased the symptom severity of gray mold disease caused by *Botrytis cinerea* on potted roses, while a reduction of disease incidence was observed on mycorrhizal roses under darkness stress. However, a large number of other studies reported a protective effect of AMF on plants against necrotrophic fungi. Fritz et al. (2006) reported fewer necrosis and chlorosis caused by *Alternaria solani* in tomato plants colonized by *R. irregularis*. Similarly, de la Noval et al. (2007) and Nair et al. (2015b) observed reduced foliar damages due to *A. solani* or *Alternaria alternata*, respectively, in tomato plants colonized by *Glomus fasciculatum*. When colonized by *F. mosseae*, tomato plants also exhibited fewer symptoms in their leaves upon *A. solani* attack, as well as a reduced disease incidence and a slower disease development (Song et al. 2015). Tomato plants colonized by *F. mosseae* also exhibited less symptoms of disease caused by *B. cinerea* (Pozo et al. 2010; Fiorilli et al. 2011). Finally, Mora-Romero et al. (2015a) reported reduced lesion diameters caused by *Sclerotinia sclerotiorum* on leaves of common bean colonized by *R. irregularis*.

Towards hemibiotrophic fungi or oomycota, initially depending on living host cells like biotrophs and switching to necrotrophic lifestyle in later stages of infection, the impact of AMF was also controversial. Depending on the reported studies, their impact varies from no effect to a significant disease reduction. Chandanie et al. (2006) observed no impact of *F. mosseae* on the protection of mycorrhizal cucumber against *Colletotrichum orbiculare*, whereas Lee et al. (2005) and Saldajeno and Hyakumachi (2011) showed that root colonization of cucumber, respectively, by *R. irregularis* and *F. mosseae*, reduced the disease severity caused by this pathogenic agent. Moreover, *Fusarium oxysporum* f.sp. *lini* infection was shown to be reduced in flax colonized by *R. irregularis* (Dugassa et al. 1996) and, more recently, in tomato plants colonized by *G. fasciculatum* (Nair et al. 2015a). Likewise, contrasting results have been obtained for the control of the hemibiotrophic oomycota *Phytophthora* spp. on various crops and vegetables colonized by *Glomus* spp. or *R. irregularis*. Thus, the presence of these AMF can either reduce (Alejo-Iturvide et al. 2008; Gallou et al. 2011; Li et al. 2013) or increase (Vestberg et al. 1994; de la Noval et al. 2007) symptoms severity. Another study demonstrated that the hemibiotrophic fungus *Magnaporthe oryzae* developed lesions on leaves of mycorrhizal rice plants at early stage of the infection process, but the lesion development appeared to be blocked at later stages, suggesting a bioprotective effect of *R. irregularis* during the necrotrophic phase of *M. oryzae* development (Campos-Soriano et al. 2012). AMF are likely to operate in different ways, depending on whether the fungal pathogen invades the host tissue in a biotrophic or necrotrophic manner. These results support the idea that the lifestyle of pathogens affects the level of protection conferred by AMF against aerial pathogens.

Most cases of lack of protection have been encountered in the case of biotrophic phytopathogens, which colonize living plant tissues and obtain nutrients from living host cells. Larsen and Yohalem (2004) did not observe any significant effect of *R. irregularis* on powdery mildew caused by *Podosphaera xanthii* on mycorrhizal cucumber plants. However, Dugassa et al. (1996) observed an increased disease index and sporulation rate of *Oidium lini* on flax colonized by *R. irregularis*. On barley colonized by *Glomus etunicatum*, the sporulation rate of *Erysiphe graminis* f. sp. *hordei* was twice as high as in control plants (Gernns et al. 2001). However,

Feldmann and Boyle (1998) reported a study where the establishment of arbuscular mycorrhizal symbiosis could protect plants against a biotrophic aerial pathogen. The authors suggested that root colonization of begonia by *G. etunicatum* could reduce the extent of infection caused by *Oidium begoniae*. More recently, Mustafa et al. (2016, 2017) have reported the bioprotective effect of *F. mosseae* and *R. irregularis* against *Blumeria graminis* f. sp. *tritici* on wheat. *F. mosseae* provided the highest protection level, and its protective effect was associated to the reduction of the number of conidia with haustoria, which are specialized infection hyphae forming invaginated structures in host cells.

Discrepancies in these studies on AMF-mediated bioprotection could be attributed to differences between the targeted plant pathosystem, tested AMF inoculum, or even to the experimental conditions.

9.2.4 Pest Insects

AMF symbioses can also impact infestation by pest insects. The effect seems to depend on the insect lifestyle and degree of specialism. The presence of AMF usually affects negatively generalist insects, able to feed on a wide range of hosts and susceptible to plant defense responses (Gange and West 1994; Vicari et al. 2002). In particular, AMF may modulate the metabolism of iridoid glycosides and volatile organic compounds, involved in direct and indirect defenses of the plant (Fontana et al. 2009). On the contrary, specialist insects, feeding on one or few hosts and showing a high degree of adaptation to their hosts, usually perform better on mycorrhizal plants, probably because of a lesser sensitivity to plant defense responses combined with an improved nutritional quality of mycorrhizal plants (Gehring and Bennett 2009; Hartley and Gange 2009). The protection level also depends on the feeding guild of the attacking pest insect. Leaf chewers and miners are usually negatively affected by the establishment of mycorrhizal symbiosis. Gange and West (1994) reported a reduced number of larvae of *Arctia caja* in mycorrhizal *Plantago lanceolata*, and Vicari et al. (2002) have shown that the colonization of *Lolium perenne* by *F. mosseae* decreased the survivorship and growth of larvae of the moth *Phlogophora meticulosa*. Such reductions are likely to be related to the rapid activation of plant defenses following their feeding. Indeed, their feeding causes massive damages on leaves tissues, which defenses are potentiated in mycorrhizal plants. However, phloem-sucking insects such as aphids produce minimal damage to the leaves while feeding and therefore avoid their detection by the host immune system. Thus, Gange and West (1994) reported that the sucking insect *Myzus persicae* performed well on mycorrhizal plants, possibly benefiting from their higher nutritional value.

9.3 Mechanisms Involved in AMF-Mediated Biocontrol Against Aerial Pathogens and Pests

Whenever an AMF induces protection against aerial pathogens or pests, the underlying mechanisms of such a protection are still poorly understood. Bioprotection through mycorrhization is likely to result from a combination of several mechanisms, instead of a single one (Vierheilig et al. 2008). Although mycorrhizal colonization takes place in roots, symbiosis establishment has systemic effects in other parts of plants. In aboveground organs of mycorrhizal plants, two main mechanisms against aerial pathogens and pests may be involved. On one hand, the increased level of nutrients in the host plant and the modifications of the source-sink relations within it may affect the susceptibility of plant leaves to attackers (Wright et al. 1998a, b). On the other hand, the modulation of plant defense mechanisms by AMF was proposed as one of the main mechanisms responsible for the control of aerial pathogens and pests (Pozo et al. 2009). However, the processes of this mycorrhiza-induced resistance (MIR) remain elusive (Song et al. 2015).

9.3.1 Changes in Host Nutrition

Although mineral nutrient uptake primarily happens in the roots, minerals must also reach aerial organs. Several studies showed that the increased mineral uptake in plants colonized by AMF may be associated with an improved resistance to pathogens (Li et al. 2013). However, it was reported that improved nutritional status alone did not account for a better protection towards pathogens. Mustafa et al. (2016, 2017) observed that the susceptibility of wheat towards the biotrophic fungus *B. graminis* f. sp. *tritici* increased with the phosphorous concentration. More colonies per leaf were observed with a higher phosphorous supply, in non-mycorrhizal as well as in mycorrhizal plants. In the case of biotrophic pathogens, which develop nutrient-absorbing structures in host cells, the increased levels of assimilates in leaves of host plant can favor their development. To that respect, mycorrhizal symbiosis, since it provides an increased nutrient uptake, can benefit to biotrophs instead of conferring bioprotection. Other studies showed that an increased phosphorous supply to tomato plants did not significantly reduce symptoms caused by the necrotrophic fungus *A. solani* or by the hemibiotrophic bacteria *Xanthomonas campestris* pv. *vesicatoria*, whereas the colonization of plants by the AMF *R. irregularis* did (Fritz et al. 2006; Cervantes-Gámez et al. 2016). These results suggest that an increased level of nutrients in mycorrhizal plant is not a key biochemical factor modifying the plant's response to aerial pathogens, independently of their lifestyle, but only indirectly contributes to protection against necrotrophic or hemibiotrophic pathogen attacks. One can assume that arbuscular mycorrhizal symbiosis results in more vigorous plants, which may be more resistant to such pathogens.

Therefore, enhanced disease protection in mycorrhizal plants, when it takes place, may be primarily triggered by AMF colonization and not determined by the nutritional status of plants.

Nonetheless, from an agronomical point of view, plant health outranks the absence of plant diseases and rather corresponds to the plant's ability to maintain its productivity even under unfavorable conditions. In this way, mycorrhizal symbiosis can still be of interest if mycorrhizal plants suffer less than non-mycorrhizal ones in terms of fresh weigh production, whatever its susceptibility level to pathogens. For example, Dugassa et al. (1996) reported a disease reduction caused by *O. lini* on mycorrhizal plants. These plants maintained a better level of production compared to non-mycorrhizal ones, thanks to the changes in host nutrition provided by the symbiosis with *R. irregularis* which compensate the damages caused by the pathogen, through an increased plant growth.

9.3.2 Modulation of Plant Defense Mechanisms by AMF

9.3.2.1 Regulation of Plant Defense Mechanisms During Mycorrhizal Establishment

Root exudation of strigolactones, sesquiterpenes serving as signals to induce hyphal branching of AMF, allows the fungus to localize host roots and facilitates the colonization process (Cameron et al. 2013). Host plants initially perceive AMF as potential pathogens, due to the recognition of MAMPs (microbe-associated molecular patterns) being conserved between beneficial and pathogenic fungi (Zamioudis and Pieterse 2012). In AMF, MAMPs triggering plant defenses are not yet known (Nair et al. 2015b), but their recognition is responsible for the induction of a local transient defense response of the plant immune system, called MTI for MAMPs-triggered immunity. MTI involves, among others, accumulation of reactive oxygen species (Fester and Hause 2005), activation of phenylpropanoid metabolism (Volpin et al. 1994), transcript accumulation of hydrolytic enzymes such as chitinases or glucanases (Salzer et al. 2000), callose deposition (Lee et al. 2005), and enhanced production of salicylic acid (SA) in plant roots (Blilou et al. 1999).

As in plant-pathogen interactions, AMF secrete effectors to manipulate host signaling and locally suppress MTI in order to allow a functional symbiosis with the plant. The first characterized secreted effector in AMF is the protein effector SP7 from *R. irregularis*, which inhibits the transcription factor ERF19 involved in ethylene signaling and thus suppresses plant defense, allowing root colonization by the AMF (Kloppholz et al. 2011). Pel and Pieterse (2013) postulated that plants cannot distinguish symbionts from pathogens. Consequently, the secretion of effectors is essential to allow interactions between plants and beneficial microorganisms. A large range of fungal effectors has been predicted, and various effectors may target different host signaling processes at specific stages of arbuscular mycorrhizal symbiosis (Selin et al. 2016).

9.3.2.2 Systemic Induction of Plant Defense Mechanisms in Response to AMF

The induction of MTI upon AMF colonization can provoke significant transcriptional and hormonal modifications in whole plants, leading to a systemic induction of defense (Pieterse et al. 2014). In order to understand how aerial plants parts respond to AMF colonization, high-throughput transcriptional profiling analysis by microarrays have been conducted and revealed, for instance, that 422 genes were differentially expressed in shoots of tomato plants colonized by *F. mosseae* (Fiorilli et al. 2011), while 599 genes with modified regulation were identified in leaves of *Medicago truncatula* colonized by *R. irregularis* (Liu et al. 2007). The first genome-wide analysis to detect the expression changes in leaves of mycorrhizal tomato plants has been conducted by Cervantes-Gámez et al. (2016). They identified 742 genes overexpressed in leaves of arbuscular mycorrhizal plants, which may be related to systemic defense induction. Among hormonal modifications found out, AMF colonization alone is likely to induce jasmonic acid (JA) accumulation in plants. Li et al. (2013) also demonstrated that mycorrhizal soybean exhibited a higher JA content compared to control plants, regardless of the pathogenic condition. Other studies have shown that AMF establishment can induce defense responses in aerial parts of plants, such as accumulation of PR proteins, heat-shock protein, and enhanced glutathione S-transferase and lipoxygenase activities before a pathogen attack (Liu et al. 2007; Gallou et al. 2011; Campos-Soriano et al. 2012; Song et al. 2015). Mustafa et al. (2017) reported an upregulation of genes encoding peroxidase, phenylalanine ammonia-lyase, and chitinase in mycorrhizal wheat plants under non-infected conditions.

9.3.2.3 Priming of Plant Defenses upon Pathogen or Pest Attack

The earliest interaction stages between AMF and host plant can prime the plant and lead to a faster and stronger systemic defense response upon a subsequent pathogen or pest attack (Jung et al. 2012). Such a systemic-induced resistance conferred by mycorrhizal symbiosis, referred as MIR, has long been demonstrated to be involved in AMF-mediated bioprotection against root pathogens. Split-root experiments, allowing physical separation between AMF and pathogens, showed a protection in non-AMF-colonized root parts and therefore confirmed a plant-mediated systemic effect (Cordier et al. 1998; Pozo et al. 2002; Khaosaad et al. 2007). However, reports of such a phenomenon about protection against aerial pathogens and pests are more recent. Gange (2006) reported an accumulation of insect antifeedant compounds in leaves of arbuscular mycorrhizal plants. Liu et al. (2007) showed that *M. truncatula* colonization by *R. irregularis* led to different systemic responses in plant roots or aerial organs. In shoots, transcription of defense-related genes encoding glycosyltransferases, calcium-binding protein, kinase, and ubiquitin-protein ligase was upregulated. Pozo et al. (2010) found out that arbuscular mycorrhizal tomato plants exhibited a lower level of *B. cinerea* infection in their leaves, as assessed by real-time polymerase chain reaction analysis, while the expression of JA-regulated defense-related genes was higher compared to control plants. Changes in phytohormone homeostasis are induced in arbuscular mycorrhizal plants, and these changes have often been associated with an enhanced tolerance to stresses (Fernández et al. 2014; Selosse et al. 2014; Pozo et al. 2015).

MIR shares common features with systemic acquired resistance (SAR) after pathogen infection and with induced systemic resistance (ISR) following root colonization by beneficial bacteria (Jung et al. 2012; Cameron et al. 2013; Pieterse et al. 2014). Indeed, MIR has been associated with SAR-like priming of SA-dependent genes, upon a pathogen attack, such as PR protein-encoding genes. For example, Song et al. (2015) showed that the pre-inoculation of tomato plants with *F. mosseae* led to significant increases in β-1,3-glucanase, chitinase, phenylalanine ammonia-lyase, and lipoxygenase activities in tomato leaves upon *A. solani* attack, as well as an induction of PR protein genes. Similarly, Gallou et al. (2011) and Campos-Soriano et al. (2012) observed a stronger induction of defense genes, including PR protein genes, in mycorrhizal potato plants challenged with *P. infestans* and tomato plants challenged with *M. oryzae*, respectively, compared to mycorrhizal plants under non-infected conditions.

This upregulation of defense responses associated with SA pathway links MIR to a SAR-like response, contrary to ISR that is assumed to occur without any accumulation of defense compounds prior to pathogen attack. However, MIR, like ISR, is often considered as a JA-dependent defense response. Several reports showed that JA-responsive genes were enhanced in leaves of mycorrhizal plants following a pathogen attack, compared to non-mycorrhizal ones (Pozo et al. 2010; Gallou et al. 2011; Campos-Soriano et al. 2012). Mora-Romero et al. (2015b) showed that common bean colonized by *R. irregularis* whose oxylipin-encoding genes involved in JA biosynthesis have been silenced lost their capacity to express a systemic-induced resistance compared to wild-type plants upon *S. sclerotiorum* infection. In addition, MIR confers a protection against a wide range of pathogens, including biotrophic and necrotrophic ones. It is generally assumed that biotrophic pathogens are more sensitive to SA-mediated responses, whereas necrotrophs are rather targeted by JA- or ethylene-mediated defenses (Pieterse et al. 2014). As discussed above (see Sect. 9.2.3), AMF-colonized plants are usually more resistant to necrotrophs than to biotrophs, and this correlates with an activation of JA-dependent defenses and a repression of SA-dependent ones in a well-established mycorrhizal symbiosis (Pozo and Azcón-Aguilar 2007), which is in accordance with an ISR-like response.

From an agronomical point of view, ISR-like response to mycorrhizal symbiosis, triggering systemic plant defenses only upon a pathogen attack, may be more promising as it is cost efficient compared to SAR-like response, where a fitness cost is potentially associated with the induction of defense even in the absence of any pathogen attack.

Despite such transcriptional upregulation of defense-related genes, repression of various systemic defense responses can also occur upon mycorrhizal colonization. Indeed, Shaul et al. (1999) reported a delay in the systemic accumulation of PR proteins in tobacco plants colonized by *R. irregularis* and challenged with *B. cinerea*. All of these changes systemically induced upon AMF colonization can affect the interactions of plants with aerial attackers.

9.3.2.4 Induction of Resistance Through Mycorrhizal Networks

In addition to the local or the priming effects of arbuscular mycorrhizal on aboveground tissues of a given plant, AMF may extend the induction of defense to neighboring plants. Indeed, AMF can spread from the root system of a given mycorrhizal plant, colonize thereafter roots from other plants and form common mycorrhizal networks (CMNs) interconnecting different plants from the same or even different species (Selosse et al. 2006). CMNs can serve as conduits for interplant signaling, influencing defense against aerial pathogens and pests. Song et al. (2010) were the first to show that CMNs may function as a defense communication conduit between infected and healthy tomato plants. Following the establishment of CMNs with the AMF and *F. mosseae* between tomato plants, the authors demonstrated that challenging one plant with *A. solani* led to an increase of disease resistance in healthy neighboring plants. Subsequent works demonstrated a similar interplant signaling through CMNs under herbivore attack with aphids (Babikova et al. 2013) or caterpillars (Song et al. 2014), where induced defense mechanisms have been observed in pest-free neighboring plants. Potential mechanisms for interplant signaling may be a direct transfer of signaling molecules within fungal hyphae and also electrical signals, which can be induced by wounding following a pathogen attack and are able to travel over relatively long distances (Johnson and Gilbert 2015).

9.4 Factors Affecting AMF Protection Efficacy Towards Aerial Pathogens and Pests

Like for other biocontrol agents, several factors can impact the level of protection conferred by AMF towards aerial phytopathogens and pests. The ultimate application of arbuscular mycorrhizal symbiosis is to be integrated within a technical management mode and used at field scale. Under field conditions, AMF will both interact with soil microorganisms and be subjected to fluctuations of environmental conditions, and its bioprotective effect could be affected.

9.4.1 *Biotic Factors*

9.4.1.1 Host Genotype

The host plant genotype can affect the protective effect induced by AMF, different genotypes of a given plant exhibiting different resistance levels. For example, Mora-Romero et al. (2015a) studied the protection conferred by *R. irregularis* towards *S. sclerotiorum* on three common bean cultivars and towards *X. campestris* pv. *vesicatoria* on two tomato cultivars. AMF failed to protect against pathogen attacks in one bean cultivar and one tomato cultivar, in spite of similar levels of mycorrhizal root colonization in all cultivars of a given plant species, indicating that mycorrhiza-induced plant resistance is genotype dependent. Mustafa et al. (2016, 2017) found out different mechanisms involved in protection towards *B. graminis* f. sp. *tritici* in two different wheat cultivars colonized by *F. mosseae*. In both cultivars, the protective effect was associated with a reduction of haustoria formation linked to the accumulation of polyphenolic compounds at *B. graminis* f. sp. *tritici* penetration sites. However, this accumulation of polyphenolic compounds was observed in the whole cells of the moderately resistant mycorrhizal cultivar Lord, whereas it was restricted to papillae in the moderately susceptible mycorrhizal cultivar Orvantis. Differences among cultivars can be linked to the well-known fact that host genotype affects the expression of induced resistance in general (Walters et al. 2013).

9.4.1.2 AMF Species

Variation in resistance to pathogens in aerial parts of plants can also be observed depending on the AMF species. De la Noval et al. (2007) have shown that colonization of tomato plants by *G. fasciculatum* conferred a protection towards *A. solani*, whereas the colonization of plants by *Glomus clarum* failed to protect against the same fungal pathogen, in spite of comparable mycorrhizal colonization levels of both AMF. Similarly, the colonization of chrysanthemum by *F. mosseae* reduced the number of plants infected by the chrysanthemum yellows phytoplasma, whereas the colonization by *R. irregularis* did not (D'Amelio et al. 2007). More recently, Mustafa et al. (2016) obtained different levels of protection towards *B. graminis* f. sp. *tritici* on wheat, depending on the inoculum used for the mycorrhizal symbiosis. The highest protection level was obtained in wheat plants colonized by *F. mosseae*. The lowest protection level was obtained with *R. irregularis*, and an intermediate level of protection was obtained with the commercially available product Solrize®, containing several *Glomus* species. Interestingly, these authors demonstrated that mycorrhizal colonization levels were the highest with *F. mosseae*, intermediate with *R. irregularis*, and the lowest with Solrize®, suggesting that a higher colonization level by AMF did not necessarily result in higher protection against a given pathogen. This lack of correlation between mycorrhizal rates and induced protection has been more deeply investigated in the case of root pathogens (St-Arnaud et al. 1997; Castellanos-Morales et al. 2012).

Literature about AMF-based bioprotection towards aerial pathogens and pests consists of 77% of studies being performed either with *F. mosseae* or *R. irregularis* (Table 9.1). Other studies mainly relied on *Glomus* species. Although these species are widely distributed at a global scale and are present in drastically different environments (Smith and Read 2008), it is also known that agricultural soils are often dominated by *Glomus* spp. (Oehl et al. 2003). The large occurrence of *F. mosseae* and *R. irregularis* in literature reflects the general use of these AMF in commercially available inocula and as model species in functional and biodiversity studies on arbuscular mycorrhizal symbiosis.

9.4.1.3 Pathogen Lifestyle

As previously mentioned, AMF-mediated protection towards aerial pathogens and pests may depend on their lifestyle or even their feeding guild concerning pest insects. The bioprotective effect of AMF would be greater towards necrotrophic or hemibiotrophic pathogens, as well as leaf chewers, than towards biotrophic pathogens or phloem-sucking insects, these latest benefiting from the higher nutritional status of mycorrhizal plants (see Sects. 9.2.3 and 9.2.4).

9.4.1.4 Interaction with Soil Microfauna

It is known that a fungus in a root system can alter the ability of another introduced AMF to further colonize plant roots (Schwartz et al. 2006). For example, Hepper et al. (1988) showed that *Glomus caledonium* was more competitive than *F. mosseae* to colonize roots of *Allium porrum*. However, most studies on the interactions between AMF and soil microfauna focused on soil bacteria. Bacteria including plant growth-promoting rhizobacteria (PGPR) and the genus *Rhizobium* are referred to as mycorrhiza helper bacteria since they promote the AMF symbiosis (Frey-Klett et al. 2007). Mycorrhiza helper bacteria mostly include *Bacillus* and *Pseudomonas* species and are usually fungal specific, but not plant specific (Miransari 2011). Such bacteria affect AMF root colonization by producing compounds that enhance AMF sporulation and hyphal growth or enhance the production of root exudates, resulting in a better arbuscular mycorrhizal establishment (Barea et al. 2002).

Some authors consider that MIR results from both direct interactions between AMF and host plants and the plant response to ISR-eliciting bacteria present in the mycorrhizosphere (Cameron et al. 2013). The current progress of studies on plant microbiome may help to confirm in the near future the importance of the mycorrhizosphere in AMF-mediated bioprotective effect.

9.4.1.5 Interaction with Other Biocontrol Agents

The use of other biocontrol agents together with AMF can improve their efficacy towards aerial attackers. Møller et al. (2009) have shown that under normal light conditions, the combined use of the AMF *F. mosseae* and the biocontrol agent *Ulocladium atrum* reduced more the disease incidence caused by *B. cinerea* in pot roses compared to the use of *F. mosseae* or *U. atrum* alone. These authors observed an enhanced colonization by *F. mosseae* in plants treated with *U. atrum* compared to control plants, and they suggested this could result from *U. atrum* inhibition of *B. cinerea*, enabling the formation and allocation of more assimilates available for *F. mosseae*. Saldajeno and Hyakumachi (2011) also reported an additive effect of the co-inoculation of *Fusarium equiseti* and *F. mossae* for the control of anthracnose caused by *C. orbiculare* in cucumber plants. Although the mechanisms underlying such a disease suppression have not been elucidated, it may result from the combined activation of plant defense responses by the AMF and *F. equiseti*.

9.4.2 *Crop Management*

9.4.2.1 Fertilization

Increased nutrient availability in soil is often detrimental to AMF. Fritz et al. (2006) showed that an additional phosphate supply increased the symptoms caused by *A. solani* in tomato plants colonized by *R. irregularis*. The lack of induced resistance with high phosphate supply can be due to a decrease in root colonization by AMF resulting from additional phosphate. Accordingly, Mustafa et al. (2016) observed a reduction of *F. mosseae* colonization on wheat under high phosphorous supply, leading to an enhanced susceptibility of plants to *B. graminis* f. sp. *tritici*. On the contrary, phosphorus deficiency can promote arbuscular mycorrhizal symbiosis by increasing strigolactone biosynthesis (Cosme et al. 2014; Fusconi 2014). Similarly, a stimulatory effect on arbuscular mycorrhizal symbiosis of nitrogen deficiency has been reported by Nouri et al. (2014).

However, it was shown that in organic systems, the use at low level of manure or compost, slowly releasing nutrients in soil, may benefit AMF (Cavagnaro 2014).

9.4.2.2 Pesticide Use

Fungicides have often detrimental effects on non-target beneficial microorganisms, including AMF. Some studies showed that AMF are targeted by sterol biosynthesis inhibitors, the most prevalent group of fungicides used in agriculture, in the same way as fungal pathogens, by the perturbation of their sterol metabolisms (Campagnac et al. 2008, 2009, 2010; Calonne et al. 2010, 2012). Thus, Calonne et al. (2012) reported the negative effect of propiconazole, from the triazole fungicide family, on

R. irregularis development in chicory roots. Zhang et al. (2006) showed that the application of chlorothalonil, a multisite fungicide from the chloronitrile family, on rice led to a reduced colonization by *F. mosseae*.

9.5 Concluding Remarks and Future Research Avenues

The potential of AMF as biocontrol agents against aerial diseases and pests has long been underestimated and is still poorly studied. But it is now recognized that they can protect plants against a large wide of pathogens and pests, mostly through changes in host nutrition and systemic-induced resistance. However, there is a clear need for further studies both on basic research and large-scale applications of AMF in agriculture.

Numerous biotic and abiotic factors have been shown to impact the mycorrhizal rate in host plants and/or the efficacy of the protection induced by AMF against aerial pathogens and pests. More specifically, host genotype and fungal species are involved in the so-called partner selection (Werner and Kiers 2015), which is crucial for the establishment of a functional symbiosis that will provide bioprotection upon a pathogen attack. In addition, crop management involves a range of various practices which can impact the arbuscular mycorrhizal symbiosis, both directly by damaging or killing AMF and indirectly by creating conditions favorable or unfavorable for the AMF. By impacting the establishment of arbuscular mycorrhizal symbiosis, these practices subsequently affect the efficacy of the protection conferred by AMF. It has been shown that practices such as fertilization or the use of pesticides can be controversial for AMF-mediated protection and therefore must be thoroughly considered for each culture system. Other management practices such as crop rotation or tillage play a role in the establishment of arbuscular mycorrhizal symbiosis, but to our knowledge no report exists about their impact on AMF-mediated bioprotection against aerial pathogens or pests.

Most studies describing AMF as potential biocontrol tool against aerial pathogens and pests have been conducted under controlled conditions, and reports of field experiments are scarce (Gange and West 1994). That is why AMF efficacy has yet to be demonstrated in large-scale crop production systems against aerial pathogens and pests. The transfer from the lab to the field scale seems to be somewhat difficult. To be effective in the field, an AMF has to be adapted to the encountered environmental conditions. Therefore, AMF with large geographic distribution such as *F. mosseae* and *R. irregularis* (Rosendahl et al. 2009) potentially rapidly adapting to environmental changes (Angelard et al. 2014) are particularly interesting. However, it would be worth investigating the bioprotective capacity of various AMF species instead of limiting the work on the few well-known species as it may be possible to breed and select more effective AMF for a given crop under specific environmental conditions. To obtain a large-scale protective effect of AMF, which would be effective over a range of environmental conditions and host plant groups, using several AMF in combination could be a solution. Furthermore, several studies have shown that using AMF in combination with other biological control agents such as mycorrhiza helper bacteria can improve their efficacy (Møller et al. 2009; Miransari 2011). AMF could also be used in combination with a low dose of fungicides or in a combination with elicitors in an integrated pest management program. This would reduce the use of pesticides which is beneficial for the durability of the resistance.

Another bottleneck for the use of AMF at field scale could be the large-scale production of AMF inocula, which is complex and challenging. AMF inocula are either produced *in vitro* in bioreactors or *in vivo* using conventional pot coculture methods. Other methods for the production of indigenous mycorrhizal inocula on the farm have been reported (Douds et al. 2010). The extent of their impact on crop yield and the economics of farming is still under debate, particularly in the case of wheat (Ryan and Angus 2003; Ryan et al. 2005; Ryan and Kirkegaard 2012; Dai et al. 2014; Rodriguez and Sanders 2015). However, the market for beneficial microbial inocula is growing continuously, and the development of these inocula for large-scale field application is moving forward (Schouteden et al. 2015); therefore, field applications of AMF all over the world might be realistic.

In parallel, the development of agricultural practices favoring local AMF populations are major challenges in the coming years, to consider these root symbionts as key players for plant productivity under a changing world. But because indigenous AMF equally or even better perform than commercial inocula, to be affordable, farmers are often encouraged to produce their own inoculum from native soils. Under profitable crop management practices, hyphal networks will remain unaltered and will spread naturally. Thus, AMF as biocontrol agents could be economically more profitable than other biocontrol agents that have to be regularly reintroduced in the field.

In addition, the industrial production of commercial AMF-based inocula requires standardized diagnosis toolkits for monitoring product quality control, assessing shelf life of formulated products and tracing inocula in soil and crop roots. Few recent studies have attempted to use molecular-based methods for detecting and quantifying the abundance of AMF. Application of quantitative real-time polymerase chain reaction (qPCR) in AMF studies demonstrated its usefulness as a rapid and sensitive technique for the enumeration of fungal propagules or gene copy numbers (Filion et al. 2003; Alkan et al. 2004; Isayenkov et al. 2004; Kiers et al. 2011; Thonar et al. 2012; Couillerot et al. 2013; Badri et al. 2016). However, it is not known how exogenous AMF interacts with local AMF communities preexisting in soils (Rodriguez and Sanders 2015), and population genomic approaches would be needed to address this point because differentiating introduced AMF from fungi naturally present in soil can be challenging. Despite the development of new technologies to track and quantify AMF in soil, an easy method for assessing the efficacy of AMF is still lacking. It would be helpful to identify arbuscular mycorrhizal inducible markers in crop fields in order to meet farmers' expectations. Next-generation sequencing may be a relevant strategy, but the studies in this area have so far been restricted to the laboratory (Salvioli and Bonfante 2013).

With respect to fundamental research, several aspects remain to be elucidated. Mechanisms of induced resistance in arbuscular mycorrhizal symbiosis interactions still need to be investigated. For instance, the nature of fungal effectors involved in the suppression of plant immunity and allowing the colonization of plant roots by AMF is still to be identified, as well as how these effectors are regulated and if host signals or environmental conditions are involved in this regulation to be addressed (Pozo et al. 2015). It is also still to understand, at molecular levels, how the suppression of local immunity, required for the establishment of arbuscular mycorrhizal symbiosis, is linked to the systemic priming of plant defenses and an enhanced tolerance to stresses (Pozo et al. 2015).

Recently, Mora-Romero et al. (2015a) showed that a shoot branch can acquire protection even if it comes from a non-mycorrhizal plant, suggesting that a signal moves from mycorrhizal roots to shoots through vascular tissue, but the nature of the moving signal has not been elucidated. One possible application of these findings would be to use mycorrhizal rootstocks from plant species easy to be colonized by AMF, as an additional strategy to improve disease protection on various crops and vegetables.

Finally, to our knowledge, the bioprotective effect of AMF against several aerial pathogens on a given plant has not been investigated. But because MIR shares common features with ISR, known to induce resistance against a broad spectrum of pathogens (Pieterse et al. 2014), it would be worth challenging mycorrhizal plants with several pathogens to assess if AMF can be used as a global control strategy against several diseases at field scale.

References

Alabouvette C, Olivain C, Steinberg C (2006) Biological control of plant diseases: the European situation. Eur J Plant Pathol 114:329–341

Alejo-Iturvide F, Márquez-Lucio MA, Morales-Ramírez I et al (2008) Mycorrhizal protection of chili plants challenged by *Phytophthora capsici*. Eur J Plant Pathol 120:13–20

Alkan N, Gadkar V, Coburn J et al (2004) Quantification of the arbuscular mycorrhizal fungus *Glomus intraradices* in host tissue using real-time polymerase chain reaction. New Phytol 161:877–885

Angelard C, Tanner CJ, Fontanillas P et al (2014) Rapid genotypic change and plasticity in arbuscular mycorrhizal fungi is caused by a host shift and enhanced by segregation. ISME J 8:284–294

Babikova Z, Gilbert L, Bruce TJ et al (2013) Underground signals carried through common mycelial networks warn neighbouring plants of aphid attack. Ecol Lett 16:835–843

Badri A, Stefani FO, Lachance G et al (2016) Molecular diagnostic toolkit for Rhizophagus irregularis isolate DAOM-197198 using quantitative PCR assay targeting the mitochondrial genome. Mycorrhiza 26:721–733

Barea J-M, Azcón R, Azcón-Aguilar C (2002) Mycorrhizosphere interactions to improve plant fitness and soil quality. Antonie Van Leeuwenhoek 81:343–351

Baum C, El-Tohamy W, Gruda N (2015) Increasing the productivity and product quality of vegetable crops using arbuscular mycorrhizal fungi: a review. Sci Hortic 187:131–141

Berendsen RL, Pieterse CM, Bakker PA (2012) The rhizosphere microbiome and plant health. Trend Plant Sci 17:478–486

Blilou I, Ocampo JA, García-Garrido JM (1999) Resistance of pea roots to endomycorrhizal fungus or *Rhizobium* correlates with enhanced levels of endogenous salicylic acid. J Exp Bot 50:1663–1668

Bonfante P, Perotto S (1995) Strategies of arbuscular mycorrhizal fungi when infecting host plants. New Phytol 130:3–21

Calonne M, Fontaine J, Debiane D et al (2010) Propiconazole toxicity on the non-target organism, the arbuscular mycorrhizal fungus, *Glomus irregulare*. In: Carisse O (ed) Fungicides. InTech Europe, pp 23–39

Calonne M, Lounès-Hadj Sahraoui A, Campagnac E et al (2012) Propiconazole inhibits the sterol 14α-demethylase in *Glomus irregulare* like in phytopathogenic fungi. Chemosphere 87:376–383

Cameron DD, Neal AL, van Wees SC et al (2013) Mycorrhiza-induced resistance: more than the sum of its parts? Trends Plant Sci 18:539–545

Campagnac E, Fontaine J, Lounès-Hadj Sahraoui A et al (2008) Differential effects of fenpropimorph and fenhexamid, two sterol biosynthesis inhibitor fungicides, on arbuscular mycorrhizal development and sterol metabolism in carrot roots. Phytochemistry 69:2912–2919

Campagnac E, Fontaine J, Lounès-Hadj Sahraoui A et al (2009) Fenpropimorph slows down the sterol pathway and the development of the arbuscular mycorrhizal fungus *Glomus intraradices*. Mycorrhiza 19:365–374

Campagnac E, Lounès-Hadj Sahraoui A, Debiane D et al (2010) Arbuscular mycorrhiza partially protect chicory roots against oxidative stress induced by two fungicides, fenpropimorph and fenhexamid. Mycorrhiza 20:167–178

Campos-Soriano L, García-martínez J, Segundo BS (2012) The arbuscular mycorrhizal symbiosis promotes the systemic induction of regulatory defence-related genes in rice leaves and confers resistance to pathogen infection. Mol Plant Pathol 13:579–592

Castellanos-Morales V, Cárdenas-Navarro R, García-Garrido J et al (2012) Bioprotection against *Gaeumannomyces graminis* in barley–a comparison between arbuscular mycorrhizal fungi. Plant Soil Environ 58:256–261

Cavagnaro TR (2014) Impacts of compost application on the formation and functioning of arbuscular mycorrhizas. Soil Biol Biochem 78:38–44

Cervantes-Gámez RG, Bueno-Ibarra MA, Cruz-Mendívil A et al (2016) Arbuscular mycorrhizal symbiosis-induced expression changes in *Solanum lycopersicum* leaves revealed by RNA-seq analysis. Plant Mol Biol Report 34:89–102

Chandanie WA, Kubota M, Hyakumachi M (2006) Interactions between plant growth promoting fungi and arbuscular mycorrhizal fungus *Glomus mosseae* and induction of systemic resistance to anthracnose disease in cucumber. Plant Soil 286:209–217

Cordier C, Pozo MJ, Barea JM et al (1998) Cell defense responses associated with localized and systemic resistance to *Phytophthora parasitica* induced in tomato by an arbuscular mycorrhizal fungus. Mol Plant-Microbe Interact 11:1017–1028

Cosme M, Franken P, Mewis I et al (2014) Arbuscular mycorrhizal fungi affect glucosinolate and mineral element composition in leaves of *Moringa oleifera*. Mycorrhiza 24:565–570

Couillerot O, Ramírez-Trujillo A, Walker V et al (2013) Comparison of prominent *Azospirillum* strains in *Azospirillum–Pseudomonas–Glomus* consortia for promotion of maize growth. Appl Microbiol Biotechnol 97:4639–4649

Dai M, Hamel C, Bainard LD et al (2014) Negative and positive contributions of arbuscular mycorrhizal fungal taxa to wheat production and nutrient uptake efficiency in organic and conventional systems in the Canadian prairie. Soil Biol Biochem 74:156–166

D'Amelio R, Massa N, Gamalero E et al (2007) Preliminary results on the evaluation of the effects of elicitors of plant resistance on chrysanthemum yellows phytoplasma infection. Bull Insectol 60:317

de la Noval B, Pérez E, Martínez B et al (2007) Exogenous systemin has a contrasting effect on disease resistance in mycorrhizal tomato (*Solanum lycopersicum*) plants infected with necrotrophic or hemibiotrophic pathogens. Mycorrhiza 17:449–460

Dehne HW (1982) Interaction between vesicular-arbuscular mycorrhizal fungi and plant pathogens. Phytopathology 72:1115–1119

Douds DD Jr, Nagahashi G, Hepperly PR (2010) On-farm production of inoculum of indigenous arbuscular mycorrhizal fungi and assessment of diluents of compost for inoculum production. Bioresour Technol 101:2326–2330

Dugassa GD, Von Alten H, Schönbeck F (1996) Effects of arbuscular mycorrhiza (AM) on health of *Linum usitatissimum* L. infected by fungal pathogens. Plant Soil 185:173–182

Elsharkawy MM, Shimizu M, Takahashi H et al (2012) The plant growth-promoting fungus *Fusarium equiseti* and the arbuscular mycorrhizal fungus *Glomus mosseae* induce systemic resistance against *Cucumber mosaic virus* in cucumber plants. Plant Soil 361:397–409

Feldmann F, Boyle C (1998) Concurrent development of arbuscular mycorrhizal colonization and powdery mildew infection on three *Begonia hiemalis* cultivars. J Plant Dis Protect 105:121–129

Fernández I, Merlos M, López-Ráez J et al (2014) Defense related phytohormones regulation in arbuscular mycorrhizal symbioses depends on the partner genotypes. J Chem Ecol 40:791–803

Fester T, Hause G (2005) Accumulation of reactive oxygen species in arbuscular mycorrhizal roots. Mycorrhiza 15:373–379

Filion M, St-Arnaud M, Jabaji-Hare SH (2003) Direct quantification of fungal DNA from soil substrate using real-time PCR. J Microbiol Methods 53:67–76

Fiorilli V, Catoni M, Francia D et al (2011) The arbuscular mycorrhizal symbiosis reduces disease severity in tomato plants infected by *Botrytis cinerea*. J Plant Pathol 93:237–242

Fontana A, Reichelt M, Hempel S et al (2009) The effects of arbuscular mycorrhizal fungi on direct and indirect defense metabolites of *Plantago lanceolata* L. J Chem Ecol 35:833–843

Frey-Klett P, Garbaye J, Tarkka M (2007) The mycorrhiza helper bacteria revisited. New Phytol 176:22–36

Fritz M, Jakobsen I, Lyngkjær MF et al (2006) Arbuscular mycorrhiza reduces susceptibility of tomato to *Alternaria solani*. Mycorrhiza 16:413–419

Fusconi A (2014) Regulation of root morphogenesis in arbuscular mycorrhizae: what role do fungal exudates, phosphate, sugars and hormones play in lateral root formation? Ann Bot 113:19–33

Gallou A, Mosquera HPL, Cranenbrouck S et al (2011) Mycorrhiza induced resistance in potato plantlets challenged by *Phytophthora infestans*. Physiol Mol Plant Pathol 76:20–26

Gange AC (2006) Insect-mycorrhizal interactions: patterns, processes, and consequences. In: Ohgushi T, Craig TP, Price PW (eds) Indirect interaction webs: nontrophic linkages through induced plant traits. Cambridge University Press, Cambridge, pp 124–144

Gange AC, West HM (1994) Interactions between arbuscular mycorrhizal fungi and foliar-feeding insects in *Plantago lanceolata* L. New Phytol 128:79–87

Garcia-Garrido JM, Tribak M, Rejon-Palomares A et al (2000) Hydrolytic enzymes and ability of arbuscular mycorrhizal fungi to colonize roots. J Exp Bot 51:1443–1448

Gehring C, Bennett A (2009) Mycorrhizal fungal–plant–insect interactions: the importance of a community approach. Environ Entomol 38:93–102

Gernns H, Alten H, Poehling HM (2001) Arbuscular mycorrhiza increased the activity of a biotrophic leaf pathogen–is a compensation possible? Mycorrhiza 11:237–243

Gutjahr C, Parniske M (2013) Cell and developmental biology of arbuscular mycorrhiza symbiosis. Annu Rev Cell Dev Biol 29:593–617

Hartley SE, Gange AC (2009) Impacts of plant symbiotic fungi on insect herbivores: mutualism in a multitrophic context. Annu Rev Entomol 54:323–342

Hepper CM, Azcon-Aguilar C, Rosendahl S et al (1988) Competition between three species of *Glomus* used as spatially separated introduced and indigenous mycorrhizal inocula for leek (*Allium porrum* L.) New Phytol 110:207–215

Isayenkov S, Fester T, Hause B (2004) Rapid determination of fungal colonization and arbuscule formation in roots of *Medicago truncatula* using real-time (RT) PCR. J Plant Physiol 161:1379–1383
Johnson D, Gilbert L (2015) Interplant signalling through hyphal networks. New Phytol 205:1448–1453
Jung SC, Martinez-Medina A, Lopez-Raez JA et al (2012) Mycorrhiza-induced resistance and priming of plant defenses. J Chem Ecol 38:651–664
Khaosaad T, Garcia-Garrido JM, Steinkellner S et al (2007) Take-all disease is systemically reduced in roots of mycorrhizal barley plants. Soil Biol Biochem 39:727–734
Kiers ET, Duhamel M, Beesetty Y et al (2011) Reciprocal rewards stabilize cooperation in the mycorrhizal symbiosis. Science 333:880–882
Kloppholz S, Kuhn H, Requena N (2011) A secreted fungal effector of *Glomus intraradices* promotes symbiotic biotrophy. Curr Biol 21:1204–1209
Lakshmanan V, Selvaraj G, Bais HP (2014) Functional soil microbiome: belowground solutions to an aboveground problem. Plant Physiol 166:689–700
Larsen J, Yohalem D (2004) Interactions between mycorrhiza and powdery mildew of cucumber. Mycol Prog 3:123–128
Lee CS, Lee YJ, Jeun YC (2005) Observations of infection structures on the leaves of cucumber plants pre-treated with arbuscular mycorrhiza *Glomus intraradices* after challenge inoculation with *Colletotrichum orbiculare*. Plant Pathol J 21:237–243
Li Y, Liu Z, Hou H et al (2013) Arbuscular mycorrhizal fungi-enhanced resistance against *Phytophthora sojae* infection on soybean leaves is mediated by a network involving hydrogen peroxide, jasmonic acid, and the metabolism of carbon and nitrogen. Acta Physiol Plant 35:3465–3475
Lingua G, D'Agostino G, Massa N et al (2002) Mycorrhiza-induced differential response to a yellows disease in tomato. Mycorrhiza 12:191–198
Liu J, Maldonado-Mendoza I, Lopez-Meyer M et al (2007) Arbuscular mycorrhizal symbiosis is accompanied by local and systemic alterations in gene expression and an increase in disease resistance in the shoots. Plant J 50:529–544
Maffei G, Miozzi L, Fiorilli V et al (2014) The arbuscular mycorrhizal symbiosis attenuates symptom severity and reduces virus concentration in tomato infected by *Tomato yellow leaf curl Sardinia virus* (TYLCSV). Mycorrhiza 24:179–186
Mendes R, Garbeva P, Raaijmakers JM (2013) The rhizosphere microbiome: significance of plant beneficial, plant pathogenic, and human pathogenic microorganisms. FEMS Microbiol Rev 37:634–663
Miozzi L, Catoni M, Fiorilli V et al (2011) Arbuscular mycorrhizal symbiosis limits foliar transcriptional responses to viral infection and favors long-term virus accumulation. Mol Plant-Microbe Interact 24:1562–1572
Miransari M (2010) Contribution of arbuscular mycorrhizal symbiosis to plant growth under different types of soil stress. Plant Biol 12:563–569
Miransari M (2011) Interactions between arbuscular mycorrhizal fungi and soil bacteria. Appl Microbiol Biotechnol 89:917–930
Møller K, Kristensen K, Yohalem D et al (2009) Biological management of gray mold in pot roses by co-inoculation of the biocontrol agent *Ulocladium atrum* and the mycorrhizal fungus *Glomus mosseae*. Biol Control 49:120–125
Mora-Romero GA, Cervantes-Gámez RG, Galindo-Flores H et al (2015a) Mycorrhiza-induced protection against pathogens is both genotype-specific and graft-transmissible. Symbiosis 66:55–64
Mora-Romero GA, Gonzalez-Ortiz MA, Quiroz-Figueroa F et al (2015b) PvLOX2 silencing in common bean roots impairs arbuscular mycorrhiza-induced resistance without affecting symbiosis establishment. Funct Plant Biol 42:18–30
Mustafa G, Randoux B, Tisserant B et al (2017) Defence mechanisms associated with mycorrhiza-induced resistance in wheat against powdery mildew. Funct Plant Biol. doi:10.1071/FP16206

Mustafa G, Randoux B, Tisserant B et al (2016) Phosphorus supply, arbuscular mycorrhizal fungal species, and plant genotype impact on the protective efficacy of mycorrhizal inoculation against wheat powdery mildew. Mycorrhiza 26:685–697

Nair A, Kolet SP, Thulasiram HV et al (2015a) Role of methyl jasmonate in the expression of mycorrhizal induced resistance against *Fusarium oxysporum* in tomato plants. Physiol Mol Plant Pathol 92:139–145

Nair A, Kolet SP, Thulasiram HV et al (2015b) Systemic jasmonic acid modulation in mycorrhizal tomato plants and its role in induced resistance against *Alternaria alternata*. Plant Biol 17:625–631

Nouri E, Breuillin-Sessoms F, Feller U et al (2014) Phosphorus and nitrogen regulate arbuscular mycorrhizal symbiosis in *Petunia hybrida*. PLoS One 9:e90841

Oehl F, Sieverding E, Ineichen K et al (2003) Impact of land use intensity on the species diversity of arbuscular mycorrhizal fungi in agroecosystems of Central Europe. Appl Environ Microbiol 69:2816–2824

Oruc HH (2010) Fungicides and their effects on animals. In: Carisse O (ed) Fungicides. InTech, Rijeka, pp 349–362

Parniske M (2008) Arbuscular mycorrhiza: the mother of plant root endosymbioses. Nat Rev Microbiol 6:763–775

Pel MJC, Pieterse CM (2013) Microbial recognition and evasion of host immunity. J Exp Bot 64:1237–1248

Pieterse CM, Zamioudis C, Berendsen RL et al (2014) Induced systemic resistance by beneficial microbes. Annu Rev Phytopathol 52:347–375

Pozo MJ, Azcón-Aguilar C (2007) Unraveling mycorrhiza-induced resistance. Curr Opin Plant Biol 10:393–398

Pozo MJ, Cordier C, Dumas-Gaudot E et al (2002) Localized versus systemic effect of arbuscular mycorrhizal fungi on defence responses to *Phytophthora* infection in tomato plants. J Exp Bot 53:525–534

Pozo MJ, Verhage A, García-Andrade J et al (2009) Priming plant defence against pathogens by arbuscular mycorrhizal fungi. In: Mycorrhizas-functional processes and ecological impact. Springer, Berlin, pp 123–135

Pozo MJ, Jung SC, López-Ráez JA et al (2010) Impact of arbuscular mycorrhizal symbiosis on plant response to biotic stress: the role of plant defence mechanisms. In: Arbuscular mycorrhizas: physiology and function. Springer, Dordrecht, pp 193–207

Pozo MJ, López-Ráez JA, Azcón-Aguilar C et al (2015) Phytohormones as integrators of environmental signals in the regulation of mycorrhizal symbioses. New Phytol 205:1431–1436

Rodriguez A, Sanders IR (2015) The role of community and population ecology in applying mycorrhizal fungi for improved food security. ISME J 9:1053–1061

Rosendahl S, Mcgee P, Morton JB (2009) Lack of global population genetic differentiation in the arbuscular mycorrhizal fungus *Glomus mosseae* suggests a recent range expansion which may have coincided with the spread of agriculture. Mol Ecol 18:4316–4329

Ryan MH, Angus J (2003) Arbuscular mycorrhizae in wheat and field pea crops on a low P soil: increased Zn-uptake but no increase in P-uptake or yield. Plant Soil 250:225–239

Ryan MH, Kirkegaard JA (2012) The agronomic relevance of arbuscular mycorrhizas in the fertility of Australian extensive cropping systems. Agric Ecosyst Environ 163:37–53

Ryan MH, Van Herwaarden AF, Angus JF et al (2005) Reduced growth of autumn-sown wheat in a low-P soil is associated with high colonisation by arbuscular mycorrhizal fungi. Plant Soil 270:275–286

Saldajeno MGB, Hyakumachi M (2011) The plant growth-promoting fungus *Fusarium equiseti* and the arbuscular mycorrhizal fungus *Glomus mosseae* stimulate plant growth and reduce severity of anthracnose and damping-off diseases in cucumber (*Cucumis sativus*) seedlings. Ann Appl Biol 159:28–40

Salvioli A, Bonfante P (2013) Systems biology and "omics" tools: a cooperation for next-generation mycorrhizal studies. Plant Sci 203:107–114

Salzer P, Bonanomi A, Beyer K et al (2000) Differential expression of eight chitinase genes in *Medicago truncatula* roots during mycorrhiza formation, nodulation, and pathogen infection. Mol Plant-Microbe Interact 13:763–777
Schouteden N, De Waele D, Panis B et al (2015) Arbuscular mycorrhizal fungi for the biocontrol of plant-parasitic nematodes: a review of the mechanisms involved. Front Microbiol 6:1280
Schwartz MW, Hoeksema JD, Gehring CA et al (2006) The promise and the potential consequences of the global transport of mycorrhizal fungal inoculum. Ecol Lett 9:501–515
Selin C, de Kievit TR, Belmonte MF et al (2016) Elucidating the role of effectors in plant-fungal interactions: progress and challenges. Front Microbiol 7:600
Selosse MA, Richard F, He X et al (2006) Mycorrhizal networks: des liaisons dangereuses? Trends Ecol Evol 21:621–628
Selosse MA, Bessis A, Pozo MJ (2014) Microbial priming of plant and animal immunity: symbionts as developmental signals. Trends Microbiol 22:607–613
Shaul O, Galili S, Volpin H et al (1999) Mycorrhiza-induced changes in disease severity and PR protein expression in tobacco leaves. Mol Plant-Microbe Interact 12:1000–1007
Sipahioglu MH, Demir S, Usta M et al (2009) Biological relationship of potato virus Y and arbuscular mycorrhizal fungus *Glomus intraradices* in potato. Pest Tech 3:63–66
Smith SE, Read DJ (2008) Mycorrhizal symbiosis. Academic, San Diego
Smith SE, Facelli E, Pope S et al (2010) Plant performance in stressful environments: interpreting new and established knowledge of the roles of arbuscular mycorrhizas. Plant Soil 326:3–20
Song YY, Zeng RS, Xu JF et al (2010) Interplant communication of tomato plants through underground common mycorrhizal networks. PLoS One 5:e13324
Song YY, Ye M, Li C et al (2014) Hijacking common mycorrhizal networks for herbivore-induced defence signal transfer between tomato plants. Sci Rep 4:3915
Song Y, Chen D, Lu K et al (2015) Enhanced tomato disease resistance primed by arbuscular mycorrhizal fungus. Front Plant Sci 6:786
St-Arnaud M, Vimard B, Fortin J et al (1997) Inhibition of *Fusarium oxysporum* f-sp. *dianthi* in the non-VAM species *Dianthus caryophyllus* by co-culture with *Tagetes patula* companion plants colonized by *Glomus intraradices*. Can J Bot 75:998–1005
Thonar C, Erb A, Jansa J (2012) Real-time PCR to quantify composition of arbuscular mycorrhizal fungal communities – marker design, verification, calibration and field validation. Mol Ecol Resour 12:219–232
Turner TR, James EK, Poole PS (2013) The plant microbiome. Genome Biol 14:209
Vestberg M, Palmujoki H, Parikka P et al (1994) Effect of arbuscular mycorrhizas on crown rot (*Phytophthora cactorum*) in micropropagated strawberry plants. Special issue of the third COST 87–8.10 joint meeting, 17–18 September 1993, Agricultural Research Centre of Finland, Laukaa, Finland
Vicari M, Hatcher PE, Ayres PG (2002) Combined effect of foliar and mycorrhizal endophytes on an insect herbivore. Ecology 83:2452–2464
Vierheilig H, Steinkellner S, Khaosaad T et al (2008) The biocontrol effect of mycorrhization on soilborne fungal pathogens and the autoregulation of the AM symbiosis: one mechanism, two effects? In: Mycorrhiza. Springer, Berlin, pp 307–320
Volpin H, Elkind Y, Okon Y et al (1994) A vesicular arbuscular mycorrhizal fungus (*Glomus intraradix*) induces a defense response in alfalfa roots. Plant Physiol 104:683–689
Walters DR, Ratsep J, Havis ND (2013) Controlling crop diseases using induced resistance: challenges for the future. J Exp Bot 64:1263–1280
Wehner J, Antunes PM, Powell JR et al (2010) Plant pathogen protection by arbuscular mycorrhizas: a role for fungal diversity? Pedobiologia 53:197–201
Werner GD, Kiers ET (2015) Partner selection in the mycorrhizal mutualism. New Phytol 205:1437–1442
Whipps JM (2004) Prospects and limitations for mycorrhizas in biocontrol of root pathogens. Can J Bot 82:1198–1227

Wright DP, Read DJ, Scholes JD (1998a) Mycorrhizal sink strength influences whole plant carbon balance of *Trifolium repens* L. Plant Cell Environ 21:881–891

Wright DP, Scholes JD, Read DJ (1998b) Effects of VA mycorrhizal colonization on photosynthesis and biomass production of *Trifolium repens* L. Plant Cell Environ 21:209–216

Zamioudis C, Pieterse CM (2012) Modulation of host immunity by beneficial microbes. Mol Plant-Microbe Interact 25:139–150

Zhang XH, Zhu YG, Lin AJ et al (2006) Arbuscular mycorrhizal fungi can alleviate the adverse effects of chlorothalonil on *Oryza sativa* L. Chemosphere 64:1627–1632

Chapter 10
Arbuscular Mycorrhiza and Reactive Oxygen Species

Rupam Kapoor and Neeraja Singh

Abstract The accrual of reactive oxygen species (ROS) is a common biochemical response to all abiotic and biotic stresses in plants. ROS are extremely lethal to biological cells causing oxidative damage to DNA, lipids and proteins. Plants have developed many strategies to overcome oxidative stress to restore the redox homeostasis. One of the strategies is to establish symbiotic association in roots with arbuscular mycorrhizal (AM) fungi to improve host resistance to stress. Initial stages of AM fungus colonization trigger intracellular ROS burst in host plant; however, this effect is transient and is overcome by enhanced activities of antioxidant enzymes and molecules such as carotenoid. Accumulation of ROS in cortical cells has also been related to arbuscule digestion. Improvement of stress resistance has usually been associated with AM-induced escalation in P acquisition and plant growth. Nevertheless, non-nutritional effects of AM on host plants have attracted increasing attention. Under stress conditions, AM plants show reduced lipid peroxidation and lower levels of hydrogen peroxide and superoxide. The formation of AM reinforces the antioxidant defence system of the plant for the prevention of oxidative damage. AM symbiosis is capable of increasing activities of enzymes involved directly in removal of ROS such as superoxide dismutase, catalase (CAT) and ascorbate (ASH)- or thiol-dependent peroxidases (POX) and indirectly by generation of two redox molecules ascorbate and glutathione such as glutathione reductase (GR), dehydroascorbate reductase and monodehydroascorbate reductase. AM helps in augmenting the concentrations of non-enzymatic antioxidants such as α-tocopherol, proline, carotenoids, glutathione and ascorbic acid. This review summarizes current knowledge on the effects of AM symbiosis on the accumulation of ROS and correspondingly on the antioxidant defence system. New perspectives and challenges in molecular studies on oxidative stress alleviation by AM symbiosis are proposed.

Keywords Lipid peroxidation • Hydrogen peroxide • Superoxide dismutase • Catalase and ascorbate peroxidase • Proline • Tocopherol • Glutathione • Ascorbate

---

R. Kapoor (✉) • N. Singh
Department of Botany, University of Delhi, Delhi 110007, India
e-mail: kapoor_rupam@yahoo.com

Q.-S. Wu (ed.), *Arbuscular Mycorrhizas and Stress Tolerance of Plants*,
DOI 10.1007/978-981-10-4115-0_10

10.1 Introduction

Plants respond to all abiotic and biotic stresses by the accumulation of reactive oxygen species (ROS) (Singh et al. 2011). ROS are a family of molecules that include highly reactive free oxygen radicals such as superoxide anion [$\bullet O_2^-$] and the hydroxyl radical [•OH] and the steady 'diffusable' nonradical oxidants such as hydrogen peroxide [H_2O_2]. ROS are extremely lethal to biological cells causing oxidative injury to DNA, lipids and proteins. At low concentration, ROS can act as signalling molecules for stress responses (Singh et al. 2011). This biological paradox implies the existence of perfected mechanisms that are important for the integrity and fitness of living organisms. Generation and elimination of ROS in plants remain in dynamic balance under unstressed condition, but such balance of ROS is interrupted under stress condition, thereby inducing an elevation in ROS concentration (Bowler et al. 1994).

Plants have developed many ways to overcome oxidative stress. One way is to induce an adaptive response, consisting of a compensatory upregulation of antioxidant systems, targeted to reinstate the redox homeostasis (Lázaro et al. 2013; Garcia-Sanchez et al. 2014). Another strategy is to establish symbiotic association in roots—arbuscular mycorrhiza (AM)—to improve host resistance to stress (Smith and Read 2010; Miransari 2011). Plant antioxidant defence system includes low-molecular-weight compounds (glutathione, ascorbate (ASH) and α-tocopherol) and antioxidant enzymes. These enzymes participate in the exclusion of ROS either directly (superoxide dismutases, catalases (CAT) and ascorbate- or thiol-dependent peroxidases (POX)) or indirectly via the restoration of the two main redox molecules in the cell, ascorbate and glutathione (glutathione reductases (GR), dehydroascorbate reductases and monodehydroascorbate reductases) (Rouhier and Jacquot 2008; Singh et al. 2011).

Arbuscular mycorrhizal symbiosis is the most ubiquitous and ancient rhizospheric interaction (Parniske 2008). During the establishment of this mutualistic association, there is a continuous cellular and molecular dialogue between the symbionts (fungus and plant root) (Bonfante-Fasolo 1984; Hause and Fester 2005) that comprises the induction of the antioxidant (Garmendia et al. 2004), phenylpropanoid (Azcon-Aguilar et al. 2002) or carotenoid (Strack and Fester 2006) synthesis. The relationship between AM fungi and plant roots develops in two operative phases (Harley and Smith 1983), the extraradical phase spreading from the root into the soil and the cortical phase with intercellular hyphae, and specialized intracellular structures called 'arbuscules'. Arbuscules are the sites where exchange of carbon to the fungus and nutrients to the host plant takes place.

Arbuscular mycorrhiza can augment plant forbearances to abiotic and biotic stresses (Wu et al. 2006; Marulanda et al. 2007), such as heavy metals (Hetrick et al. 1994; Schützendübel and Polle 2002), salinity (Garg and Manchanda 2009a; Hajiboland et al. 2010; Abdel Latef and Chaoxing 2011; Evelin et al. 2012, 2013) and phytopathogens (Azcón-Aguilar and Barea 1996; Smith and Read 1997). AM improvement of stress resistance has usually been associated with AM-induced increase in P acquisition and plant growth. Nevertheless, non-nutritional effects of AM fungi on host plants have attracted increasing attention. The amelioration of stresses by AM has been attributed to a range of physiological, biochemical and molecular mechanisms. Several studies have shown that the formation of AM reinforces the antioxidant defence system of the plant for the prevention of oxidative damage (Ruiz-Lozano et al. 2001a, b; Schützendübel and Polle 2002; Lambais et al. 2003; Porcel et al. 2003; Ruiz-Lozano 2003; Wu et al. 2006). In this review, we describe the accumulation of ROS during formation of AM and its effect on ROS-scavenging antioxidant defence machinery with special reference to abiotic stress.

10.2 Oxidative Stress

10.2.1 Hydrogen Peroxide (H_2O_2)

H_2O_2 is one of the most abundant ROS, with a relatively high half-life, as compared with other ROS (Bárzana et al. 2015). At low concentration, H_2O_2 acts as a signalling molecule in defence responses and other physiological processes (Potters et al. 2010), but at high concentrations, it is harmful as other ROS (Quan et al. 2008; Zhang et al. 2013). H_2O_2 and superoxide have been the main emphases of ROS biology investigations in recent years. Since superoxide is rapidly converted to H_2O_2 in the cell, most studies concentrate on H_2O_2 as the principal ROS member. H_2O_2 is an important signalling compound implicated in the interactions of plants with pathogenic microorganisms (Apel Laloi and Hirt 2004; Laloi et al. 2004) as well as symbiotic organisms such as rhizobium (Matamoros et al. 2003) and many other processes referring to root biology, such as gravitropism (Joo et al. 2001), root elongation and root hair elongation (Foreman et al. 2003; Kwak et al. 2003; Liszkay et al. 2004).

10.2.1.1 H_2O_2 and AM Fungus Colonization

The AM symbiosis is the result of complicated and perfected signalling events. It involves complex morphological and physiological alterations of symbiotic partners, molecular recognition, response dialogue mechanism and the signalling molecules regulating transcription in mycorrhizal fungi and plant cells (Dong et al. 2004; Parniske 2008). The elicitor chitin secreted by AM fungal elicits the outburst of intracellular reactive oxygen species and the accumulation of salicylic acid and jasmonic acid, which trigger plant defence gene expression in the nucleus (Fig. 10.1a) (Song et al. 2011). The first evidence for accumulation of H_2O_2 in the course of the arbuscular mycorrhizal (AM) symbiosis was provided in roots of *Medicago truncatula* stained by diaminobenzidine (Salzer et al. 1999). Recently, Zhang et al. (2013) reported H_2O_2 accumulation in clover roots in response to the

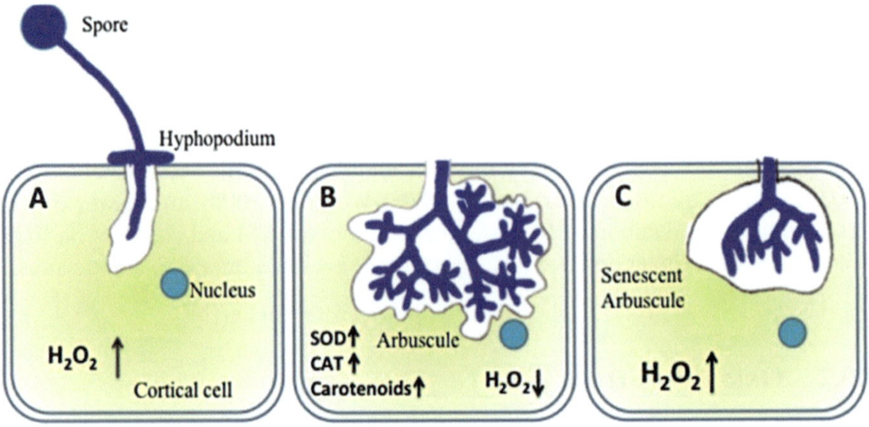

Fig. 10.1 (**a–c**) Reactive oxygen species accumulation in AM fungus-colonized root cells. (**a**) Initial stages of AM fungus colonization triggers intracellular ROS burst in host plant; (**b**) however, this effect is transient and is overcome by enhanced activities of antioxidant enzymes and molecules such as carotenoid. (**c**) Accumulation of ROS in cortical cells has also been related to clumped or less branched or senescent arbuscule

colonization by AM fungus that gradually decreased with plant growth (Fig. 10.1b). The accumulation of H_2O_2 has been observed in the plant cytoplasm close to fungal structure and in clumped or less branched or senescenced arbuscules and intercellular fungal hyphae in AM-colonized roots (Fig. 10.1c) (Salzer et al. 1999; Fester and Hause 2005), but never in cells with very branched arbuscules (Salzer et al. 1999).

Generation of H_2O_2 has also been observed in the early and late phases of the rhizobial interaction (Matamoros et al. 2003). The accumulation of ROS in later stage is associated with nodule senescence (Puppo et al. 2005). The correlation of H_2O_2 with arbuscular degradation rather suggests interesting similarities with root nodule senescence, which also involves the generation of H_2O_2. In rhizobial interaction, an increase in cytoplasmic H_2O_2 levels is triggered by endogenous or by plant-derived signals. Accumulation of H_2O_2 or a reduced activity of antioxidant systems induces the degradation of symbiotic structures and the programmed cell death of nodules (Puppo et al. 2005). Although plant cells are not dying upon arbuscular disintegration, H_2O_2 in AM roots presumed to be generated in a similar way. The H_2O_2 diffuses across the thin hyphal wall of arbuscular branches may initiate the fungal programme for senescence (Fester and Hause 2005). The accretion of ROS in the cytoplasm of arbuscule-containing cells might finally lead to arbuscular disintegration.

Some studies have reported suppression of defence signalling molecules in the early stages of symbiosis (Singh 2007) to evade plant defence gene activation against the AM symbiosis. In plants, synthesis of H_2O_2 is also known as an effective defence mechanism against microbial invasion (Singh 2007). Transient increases in catalase and peroxidase activities have been reported during the early stage of colonization of *G. intraradices* in bean root (Blee and Anderson 2000).

There are several features characteristic of AM roots that might be induced by the accumulating H_2O_2. One example is the stimulation of carotenoid biosynthesis in the late phase of arbuscular development (Bouvier et al. 1998; Fester et al. 2002; Hans et al. 2004). Another example is of AM plants against root pathogens and augmented tolerance to abiotic stress (Dumas-Gaudot et al. 2000; Linderman 2000). This represents cross-tolerance in AM plants due to the accumulation of ROS (Bowler and Fluhr 2000). Additional findings confirmed that cortical cells with arbuscule digestion have more intense defence response (Song et al. 2011). This may be due to lower antioxidant enzyme activities in the cells that weaken decomposition of H_2O_2 and consequently reinforce the plant defence response (Song et al. 2011).

Further, it has been suggested that enhanced ROS accumulation in AM roots includes increased concentration of jasmonates at later stages of the symbiosis (Hause et al. 2002) and the induction in activities of various antioxidative enzymes (Arines et al. 1994; Blilou et al. 2000; Lambais et al. 2003). Besides these studies, no molecular responses of plants or fungi to the accumulation of H_2O_2 in the AM symbiosis have been reported.

10.2.1.2 H_2O_2 and Stress

H_2O_2 signalling pathway is normally reported in relation to diverse biotic and abiotic stresses (Cheeseman 2007). Generally, H_2O_2 concentration in unstressed roots is much lower than that in stressed roots (Huang et al. 2008; Hajiboland et al. 2010; Ruiz-Sánchez et al. 2010; Garg and Bhandari 2012).

A large number of experiments have been conducted in stressed condition, and it is widely accepted that diminishing H_2O_2 levels is one of mechanisms by which AM protect plants against diverse stresses. Under normal (unstressed) conditions, however, results varied with individual studies. While most of the studies reported decreased H_2O_2 level in both stressed and unstressed roots in mycorrhizal plants (Hajiboland et al. 2010; Ruiz-Sánchez et al. 2010; Garg and Bhandari 2012), few demonstrate that AM increased H_2O_2 in unstressed roots but decreased in stressed roots (Huang et al. 2008). Under salinity stress, AM significantly decreased H_2O_2 by up to 40% in the roots of tomato plants (Hajiboland et al. 2010) and by up to 22% in the roots of pigeon pea plants in Cd-polluted soil (Garg and Bhandari 2012). Similarly, *G. intraradices* sharply decreased H_2O_2 by about 60% in the roots of rice plants (Ruiz-Sánchez et al. 2010). A lower oxidative burst in AM plants might be associated with intensive growth of arbuscules and hyphae within host roots (Salzer et al. 1999; Fester and Hause 2005).

Recently, Zou et al. (2015) showed that H_2O_2 is effusive in roots. Seedlings of trifoliate orange exhibited considerably higher net H_2O_2 effluxes in roots when colonized by *Rhizophagus intraradices*, especially in the root meristem zone. The H_2O_2 dispersion occurs over short distances, bypassing membrane transit (Cheeseman

2007). Hyphae of AM fungi are aseptate (Allen 2006); therefore, the hyphal symplast is continuous starting from the arbuscules in the cortical cells till the tips of extraradical hyphae. Extraradical hyphae of AM fungi possess functional aquaporins, which are membrane proteins involved in water uptake (Aroca et al. 2009; Li et al. 2013). AM fungal hyphae could provide a pathway for H_2O_2 effluxes from the root to the rhizosphere by specific transport of H_2O_2 through aquaporin channels (Bienert et al. 2007); further studies are needed to clarify the functioning of AM fungi in H_2O_2 effluxes.

10.2.2 Lipid Peroxidation

High concentration of ROS generated in cells as a consequence of stress can cause oxidative damage, such as lipid peroxidation, protein oxidation and DNA disintegration (Sharma et al. 2010, 2012). The mechanism of lipid peroxidation is explained in many reviews (Singh et al. 2011). The extent of lipid peroxidation is a reliable indicator of ROS production leading to oxidative stress. Several studies have shown that when subjected to stress, lipid peroxidation is lesser in mycorrhizal plants than the non-mycorrhizal plants indicating that AM fungal inoculation helps in reduction of oxidative damage (Evelin and Kapoor 2014; Tan et al. 2015; Yang et al. 2015; Jiang et al. 2016). This decrease in lipid peroxidation prevents cell membrane injury and maintains integrity and stability of the plasma membrane evident by decrease in electrolyte leakage in AM fungus-inoculated plants (Evelin et al. 2012; Garg and Chandel 2015). The lower leakage value of macromolecules, malondialdehyde content displayed as lower cell membrane damage, is often correlated with the enhanced enzymatic and non-enzymatic components of antioxidative system stimulated by AM fungal inoculation (Evelin and Kapoor 2014; Bárzana et al. 2015; Tan et al. 2015; Yang et al. 2015; Jiang et al. 2016).

10.3 AM and Antioxidant Defence Systems

10.3.1 Antioxidative Enzymes

10.3.1.1 Superoxide Dismutases, Catalases and Ascorbate Peroxidases

Superoxide dismutases (SODs) are metalloenzymes that are considered as the leading defence against ROS catalyzing dismutation reaction via conversion of superoxide radicals to oxygen and H_2O_2 (Fridovich 1995), which is then disproportioned to water by catalases (CAT) or peroxidases (POX) (Ruis and Koller 1997). In eukaryotes, there are diverse forms of SODs, containing Cu/Zn, Mn or Fe at their active sites and with distinctive cellular localizations. CuZnSOD is predominantly present in the cytoplasm, while MnSOD and FeSOD are primarily located in the mitochondria and

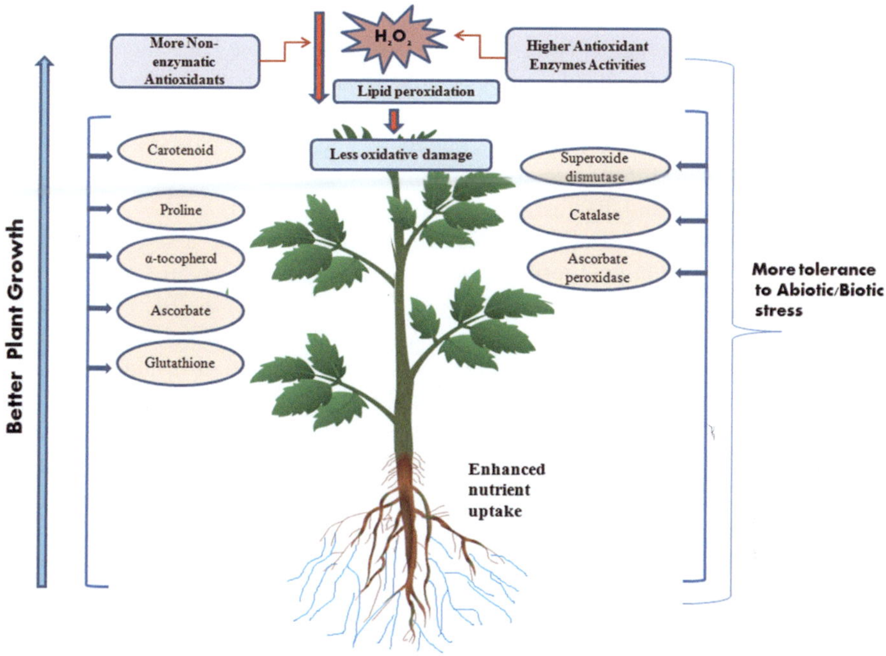

Fig. 10.2 AM plants showing reduced lipid peroxidation and lower levels of hydrogen peroxide and superoxide (Fig. 10.2). The formation of AM reinforces the antioxidant defence system of the plant for the prevention of oxidative damage. AM symbiosis is capable of increasing activities of enzymes involved directly in the removal of ROS such as superoxide dismutase, catalase and ascorbate peroxidase. AM helps in augmenting the concentrations of non-enzymatic antioxidants such as α-tocopherol, proline, carotenoids, glutathione and ascorbic acid—resulting in more tolerance to abiotic and biotic stresses

chloroplasts, respectively (Fridovich 1997). CuZnSOD plays an important role to safeguard against metabolically and externally produced ROS (Zyracka et al. 2005).

Catalases (CAT) are tetrameric heme-containing enzymes with the potential to directly dismutate H_2O_2 into H_2O and O_2 and are imperative for ROS clearing during stressed conditions (Garg and Manchanda 2009b). Ascorbate peroxidase (APX) is involved in ASH–GSH cycles, in conversion of H_2O_2 into water, and utilizes ASH as the electron donor. APX has a higher affinity for H_2O_2 than CAT and POD, and it may have a more important role in the management of ROS under stress conditions (Gill and Tuteja 2010).

AM-associated changes in antioxidant enzymes are widely reported (He et al. 2007; Ni et al. 2013; Huang et al. 2014). Effects of AM fungus colonization on antioxidant enzyme activities of host plants exposed to various abiotic stresses are comprehensively summarized by Wu et al. (2014). It is proposed that AM symbiosis can escalate antioxidative defence systems of plants through higher SOD activity in AM fungi-inoculated plants, accelerating dismutation of superoxide to H_2O_2 and subsequently impeding H_2O_2 buildup by higher activities of CAT, POX and APX (Fig. 10.2)

(Abdel Latef and He 2011; Evelin and Kapoor 2014; Garg and Chandel 2015; Liu et al. 2016). On the contrary, no significant correlations have been found between activities of SOD, CAT and H_2O_2 and $\bullet O_2^-$ concentration in some studies (Porcel and Ruiz-Lozano 2004; Azcón et al. 2009; Zou et al. 2015), suggesting that low oxidative levels in AM plants are not always due to the enhanced SOD and CAT activity. Thus, the contribution of the individual enzyme has been shown to differ with regard to the fungal species and the host plant (Ruiz-Lozano et al. 2012).

The inductive effect of AM symbiosis on activities of the above antioxidant enzymes may be the indirect result of the mycorrhizal effects on host plant growth and procurement of phosphorus and nitrogen (Ruiz-Lozano et al. 2012). CAT, APX and SOD are metalloenzymes whose activities can be determined by the availability of the metals they utilize. Increase in the absorption of slowly diffusing Zn and Cu that serve as cofactors for SODs in mycorrhizal plants results in their higher activity (Alguacil et al. 2003; Subramanian et al. 2011; Evelin and Kapoor 2014). Further, stress conditions such as salt stress result in decrease in Fe concentrations (Shim et al. 2003; Foyer and Noctor 2005) and provoke CAT degradation, resulting in direct functional or structural effects on CAT protein and prevention of synthesis of new enzyme (Feierabend and Engel 1986). It has been proposed that formation of AM overcomes this effect and hence prevents decline in CAT activity (Evelin and Kapoor 2014). The AM fungi species-specific effect on individual antioxidative enzyme may also depend on the micronutrients available to some of the enzymes (Ruiz-Lozano et al. 2012). In addition, Lambais et al. (2003) proposed CAT activity to be an important regulating factor in intraradical fungal growth and mycorrhizal development. The induction of SOD on AM formation may be a defence response for quenching ROS. AM fungal root colonization can activate the expression of some plant SOD genes under abiotic stress (Palma et al. 1993; Li et al. 2012).

10.3.1.2 Ascorbate Peroxidase, Monodehydroascorbate Reductase, Dehydroascorbate Reductase, and Glutathione Reductase

AM-induced stress tolerance is also associated with efficient neutralization of H_2O_2 by regulating the ascorbate–glutathione cycle (Liu et al. 2016). In the ascorbate–glutathione cycle, ascorbate peroxidase (APX) utilizes AsA as an electron donor for reduction of H_2O_2, monodehydroascorbate is reduced to AsA by monodehydroascorbatereductase (MDHAR), and dehydroascorbate is reduced to AsA by dehydroascorbatereductase (DHAR) (Ordoñez et al. 2014). Glutathione reductase (GR) is a flavoprotein oxidoreductase and a potential enzyme of the ASH–GSH cycle. GR catalyzes the reduction of GSH, a molecule involved in several regulatory metabolic and antioxidative activities in plants where GR catalyzes the NADPH-dependent reaction of disulphide bond of GSSG and is thus important for upholding the GSH pool (Reddy and Raghavendra 2006; Rao and Reddy 2008).

AM significantly promoted APX, GR and MDHAR activities as compared to control during different abiotic stresses in different studies (Alqarawi et al. 2014; Kumar et al. 2014; Evelin and Kapoor 2014; Garg and Chandel 2015; Liu et al. 2016). Further, Liu et al. (2016) observed that the expression levels of *APX*,

MDHAR, *GR* and *DHAR* genes in the AM plants were higher compared with non-mycorrhizal in low temperature.

10.3.2 Antioxidant Molecules

10.3.2.1 Glutathione and Ascorbate

Glutathione is a cysteine-containing tripeptide found amply in all cell compartments in its reduced form (Foyer and Noctor 2005). The reduced form (GSH) in combination with its oxidized state disulphide glutathione (GSSG) is required to maintain cellular redox system in plants (Gill and Tuteja 2010). Stress conditions often result in depletion of GSH pool, and redox state becomes more oxidized, leading to deterioration of the system. GSH is crucial to preserve the normal reduced state of cells so as to neutralize the adverse effects of ROS-induced oxidative stress (Meyer 2008; Noctor et al. 2012). It is a potential scavenger of $\bullet O_2^-$, H_2O_2 (Briviba et al. 1997) and most dangerous ROS like •OH (Larson 1998). Additionally, GSH plays a crucial role in the antioxidative defence system by regenerating additional water-soluble antioxidant like ascorbate (ASH), via the ASH–GSH cycle (Foyer and Halliwell 1976). Ascorbate is an important non-enzymatic antioxidant compound involved in the removal of H_2O_2 by ascorbate peroxidases that use ascorbate as electron donor (Foyer and Noctor 2011).

Liu et al. (2016) observed that the contents of AsA and GSH and the redox ratio of ascorbate (AsA/DHA) and glutathione (GSH/GSSG) were all increased in the AM fungus-colonized roots compared with non-inoculated tomato plants (Fig. 10.2). AM fungus inoculation under chilling stress almost preserved the content of AsA and GSH to a level near to the control. More GSH in mycorrhizal plants enables them (1) to directly scavenge more $\bullet O_2^-$ and H_2O_2 as well as other ROS-like hydroxyl radicals and (2) to reproduce more ascorbic acid via the ascorbate–glutathione cycle compared to NM plants (Evelin and Kapoor 2014). Higher level of ascorbate was observed in maize (Bárzana et al. 2015), citrus (Wu and Zou 2009) and rice (Ruiz-Sánchez et al. 2011). Garg and Chandel (2015) ascertained the role of AM in scavenging ROS and reinforcement of antioxidant defence in *Cajanus cajan* nodules under salinity (NaCl) and cadmium (Cd) stress. They proposed that higher redox ratio of glutathione in mycorrhizal plants served as an important defence line against ROS to protect nodules from oxidative stress damage. Moreover, mycorrhizal symbiosis remarkably enhanced GR activity that restored the depleted reduced glutathione (GSH) pool by concomitant reduction of GSSG under NaCl and Cd stress. Though contributions of AM in sustaining high levels of ascorbate and glutathione have been realized, however, the mechanisms responsible for this effect are largely unknown. More experiments are required to analyse the transcription levels of enzymes involved in the biosynthesis of ascorbate and glutathione. For this, the genes encoding enzymes GDP-D-mannose 3,5-epimerase (GME), vital in the biosynthetic pathway of ascorbic acid (Watanabe et al. 2006), and cysteine synthase (CS), central in the biosynthesis of cysteine and glutathione (Harada et al. 2001; Choi et al. 2007), are proposed.

10.3.2.2 Proline

In addition to being an osmolyte, proline is considered a potent antioxidant and a potential inhibitor of programmed cell death. Free proline has been proposed to act as an osmoprotectant, a protein stabilizer, a metal chelator, an inhibitor of lipid peroxidation and •OH and $\bullet O_2^-$ scavenger (Ashraf and Foolad 2007; Trovato et al. 2008). High concentration of proline has been correlated with improved tolerance to various abiotic stresses especially salt and drought. More synthesis of proline under drought or salt stress has been implicated as a mechanism to alleviate cytoplasmic acidosis and maintain $NADP^+$/NADPH at values compatible with metabolism (Hare and Cress 1997). An additional advantage of the refilling of $NADP^+$ supply by proline synthesis is to assist redox cycling, which is especially important in plant antioxidant defence mechanisms during stress (Babiychuk et al. 1995).

Several studies have reported more accumulation of proline under abiotic stress in AM plants (Sharifi et al. 2007; Garg and Manchanda 2009a). Garg and Baher (2013) observed that AM fungi stimulated the salinity-induced buildup of proline by restraining proline dehydrogenase in AM plants. Additionally, the proline biosynthetic enzymes pyrroline-5-carboxylate synthetase and glutamate dehydrogenase activities increased in stressed mycorrhizal plants. Further, Abo-Doma et al. (2011) indicated that mycorrhiza increased transcript expression of pyrroline-5-carboxylate synthetase. In contrast there are reports on decline in proline concentration in mycorrhizal plants (Wu and Xia 2006; Manoharan et al. 2010; Doubková et al. 2013; Zou et al. 2013). This may be explained by the fact that proline is also an indicator of level of stress (Wang et al. 2004). Therefore, lower proline concentration in AM plants observed in these study could be due to reduced stress on plants (Evelin et al. 2013; Zou et al. 2013).

10.3.2.3 Carotenoids

Some isoprenoids (such as some carotenoids and α-tocopherols) play an effective role in photoprotection (Peñuelas and Munné-Bosch 2005). Furthermore, it has been proved that monoterpene have a protecting role against oxidative stress (Loreto et al. 2004). Carotenoids are pigments that are found in plants and microorganisms. Carotenoid a lipid soluble antioxidant plays a multitude of functions in plant metabolism including oxidative stress tolerance. They protect the photosynthetic apparatus by reducing a triplet sensitizer (Chl_3), $^1O_2$ and other harmful free radicals that are naturally formed during photosynthesis (an antioxidant function) (Collins 2001). Activation of carotenoid biosynthesis in AM roots is a general phenomenon (Fester et al. 2002).

10.3.2.4 α-Tocopherols

Tocopherols, lipid soluble antioxidants are considered as potential scavengers for protection of membrane stability, including quenching or scavenging ROS (Hollander-Czytko et al. 2005). Tocopherols are present in plants in the thylakoid membrane of chloroplasts. Among the four isomers of tocopherols synthesized in plants, α-tocopherol has the maximum antioxidative activity due to the existence of three methyl groups in its molecular structure (Serbinonva and Packer 1994). The α-tocopherol disrupts the chain propagation step in lipid autoxidation (Serbinonva and Packer 1994). Thus, lower lipid peroxidation in AM plants can be explained by higher α-tocopherol in them as they prevent peroxidation of membrane lipids by disrupting chain propagation step in lipid autoxidation (Evelin et al. 2013). It was observed that AM fungus colonization resulted in increase in number and size of plastoglobules (Evelin et al. 2013). Plastoglobules are lipoprotein bodies generated in plastids and are sites for synthesis of tocopherols under oxidative stress (Brehelin et al. 2007). The reduced damage in membrane system of the cell in AM plants may be due to higher number of plastoglobules that averted oxidative damage of the membrane. Further, more α-tocopherol in AM plants may be due to increased production from tocopheroxyl radicals, enabled by the higher concentration of ascorbic acid in them (Thomas et al. 1992; Noctor and Foyer 1998).

The influence of AM symbiosis on the levels of non-enzymatic antioxidants such as ascorbic acid, glutathione, carotenoids or α-tocopherols in the host plant has been very less reviewed. Therefore, this aspect requires an in-depth investigation.

10.4 Antioxidation Defence Arsenal in AM Fungi

There is evidence that antioxidant systems, including SODs, from the plant and/or the fungus play a role in symbiosis (Pauly et al. 2006; Takemoto et al. 2007; Abbá et al. 2009); however, far less is known about the contribution of each partner. Very few attempts have been made to reveal the existence of different enzymes in AM fungi that may be implicated in cellular defence against oxidative stress. This includes genes encoding SOD (Lanfranco et al. 2005; González-Guerrero et al. 2010), glutaredoxins (GRXs), that are small proteins with glutathione-dependent disulphide oxidoreductase activity (Benabdellah et al. 2009b), pyridoxine 5′-phosphate (PNP) synthase that is involved in the biosynthesis of vitamin B_6, a vital metabolite for defence against cellular oxidative stress (Benabdellah et al. 2009a) and a metallothionein (MT)-encoding gene that was proposed to play a role in regulation of the redox status of the extraradical mycelium of *Rhizophagus intraradices* (González-Guerrero et al. 2007). Germinated spores of the AM fungus *Gigaspora margarita* possess specific SOD genes to tolerate oxidative stress (Lanfranco et al. 2005; Corradi et al. 2009). As the peak transcription levels are perceived in intraradical fungal structures, it has been mooted that CuZnSOD might be a crucial component in the plant/fungus dialogue crucial to reach structural and functional harmony between the symbiotic partners. Gene encoding a CuZnSOD in *R. intraradices* has been characterized to postulate its role in ROS homeostasis in AM (González-Guerrero et al. 2010). Further analysis of transcriptomic data of *R. intraradices* provided by Tisserant et al. (2012) would reveal more enzymes in the AM fungi implicated in scavenging ROS (Ruiz-Lozano et al. 2012).

10.5 Conclusion

Elucidating the mechanisms that augment antioxidant capacity of AM plants during abiotic stress could provide a powerful strategy to improve the tolerance of crops to these environmental stress conditions. There is a need to discover efficient AM fungal strains for abiotic stress tolerance through assessing the species found in the nature (different stressed habitats), because it has been assumed that the ecological roles of these fungi are not fully understood. Most studies on AM are performed at experimental phase; field experiments and trials must be promoted to evaluate the efficacy of the symbiosis under natural conditions as, eventually, both the host and fungal symbionts have to deal with the natural environment and survive.

References

Abbá S, Khouja HR, Martino E et al (2009) SOD1-targeted gene disruption in the ericoid mycorrhizal fungus *Oidiodendron maius* reduces conidiation and the capacity for mycorrhization. Mol Plant-Microbe Interact 22(11):1412–1421

Abdel Latef AAH, Chaoxing H (2011) Effect of arbuscular mycorrhizal fungi on growth, mineral nutrition, antioxidant enzymes activity and fruit yield of tomato grown under salinity stress. Sci Hortic 127:228–233

Abdel Latef AAH, He CX (2011) Arbuscular mycorrhizal influence on growth, photosynthetic pigments, osmotic adjustment and oxidative stress in tomato plants subjected to low temperature stress. Acta Physiol Plant 33(4):1217–1225

Abo-Doma A, Edrees S, Abdel-Aziz SH (2011) The effect of mycorrhiza growth and expression of some genes in barley. Egypt J Genet Cytol 40:301–313

Alguacil MM, Herna´ndez JA, Caravaca F et al (2003) Antioxidant enzyme activities in shoots from three mycorrhizal shrub species afforested in a degraded semi-arid soil. Physiol Plant 118(4):562–570

Allen MF (2006) Water dynamics of mycorrhiza in arid soils. In: Gadd GM (ed) Fungi in biogeochemical cycles. Cambridge University Press, London, pp 74–97

Alqarawi AA, Abd Allah EF, Hashem A (2014) Alleviation of salt-induced adverse impact via mycorrhizal fungi in *Ephedra aphylla* Forssk. J Plant Interact 9(1):802–810

Apel Laloi K, Hirt H (2004) Reactive oxygen species: metabolism, oxidative stress, and signal transduction. Annu Rev Plant Biol 55:373–399

Arines J, Quintela M, Vilarino A et al (1994) Protein patterns and superoxide dismutase activity in non-mycorrhizal and arbuscular mycorrhizal *Pisum sativum* L. plants. Plant Soil 166(1):37–45

Aroca R, Bago A, Sutka M et al (2009) Expression analysis of the first arbuscular mycorrhizal fungi aquaporin described reveals concerted gene expression between salt-stressed and non-stressed mycelium. Mol Plant-Microbe Interact 22(9):1169–1178
Ashraf M, Foolad MR (2007) Roles of glycine betaine and proline in improving plant abiotic stress resistance. Environ Exp Bot 59(2):206–216
Azcón-Aguilar C, Barea JM (1996) Arbuscular mycorrhizas and biological control of soil-borne plant pathogens – an overview of the mechanisms involved. Mycorrhiza 6:457–464. doi:10.1007/s005720050147
Azcón-Aguilar C, Jaizme-Vega MC, Calvet C (2002) The contribution of arbuscular mycorrhizal fungi for bioremediation. In: Gianinazzi S, Schuepp H, Barea JM, Haselwandter K (eds) Mycorrhizal technology in agriculture, From genes to bioproducts. Birkhauser Verlag, Berlin, pp 187–197
Azcón R, Pera´lvarez MC, Biró B et al (2009) Antioxidant activities and metal acquisition in mycorrhizal plants growing in a heavy-metal multi-contaminated soil amended with treated lignocellulosic agrowaste. Appl Soil Ecol 41(2):168–177
Babiychuk E, Kushnir S, Belles-Boix E et al (1995) *Arabidopsis thaliana* NADPH oxidoreductase homologs confer tolerance of yeasts toward the thiol-oxidizing drug diamide. J Biol Chem 270(3):26224–26231
Bárzana G, Aroca R, Ruiz-Lozano JM (2015) Localized and non-localized effects of arbuscular mycorrhizal symbiosis on accumulation of osmolytes and aquaporins and on antioxidant systems in maize plants subjected to total or partial root drying. Plant Cell Environ 38:1613–1627
Benabdellah K, Azcón-Aguilar C, Valderas A et al (2009a) GintPDX1 encodes a protein involved in vitamin B6 biosynthesis that is up-regulated by oxidative stress in the arbuscular mycorrhizal fungus *Glomus intraradices*. New Phytol 184(3):682–693
Benabdellah K, Merlos MA, Azcón-Aguilar C et al (2009b) GintGRX1, the first characterized glomeromycotan glutaredoxin, is a multifunctional enzyme that responds to oxidative stress. Fungal Genet Biol 46(1):94–103
Bienert GP, Møller ALB, Kristiansen KA et al (2007) Specific aquaporins facilitate the diffusion of hydrogen peroxide across membranes. J Biol Chem 282(2):1183–1192
Blee KA, Anderson AJ (2000) Defense responses in plants to arbuscular mycorrhizal fungi. In: Podila GK, Douds DD (eds) Current advances in mycorrhizae research. American Phytopathological Society Press, St. Paul, pp 27–44
Blilou I, Bueno P, Ocampo JA et al (2000) Induction of catalase and ascorbate peroxidase activities in tobacco roots inoculated with the arbuscular mycorrhizal fungus *Glomus mosseae*. Mycol Res 104(6):722–725
Bonfante-Fasolo P (1984) Anatomy and morphology of VA mycorrhizae. In: Powell CL, Bagyaraj DJ (eds) VA mycorrhiza. CRC Press, Boca Raton, pp 35–46
Bouvier F, Backhaus RA, Camara B (1998) Induction and control of chromoplast-specific carotenoid genes by oxidative stress. J Biol Chem 273:30651–30659
Bowler C, Fluhr R (2000) The role of calcium and activated oxygen as signals for controlling cross-tolerance. Trends Plant Sci 5(6):241–246
Bowler C, Van Camp W, Van Montagu M et al (1994) Superoxide dismutase in plants. Crit Rev Plant Sci 13:199–218. doi:10.1080/07352689409701914
Brehelin C, Kessler F, van Wijk KJ (2007) Plastoglobules: versatile lipoprotein particles in plastids. Trends Plant Sci 12(6):260–266
Briviba K, Klotz LO, Sies H (1997) Toxic and signaling effects of photochemically or chemically generated singlet oxygen in biological systems. Biol Chem 378(11):1259–1265
Cheeseman JM (2007) Hydrogen peroxide and plant stress: a challenging relationship. Plant Stress 1(1):4–15
Choi YE, Kwom KW, Lee JC et al (2007) Expression of the rice cytoplasmic cysteine synthase gene in tobacco reduces ozone-induced damage. Plant Biotechnol Rep 1(2):93–100
Collins A (2001) Carotenoids and genomic stability. Mutat Res 475(1–2):1–28

Corradi N, Buffner B, Croll D et al (2009) High-level molecular diversity of copper-zinc superoxide dismutase genes among and within species of arbuscular mycorrhizal fungi. Appl Environ Microbiol 75(7):1970–1978

Dong CJ, Zhou Y, Zhao B (2004) Effect of flavonoids on AM fungi and mycorrhizal plant. Mycosystema 23(2):294–300

Doubková P, Vlasáková E, Sudová R (2013) Arbuscular mycorrhizal symbiosis alleviates drought stress imposed on *Knautia arvensis* plants in serpentine soil. Plant Soil 370(1):149–161

Dumas-Gaudot E, Gollotte A, Cordier C et al (2000) Modulation of host defence systems. In: Kapulnik Y, Douds DD (eds) Arbuscular mycorrhizas: physiology and function. Kluwer, Dordrecht, pp 173–200

Evelin H, Kapoor R (2014) Arbuscular mycorrhizal symbiosis modulates antioxidant response in salt-stressed *Trigonella foenum-graecum* plants. Mycorrhiza 24(3):197–208

Evelin H, Giri B, Kapoor R (2012) Contribution of *Glomus intraradices* inoculation to nutrient acquisition and mitigation of ionic imbalance in NaCl-stressed *Trigonella foenum-graecum*. Mycorrhiza 22(3):203–217

Evelin H, Giri B, Kapoor R (2013) Ultrastructural evidence for AMF mediated salt stress mitigation in *Trigonella foenum-graecum*. Mycorrhiza 23(1):71–86

Feierabend J, Engel S (1986) Photoinactivation of catalase in vitro and in leaves. Arch Biochem Biophys 251(2):567–576

Fester T, Hause T (2005) Accumulation of reactive oxygen species in arbuscular mycorrhizal roots. Mycorrhiza 15(5):373–379

Fester T, Hause B, Schmidt D et al (2002) Occurrence and localization of apocarotenoids in arbuscular mycorrhizal plant roots. Plant Cell Physiol 43(3):256–265

Foreman J, Demidchik V, Bothwell JH et al (2003) Reactive oxygen species produced by NADPH oxidase regulate plant cell growth. Nature 422(6930):442–446

Foyer CH, Halliwell B (1976) The presence of glutathione and glutathione reductase in chloroplasts: a proposed role in ascorbic acid metabolism. Planta 133(1):21–25

Foyer CH, Noctor G (2005) Redox homeostasis and antioxidant signaling: a metabolic interface between stress perception and physiological responses. Plant Cell 17(7):1866–1875

Foyer CH, Noctor G (2011) Ascorbate and glutathione: the heart of the redox hub. Plant Physiol 155(1):2–18

Fridovich I (1995) Superoxide radical and superoxide dismutases. Annu Rev Biochem 64:97–112

Fridovich I (1997) Superoxide anion radical superoxide dismutases and related matters. J Biol Chem 272(30):18515–18517

Garcia-Sanchez M, Palma JM, Ocampo JA et al (2014) Arbuscular mycorrhizal fungi alleviate oxidative stress induced by ADOR and enhance antioxidant responses of tomato plants. J Plant Physiol 171(6):421–428

Garg N, Baher N (2013) Role of arbuscular mycorrhizal symbiosis in proline biosynthesis and metabolism of *Cicer arietinum* (L.) (Chickpea) genotypes under salt stress. J Plant Growth Regul 32(4):767–778

Garg N, Bhandari P (2012) Influence of cadmium stress and arbuscular mycorrhizal fungi on nodule senescence in *Cajanus cajan* (L.) Millsp. Int J Phytoremediation 14(1):62–74

Garg N, Chandel S (2015) Role of arbuscular mycorrhiza in arresting reactive oxygen species (ROS) and strengthening antioxidant defense in *Cajanus cajan* (L.) Millsp. nodules under salinity (NaCl) and cadmium (Cd) stress. Plant Growth Regul 75(2):521–534

Garg N, Manchanda G (2009a) Role of arbuscular mycorrhizae in the alleviation of ionic, osmotic and oxidative stresses induced by salinity in *Cajanus cajan* (L.) Millsp. J Agron Crop Sci 195(2):110–123

Garg N, Manchanda G (2009b) ROS generation in plants: boon or bane? Plant Biosyst 143:8–96

Garmendia I, Goicoechea N, Aguirreolea J (2004) Antioxidant metabolism in asymptomatic leaves of *Verticillium*-infected pepper associated with an arbuscular mycorrhizal fungus. J Phytopathol 152(11–12):593–599

Gill SS, Tuteja N (2010) Reactive oxygen species and antioxidant machinery in abiotic stress tolerance in crop plants. Plant Physiol Biochem 48(12):909–930
González-Guerrero M, Cano C, Azcón-Aguilar C et al (2007) GintMT1 encodes a functional metallothionein in *Glomus intraradices* that responds to oxidative stress. Mycorrhiza 17:327–335
González-Guerrero M, Oger E, Benabdellah K et al (2010) Characterization of a CuZn superoxide dismutase gene in the arbuscular mycorrhizal fungus *Glomus intraradices*. Curr Genet 56(3):265–274
Hajiboland R, Aliasgharzadeh N, Laiegh SF et al (2010) Colonization with arbuscular mycorrhizal fungi improves salinity tolerance of tomato (*Solanum lycopersicum* L.) plants. Plant Soil 331(1):313–327
Hans J, Hause B, Strack D et al (2004) Cloning, characterization, and immunolocalization of a mycorrhiza-inducible 1-deoxy-D-xylulose 5-phosphate reductoisomerase in arbuscule-containing cells of maize. Plant Physiol 134(2):614–624
Harada E, Choi YE, Tsuchisaka A et al (2001) Transgenic tobacco plants expressing a rice cysteine synthase gene are tolerant to toxic levels of cadmium. J Plant Physiol 158:655–661
Hare PD, Cress WA (1997) Metabolic implications of stress-induced proline accumulation in plants. Plant Growth Regul 21(2):79–102
Harley JL, Smith SE (1983) Mycorrhizal symbiosis. Academic, London
Hause B, Fester T (2005) Molecular and cell biology of arbuscular mycorrhizal symbiosis. Planta 221(2):184–196
Hause B, Maier W, Miersch O et al (2002) Induction of jasmonate biosynthesis in arbuscular mycorrhizal barley roots. Plant Physiol 130(3):1213–1220
He ZQ, He CX, Zhang ZB et al (2007) Changes of antioxidative enzymes and cell membrane osmosis in tomato colonized by arbuscular mycorrhizae under NaCl stress. Colloid Surf B 59(2):128–133
Hetrick BA, Wilson GW, Figge DA (1994) The influence of mycorrhizal symbiosis and fertilizer amendments on establishment of vegetation in heavy metal mine spoil. Environ Pollut 86:171–179. doi:10.1016/0269-7491(94)90188-0
Hollander-Czytko H, Grabowski J, Sandorf I et al (2005) Tocopherol content and activities of tyrosine aminotransferase and cystine lyase in *Arabidopsis* under stress conditions. J Plant Physiol 162(7):767–770
Huang LL, Yang C, Zhao Y et al (2008) Antioxidant defenses of mycorrhizal fungus infection against SO_2-induced oxidative stress in *Avena nuda* seedlings. Bull Environ Contam Toxicol 81:440–444
Huang YM, Srivastava AK, Zou YN et al (2014) Mycorrhizal-induced calmodulin mediated changes in antioxidant enzymes and growth response of drought-stressed trifoliate orange. Front Microbiol 5:682. doi:10.3389/fmicb.2014.00682
Jiang QY, Zhuo F, Long SH et al (2016) Can arbuscular mycorrhizal fungi reduce Cd uptake and alleviate Cd toxicity of *Lonicera japonica* grown in Cd added soils? Sci Rep 6:21805
Joo JH, Bae YS, Lee JS (2001) Role of auxin-induced reactive oxygen species in root gravitropism. Plant Physiol 126(3):1055–1060
Kumar A, Dames JF, Gupta A et al (2014) Current developments in arbuscular mycorrhizal fungi research and its role in salinity stress alleviation: a biotechnological perspective. Crit Rev Biotechnol. doi:10.3109/07388551.2014.899964
Kwak JM, Mori IC, Pei ZM et al (2003) NADPH oxidase AtrbohD and AtrbohF genes function in ROS-dependent ABA signaling in Arabidopsis. EMBO J 22(11):2623–2633
Laloi C, Apel K, Danon A (2004) Reactive oxygen signalling: the latest news. Curr Opin Plant Biol 7(3):323–328
Lambais MR, Ríos-Ruiz WF, Andrade RM (2003) Antioxidant responses in bean (*Phaseolus vulgaris*) roots colonized by arbuscular mycorrhizal fungi. New Phytol 160(2):421–428

Lanfranco L, Novero M, Bonfante P (2005) The mycorrhizal fungus *Gigaspora margarita* possesses a CuZn superoxide dismutase that is up-regulated during symbiosis with legume hosts. Plant Physiol 137(4):1319–1330

Larson A (1998) The antioxidants of higher plants. Phytochemistry 27:969–978

Lázaro JJ, Jiménez A, Camejo D et al (2013) Dissecting the integrative antioxidant and redox systems in plant mitochondria. Effect of stress and S-nitrosylation. Front Plant Sci 4:460

Li T, Liu RJ, He XH et al (2012) Enhancement of superoxide dismutase and catalase activities and salt tolerance of euhalophyte *Suaeda salsa* L. by mycorrhizal fungus *Glomus mosseae*. Pedosphere 22(2):217–224

Li T, Hu YJ, Hao ZP et al (2013) First cloning and characterization of two functional aquaporin genes from an arbuscular mycorrhizal fungus *Glomus intraradices*. New Phytol 197(2):617–630

Linderman RG (2000) Effects of mycorrhizas on plant tolerances to diseases. In: Kapulnik Y, Douds DD (eds) Arbuscular mycorrhizas: physiology and function. Kluwer, Dordrecht, pp 345–365

Liszkay A, van der Zalm E, Schopfer P (2004) Production of reactive oxygen intermediates (O_2^-, H_2O_2, and ·OH) by maize roots and their role in wall loosening and elongation growth. Plant Physiol 136(2):3114–3123

Liu A, Chen S, Wang M et al (2016) Arbuscular mycorrhizal fungus alleviates chilling stress by boosting redox poise and antioxidant potential of tomato seedlings. J Plant Growth Regul 35(1):109–120

Loreto F, Pinelli P, Manes F et al (2004) Impact of ozone on monoterpene emissions and evidence for an isoprene-like antioxidant action of mono-terpenes emitted by *Quercus ilex* leaves. Tree Physiol 24:361–367

Manoharan PT, Shanmugaiah V, Balasubramanian N et al (2010) Influence of AM fungi on the growth and physiological status of *Erythrina variegata* L. grown under different water stress conditions. Eur J Soil Biol 46:151–156

Marulanda A, Porcel R, Barea JM et al (2007) Drought tolerance and antioxidant activities in lavender plants colonized by native drought-tolerant or drought-sensitive *Glomus* species. Microb Ecol 54:543–552. doi:10.1007/s00248-007-9237-y

Matamoros MA, Dalton DA, Ramos J et al (2003) Biochemistry and molecular biology of antioxidants in the rhizobia–legume symbiosis. Plant Physiol 133(2):499–509

Meyer AJ (2008) The integration of glutathione homeostasis and redox signaling. J Plant Physiol 165(13):1390–1403

Miransari M (2011) Hyperaccumulators, arbuscular mycorrhizal fungi and stress of heavy metals. Biotechnol Adv 29(6):645–653

Ni QD, Zou YN, Wu QS et al (2013) Increased tolerance of citrus (*Citrus tangerina*) seedlings to soil water deficit after mycorrhizal inoculation: changes in antioxidant enzyme defense system. Not Bot Horti Agrobot 41(2):524–529

Noctor G, Foyer CH (1998) Ascorbate and glutathione: keeping active oxygen under control. Annu Rev Plant Physiol Plant Mol Biol 49:249–279

Noctor G, Mhamdi A, Chaouch S et al (2012) Glutathione in plants: an integrated overview. Plant Cell Environ 35(2):454–484

Ordoñez NM, Marondedze C, Thomas L et al (2014) Cyclic mononucleotides modulate potassium and calcium flux responses to H_2O_2 in Arabidopsis roots. FEBS Lett 588(6):1008–1015

Palma JM, Longa MA, Arines J et al (1993) Superoxide dismutase in vesicular arbuscular mycorrhizal red clover plants. Physiol Plant 87(1):77–83

Parniske M (2008) Arbuscular mycorrhiza: the mother of plant root endosymbiosis. Nat Rev Microbiol 6:763–775

Pauly N, Pucciariello C, Mandon K et al (2006) Reactive oxygen and nitrogen species and glutathione: key players in the legume-Rhizobium symbiosis. J Exp Bot 57(8):1769–1776

Peñuelas J, Munné-Bosch S (2005) Isoprenoids: an evolutionary pool for photo-protection. Trends Plant Sci 10(4):166–169

Porcel R, Ruiz-Lozano JM (2004) Arbuscular mycorrhizal influence on leaf water potential, solute accumulation, and oxidative stress in soybean plants subjected to drought stress. J Exp Bot 55(403):1743–1750

Porcel R, Barea JM, Ruiz-Lozano JM (2003) Antioxidant activities in mycorrhizal soybean plants under drought stress and their possible relationship to the process of nodule senescence. New Phytol 157:135–143. doi:10.1046/j.1469-8137.2003.00658.x

Potters G, Horemans N, Jansen MAK (2010) The cellular redox state in plant stress biology – a charging concept. Plant Physiol Biochem 48(5):292–300

Puppo A, Groten K, Bastian F et al (2005) Legume nodule senescence: roles for redox and hormone signalling in the orchestration of the natural aging process. New Phytol 165(3):683–701

Quan LJ, Zhang B, Shi WW et al (2008) Hydrogen peroxide in plants: a versatile molecule of the reactive oxygen species network. J Integr Plant Biol 50(1):2–18

Rao ASVC, Reddy AR (2008) Glutathione reductase: a putative redox regulatory system in plant cells. In: Khan NA, Singh S, Umar S (eds) Sulfur assimilation and abiotic stresses in plants. Springer, Dordrecht, pp 111–147

Reddy AR, Raghavendra AS (2006) Photooxidative stress. In: Madhava Rao KV, Raghavendra AS, Reddy KJ (eds) Physiology and molecular biology of stress tolerance in plants. Springer, Dordrecht, pp 157–186

Rouhier N, Jacquot JP (2008) Getting sick may help plants overcome abiotic stress. New Phytol 180(4):738–741

Ruis H, Koller F (1997) Biochemistry, molecular biology, and cell biology of yeast and fungal catalases. In: Scandalios JG (ed) Oxidative stress and the molecular biology of antioxidant defenses. Cold Spring Harbor Laboratory Press, Cold Spring Harbor, pp 309–342

Ruiz-Lozano JM (2003) Arbuscular mycorrhizal symbiosis and alleviation of osmotic stress, new perspectives for molecular studies. Mycorrhiza 13:309–317. doi:10.1007/s00572-003-0237-6

Ruiz-Lozano JM, Collados C, Barea JM et al (2001a) Cloning of cDNAs encoding SODs from lettuce plants which show differential regulation by arbuscular mycorrhizal symbiosis and by drought stress. J Exp Bot 52(364):2241–2242

Ruiz-Lozano JM, Collados C, Barea JM et al (2001b) Arbuscular mycorrhizal symbiosis can alleviate drought-induced nodule senescence in soybean plants. New Phytol 151(2):493–502. doi:10.1046/j.0028-646x.2001.00196.x

Ruiz-Lozano JM, Porcel R, Azcón C et al (2012) Regulation by arbuscular mycorrhizae of the integrated physiological response to salinity in plants: new challenges in physiological and molecular studies. J Exp Bot 63(11):4033–4044

Ruiz-Sánchez M, Aroca R, Mun˜oz Y et al (2010) The Arbuscular mycorrhizal symbiosis enhances the photosynthetic efficiency and the antioxidative response of rice plants subjected to drought stress. J Plant Physiol 167(11):862–869

Ruiz-Sánchez M, Armada E, Munoz Y et al (2011) *Azospirillum* and arbuscular mycorrhizal colonization enhanced rice growth and physiological traits under well-watered and drought conditions. J Plant Physiol 168(10):1031–1037

Salzer P, Corbiere H, Boller T (1999) Hydrogen peroxide accumulation in *Medicago truncatula* roots colonized by the arbuscular mycorrhiza forming fungus *Glomus intraradices*. Planta 208(3):319–325

Schützendübel A, Polle A (2002) Plant responses to abiotic stresses: heavy metal-induced oxidative stress and protection by mycorrhization. J Exp Bot 53(372):1351–1365. doi:10.1093/jexbot/53.372.1351

Serbinonva EA, Packer L (1994) Antioxidant properties of tocopherol and tocotrienol. Methods Enzymol 234:354–367

Sharifi M, Ghorbanli M, Ebrahimzadeh H (2007) Improved growth of salinity-stressed soybean after inoculation with pre-treated mycorrhizal fungi. J Plant Physiol 164(9):1144–1151

Sharma P, Jha AB, Dubey RS (2010) Oxidative stress and antioxidative defense system in plants growing under abiotic stresses. In: Pessarakli M (ed) Handbook of plant and crop stress, 3rd edn. Taylor and Francis, Boca Raton, pp 89–138

Sharma P, Jha AB, Dubey RS et al (2012) Reactive oxygen species, oxidative damage and antioxidative defense mechanism in plants under stressful conditions. J Bot. doi:10.1155/2012/217037
Shim IS, Momose Y, Yamamoto A (2003) Inhibition of catalase activity by oxidative stress and its relationship to salicylic acid accumulation in plants. Plant Growth Regul 39(3):285–292
Singh A (2007) Molecular basis of plant–symbiotic fungi interaction: an overview. Sci World J 5(5):115–131
Singh LP, Gill SS, Tuteja N (2011) Unraveling the role of fungal symbionts in plant abiotic stress tolerance. Plant Signal Behav 6(2):175–191
Smith SE, Read DJ (1997) Mycorrhizal symbiosis. Academic, San Diego
Smith SE, Read DJ (2010) Mycorrhizal symbiosis. Academic, San Diego
Song F, Song G, Airong Dong A et al (2011) Regulatory mechanisms of host plant defense responses to arbuscular mycorrhiza. Acta Ecol Sin 31:322–327
Strack D, Fester T (2006) Isoprenoid metabolism and plastid reorganization in arbuscular mycorrhizal roots. New Phytol 172(1):22–34
Subramanian KS, Virgine Tenshia JS, Jayalakshmi K et al (2011) Antioxidant enzyme activities in arbuscular mycorrhizal (*Glomus intraradices*) fungus inoculated and non-inoculated maize plants under zinc deficiency. Indian J Microbiol 51(1):37–43
Takemoto D, Tanaka A, Scott B (2007) NADPH oxidases in fungi: diverse roles of reactive oxygen species in fungal cellular differentiation. Fungal Genet Biol 44:1065–1076
Tan SY, Jiang QY, Zhuo F et al (2015) Effect of inoculation with *Glomus versiforme* on cadmium accumulation, antioxidant activities and phytochelatins of *Solanum photeinocarpum*. PLoS One 10(7):e0132347. doi:10.1371/journal.pone.0132347
Thomas CE, McLean R, Parker RA et al (1992) Ascorbate and phenolic antioxidant interactions in prevention of liposomal oxidation. Lipids 27(7):543–550
Tisserant E, Kohler A, Dozolme-Seddas P et al (2012) The transcriptome of the arbuscular mycorrhizal fungus *Glomus intraradices* (DAOM 197198) reveals functional tradeoffs in an obligate symbiont. New Phytol 193(3):755–769
Trovato M, Mattioli R, Costantino P (2008) Multiple roles of proline in plant stress tolerance and development. Rend Lincei 19(4):325–346
Wang S, Wan C, Wang Y et al (2004) The characteristics of Na^+, K^+ and free proline distribution in several drought-resistance plants of the Alxa Desert, China. J Arid Environ 56:525–539
Watanabe K, Suzuki K, Kitamura S (2006) Characterization of a GDP-D-mannose 3″, 5″-epimerase from rice. Phytochemistry 67(4):338–346
Wu QS, Xia RX (2006) Arbuscular mycorrhizal fungi influence growth, osmotic adjustment and photosynthesis of citrus under well-watered and water stress conditions. J Plant Physiol 163(4):417–425
Wu QS, Zou YN (2009) Mycorrhiza has a direct effect on reactive oxygen metabolism of drought-stressed citrus. Plant Soil Environ 55:436–442
Wu QS, Xia RX, Zou YN (2006) Reactive oxygen metabolism in mycorrhizal and non-mycorrhizal citrus (*Poncirus trifoliata*) seedlings subjected to water stress. J Plant Physiol 163:1101–1110. doi:10.1016/j.jplph.2005.09.001
Wu QS, Zou YN, Abd-Allah EF (2014) Mycorrhizal association and ROS in plants. In: Ahmad P (ed) Oxidative damage to plants. Elsevier Inc., Amsterdam, pp 453–475. doi:http://dx.doi.org/10.1016/B978-0-12-799963-0.00015-0
Yang Y, Han X, Liang Y et al (2015) The combined effects of Arbuscular mycorrhizal fungi (AMF) and lead (Pb) stress on Pb accumulation, plant growth parameters, photosynthesis, and antioxidant enzymes in *Robinia pseudoacacia* L. PLoS One 10(12):e0145726. doi:10.1371/journal.pone.0145726
Zhang RQ, Zhu HH, Zhao HQ et al (2013) Arbuscular mycorrhizal fungal inoculation increases phenolic synthesis in clover roots via hydrogen peroxide, salicylic acid and nitric oxide signaling pathways. J Plant Physiol 170(1):74–79
Zou YN, Wu QS, Huang YM et al (2013) Mycorrhizal-mediated lower proline accumulation in *Poncirus trifoliata* under water deficit derives from the integration of inhibition of proline

synthesis with increase of proline degradation. PLoS One 8(11):e80568. doi:10.1371/journal.pone.0080568

Zou YN, Huang YM, Wu QS et al (2015) Mycorrhiza-induced lower oxidative burst is related with higher antioxidant enzyme activities, net H_2O_2 effluxes, and Ca^{2+} influxes in trifoliate orange roots under drought stress. Mycorrhiza 25(2):143–152

Zyracka E, Zadrag R, Koziol S et al (2005) Yeast as a biosensor for antioxidants: simple growth tests employing a *Saccharomyces cerevisiae* mutant defective in superoxide dismutase. Acta Biochim Pol 52(3):679–684

Chapter 11
Arbuscular Mycorrhizas and Ecosystem Restoration

Fayuan Wang

Abstract Human unreasonable activities have led to a series of environmental problems, including ecological imbalance and ecological degradation, resulting in the decay or loss of the service function of ecosystems and the reduction of biodiversity and soil productivity. Ecosystem restoration using pioneer plants and beneficial microorganisms is considered to be necessary and useful and has achieved great progress in recent years. Arbuscular mycorrhizae (AM) are ubiquitous symbiotic associations formed between arbuscular mycorrhizal fungi (AMF) and more than 90% of surveyed higher plants in terrestrial ecosystems, including fragile and degraded environments. It's well known that AMF improve acquisition of mineral nutrients (notably P) of host plants, contribute to processes associated with soil aggregation, and help to maintain plant community stability and productivity and ecosystem functioning. AMF can also benefit host plants' growth via alleviating various environmental stresses, such as heavy metals, drought, salinity, and soil compaction, allowing some plants to grow in adverse conditions. Hence, AMF may play potential roles in ecosystem restoration at different scales. In this chapter, AMF improvements in plant nutrition and growth, plant resistance, soil structure, and ecosystem processes and the related mechanisms employed by AMF are briefly discussed. The current status of using AMF in ecosystem restoration of mine soils/spoils, desertified areas, salt-affected soils, and degraded grasslands is comprehensively summarized. The future prospects regarding selecting efficient AMF-plants associations and enhancing AMF-assisted restoration are discussed.

Keywords Arbuscular mycorrhizal fungi • Restoration • Revegetation • Degraded ecosystem • Disturbed ecosystem

---

F. Wang (✉)
College of Environment and Safety Engineering, Qingdao University of Science and Technology, Qingdao, Shandong Province 266042, People's Republic of China

Agricultural College, Henan University of Science and Technology, Luoyang, Henan Province 471003, People's Republic of China
e-mail: wfy1975@163.com

Q.-S. Wu (ed.), *Arbuscular Mycorrhizas and Stress Tolerance of Plants*,
DOI 10.1007/978-981-10-4115-0_11

11.1 Introduction

Ecosystems provide a variety of goods and services upon which human beings depend, and generally they have a self-regulation mechanism which modify external disturbance and maintain them healthy and sustainable. However, in recent years, natural and anthropogenic activities, such as global warming, weather disasters, overgrazing, mining, artificial pollution, and desertification, have caused a series of environmental problems, including ecological imbalance and ecological degradation, leading to the decay or loss of the service function of ecosystems and the reduction of biodiversity and soil productivity. Ecosystem restoration means to return degraded or polluted ecosystem to their original state using ecological principles and techniques. Using pioneer plants and beneficial microorganisms for ecological restoration is considered to be useful and necessary and has achieved great progress in recent years. However, successful restoration is very difficult, and the ecological benefits are still quite low, because of various abiotic or biotic stresses making plants grow hardly. Especially in fragile and extreme ecosystems, environmental stresses generally threaten plant growth and survivorship: in rocky desertified land, water shortage and low soil fertility inhibit plant growth, while in mine sites, topsoil loss, toxic heavy metals, and extreme soil pH generally hinder vegetation to survive. More efforts should be made to substantially enhance ecosystem restoration in future.

Arbuscular mycorrhizae (AM) are ubiquitous symbiotic associations formed between arbuscular mycorrhizal fungi (AMF) and more than 90% of surveyed higher plants in terrestrial ecosystems (Smith and Read 2008). In fact, AMF have a global distribution (Kivlin et al. 2011; Treseder and Cross 2006). They occurred widely in diverse environments including fragile and degraded ecosystems. It's well known that AMF facilitate acquisition of mineral nutrients (notablly P) by host plants, contribute to processes associated with soil aggregation, and maintain plant community stability and productivity and ecosystem functioning (van der Heijden et al. 1998a, b). AMF can also benefit host plants' growth via alleviating various environmental stresses, such as heavy metal toxicity, drought, salinity, and soil compaction, allowing some plants to grow in stressful conditions (Meier et al. 2012; Miransari 2010). A great number of studies have indicated that the AMF are one of the main mediators of the disturbed communities and ecosystem succession, and hence they may play potential roles in ecosystem restoration at different scales. In this chapter, the roles of AMF in plant nutrition and growth, soil structure, and plant community and ecosystems and the related mechanisms employed by AMF are briefly reviewed, the current status of using AMF in ecosystem restoration are comprehensively summarized, and the future prospects are also suggested.

11.2 AMF and Plant Nutrition and Resistance

11.2.1 AMF and Plant Nutrient Acquisition

It has been widely confirmed that, AMF can improve the nutritional status of the host plants, especially P nutrition (Li et al. 1991, 1997b; Yao et al. 2001; Feng et al. 2003; Smith et al. 2003). Four phosphate transporter (PT) genes have been isolated from AMF hyphae. Sokolski et al. (2011) identified PT genes from 25 strains of 10 AMF species in *Glomus*, indicating that the PT genes are likely to widely occur in AMF. In addition, PT genes in plant tissues are also regulated by AMF (Table 11.1). AMF are believed to provide the dominant route for P supply of host plants, even when overall growth or P uptake remains unchanged (Smith et al. 2003). AM-specific plant PT transporters play a key role in plant P uptake irrespective of plant growth by taking up more P or by increased dry weight, compared with nonmycorrhizal

Table 11.1 Some reported phosphate transporter (PT) genes in AMF and plants

Phosphate transporter gene	Source	References
GvPT	*Glomus versiforme*	Harrison and van Buuren (1995)
GiPT	*Glomus intraradices*	Maldonado-Mendoza et al. (2001)
GmosPT	*Glomus mosseae*	Benedetto et al. (2005)
TPT	*Rhizophagus irregularis*	Halary et al. (2013)
LePT1	*Lycopersicon esculentum*	Rosewarne et al. (1999)
StPT3	*Solanum tuberosum*	Rausch et al. (2001)
MtPT4	*Medicago truncatula*	Harrison et al. (2002)
OsPT11	*Oryza sativa*	Paszkowski et al. (2002)
StPT4, StPT5	*Solanum tuberosum*	Nagy et al. (2005)
HORvu	*Hordeum vulgare*	Glassop et al. (2005)
TRIae	*Triticum aestivum*	Glassop et al. (2005)
ZEAma; PhT1;6	*Zea mays*	Glassop et al. (2005)
OsPT1–13	*Oryza sativa*	Chen et al. (2013)
LjPT3	*Lotus japonicus*	Maeda et al. (2006)
CfPT3, CfPT4, CfTT5	*Capsicum frutescens*	Chen et al. (2007a)
SmPT3, SmPT4, SmPT5	*Solanum melongena*	Chen et al. (2007a)
NtPT3, NtPT4, NtPT5	*Nicotiana tabacum*	Chen et al. (2007a)
LePT4	*Lycopersicon esculentum*	Xu et al. (2007)
AsPT1, AsPT4	*Astragalus sinicus*	Xie et al. (2013)
PtPT8, PtPT9, PtPT10	*Populus trichocarpa*	Loth-Pereda et al. (2011)
GmPT10, GmPT11	*Glycine max*	Tamura et al. (2011)
PhPT1, PhPT2, PhPT3, PhPT4, PhPT5, PhPT7	*Petunia hybrida*	Breuillin et al. (2010)
PhPT4	*Petunia hybrida*	Tan et al. (2012)

plants. In another words, the roles of AMF in agricultural and natural ecosystems must be underestimated.

The direct and indirect mechanisms underlying P improvement by AMF are summarized as follow:

1. AMF extraradical hyphae can absorb P directly and translocate it more rapidly to AM structures within the roots (Smith et al. 2003).
2. AM have a high affinity toward P, and the transport of P is much faster in hyphae than in roots.
3. AMF hyphae are much thinner than root hairs and can absorb P from outside the P depletion zone where roots can't reach, and at the same time, they have a relatively larger surface area, hence expand the total absorption area of host root.
4. AMF can acquire P from compacted soil (Li et al. 1997a).
5. AMF secrete organic acid which help to activate poorly soluble P.
6. AMF secrete phosphatase which hydrolyzes organic sources of P into available forms.
7. AMF may upregulate PT genes in plants and themselves under P-deficient conditions (Karandashov and Bucher 2005).
8. AMF stimulate plant growth-promoting rhizobacteria especially other P cycle-related microorganisms.

In AM associations, AMF can also assists their plant hosts by providing N in addition to P, (Suzuki et al. 1999; Feng et al. 2002). AMF can capture inorganic or organic forms of N and translocate them via amino acids (arginine) from the extra- to the intraradical mycelium and then transfer them to the plants without any carbon skeleton (Guether et al. 2009). Ammonium transporter genes *GintAMT1* and *GintAMT2* were identified from *Glomus intraradices* extraradical hyphae (López-Pedrosa et al. 2006; Pérez-Tienda et al. 2011). A high-affinity ammonium transporter gene *LjAMT2;2* has been characterized from *Lotus japonicus* roots, which is exclusively expressed in the mycorrhizal roots and can be upregulated by colonization with *Gigaspora margarita* (Guether et al. 2009). In the soybean genome, five ammonium transporter genes are found to be AM inducible, and among them, *GmAMT4.1* showed specific expression in arbusculated cortical cells (Kobae et al. 2010). AMF hyphae can produce some hydrolytic enzymes, such as phosphatase, pectinase, cellulase, xylanase, and chitinase (Varma 1999), which may help the mineralization of organic N into available forms. AMF has been proved to both promote decomposition of and increase N capture from complex organic material in soil (Hodge et al. 2001). Also, AMF influence soil microorganisms and, hence, they may influence biochemical reactions in soil including mineralization of organic forms of N and nitrification (Hamel 2004).

In addition, AMF are able to absorb almost all other macronutrients (K, Ca, Mg, and S) and micronutrients (Zn, Fe, Cu, and Mn) (Miransari 2013; Smith and Read 2008). Most related mechanisms are believed similar to the abovementioned, but the details still need more investigation.

11.2.2 AMF and Plant Resistance

Numerous studies have shown that AMF help to enhance plant resistance to abiotic and biotic stresses, such as drought, salinity, acidity, heavy metals, disease, soil compaction, extreme temperature, and waterlogging (Meier et al. 2012; Miransari 2010; Pozo and Azcón-Aguilar 2007; Madiba 2014). Under stressful conditions, plants generally grow poorly because the nutrients are low or unavailable or some factors limit them to take up nutrients. In most cases, AMF help their host plants absorb nutrients, and thus AM plants are able to grow much better than the nonmycorrhizal ones. Not surprisingly, the most common mechanism involved in mycorrhizal plant resistance to different stresses is the improved nutrient uptake. Some stresses such as salinity and drought all can produce osmotic stress and limit plants to absorb water. AMF can adjust osmotic potential of mycorrhizal plants through the higher accumulation of organic products, e.g., proline, glycine betaine, carbohydrates such as sucrose and mannitol, and nonorganic ions including K and Cl, and help the host plants to grow better under water deficiency (Porcel et al. 2012). Glomalin-related soil protein (GRSP, formerly named as glomalin), a glycoprotein produced by AMF, is involved in inducible stress responses in AMF for salinity and heavy metals. GRSP has shown a potential role in salt (Na^+) and heavy metal immobilization (González-Chavez et al. 2004; Vodnik et al. 2008; Hammer and Rillig 2011). In soils, GRSP has been proven to bind up to 0.08 mg g^{-1} Cd and 1.12 mg g^{-1} Pb (González-Chavez et al. 2004). AM hyphae can enhance soil structure by binding soil particles and through production of GRSP, which affect soil moisture retention (Miransari 2010). In general, GRSP production in AMF may be a stress response involved in various stresses. AMF colonization generally increases antioxidant activities (such as SOD and POD) under stresses of heavy metals, drought, or salinity, thus helping plants scavenge ROS and reduce oxidative damage to biomolecules (Azcón et al. 2009; Ruiz-Lozano 2003; Ruiz-Lozano et al. 2006).

At the molecular level, AMF enhances plant tolerance to drought stress through altering plant gene expression (Ruiz-Lozano et al. 2006). AMF regulate the expression of drought-related genes in host plants, including *p5cs* genes encoding a rate-limiting enzyme in the biosynthesis of proline (Porcel et al. 2004), genes encoding aquaporins (Porcel et al. 2006), and *nced* genes encoding a key enzyme in the biosynthesis of ABA (Aroca et al. 2008), allowing mycorrhizal plants to maintain a better water status in their tissues (Porcel et al. 2012). Aquaporin is a kind of protein regulating passive transport of water, existing in cell membrane of plants and AMF. Two functional aquaporin genes, namely, *GintAQPF1* and *GintAQPF2*, were first studied using *Glomus intraradices* (Li et al. 2013b), supporting potential water transport via AMF to host plants. A series of genes involved in heavy metal detoxification have been characterized. Three metallothionein (MT) genes have been isolated from AMF, i.e., *GrosMT1* from *Gi. rosea* spores (Stommel et al. 2001), *GmarMT1* from *Gi. margarita* (BEG34) spores (Lanfranco et al. 2002), and *GintMT1* from *G. intraradices* extraradical mycelium (González-Guerrero et al. 2007). Further studies indicated that *GintMT1* and *GmarMT1* are able to restore Cu or Cd tolerance to MT-deficient yeasts and that their transcription is at least transiently induced by the presence of these metals. In addition, Lanfranco et al. (2005) also isolated SOD-encoding gene (*GmarCuZnSOD*) from *Gi. margarita*, and González-Guerrero (2010b) identified a Cu/Zn-SOD gene (*GintSOD1*) in *Glomus intraradices*. Another Zn transporter *GintZnT1* has been identified from *G. intraradices*, which has been suggested playing a role in Zn storage or efflux within hyphae (e.g., from an internal storage compartment involved in long-distance Zn translocation) or the efflux of Zn into to the plant/fungal interfacial apoplast (González-Guerrero et al. 2005). An ATP-binding cassette (ABC) transporter (*GintABC1*) was isolated from the extraradical mycelia of *Glomus intraradices*, which is upregulated by Cd and Cu (González-Guerrero et al. 2010a). These genes are likely to play an important role in AMF resistance to heavy metal toxicity and other stresses. AMF also regulate the expression of the genes in plants related to heavy metal detoxification. A plasma membrane Zn transporter *MtZIP2* has been isolated from *Medicago truncatula*, and it is upregulated in roots by Zn fertilization yet downregulated by AMF colonization (Burleigh et al. 2003). Inoculation with *Glomus mosseae* or *G. intraradices* caused an overall induction of the expression of MT genes (*PaMT1*, *PaMT2*, and *PaMT3*) and PA biosynthetic genes (*PaSPDS1*, *PaSPDS2*, and *PaADC*), together with increased free and conjugated polyamine levels, in leaves of *Populus alba* grown on a soil polluted with Cu and Zn, but not in those grown on non-polluted soil (Cicatelli et al. 2010). In leaves of *Populus alba* inoculated with *Glomus* spp., transcript level genes known to be related to heavy metal stress (metallothioneins, phytochelatin synthase, glutathione synthase, arginine decarboxylase) were generally downregulated or unaffected, in Cu-/Zn-polluted soil compared with controls, while phytochelatin synthase and clathrin were strongly upregulated in the presence of AMF, especially *Glomus mosseae* (Cicatelli et al. 2012). Pallara et al. (2013) found that *G. mosseae* reduced transcript abundance of genes involved in antioxidant defense in leaves and roots of white poplar clone "AL35" plants grown on the Cu-/Zn-polluted soil. Moreover, AMF induced the gene coding for phytochelatin synthase in leaves. Arsenate can induce the expression of the high-affinity phosphate transporter *GiPT* and the arsenite efflux pump gene *GiArsA* in the extraradical mycelium but also in arbuscules of *Glomus intraradices* (González-Chavez et al. 2011), suggesting that As tolerance mediated by AMF may be caused by an As exclusion mechanism.

AMF also change the expression of heavy metal-induced proteins in plants. Inoculation with *G. mosseae* significantly reduced Cd-induced growth inhibition on sensitive genotype of pea, and AMF modulated the expression of several proteins (Repetto et al. 2003), which was considered one of the mechanisms for Cd detoxification. Proteomic analysis revealed that 9 out of the 15 Cd-induced changes in nonmycorrhizal *M. truncatula* roots were absent or inverse in those Cd treated and colonized by *G. intraradices*, including the *G. intraradices*-dependent down-accumulation of Cd stress-responsive proteins, and 6 out of 26 mycorrhiza-related proteins displayed changes in abundance upon Cd exposure (Aloui et al. 2009). Inoculation of *G. irregulare* alleviated Cd stress on *Medicago truncatula* and induced an increase in photosynthesis-related protein in shoot of mycorrhizal plants

(Aloui et al. 2011). These results suggest that AMF play a substantial role in alleviating Cd toxicity. Under As treatment, inoculation with *G. mosseae* and *Gi. margarita* induced the differential expression of 130 proteins with specific responses in As hyperaccumulator *Pteris vittata* leaves, and multiple forms of glyceraldehyde-3-phosphate dehydrogenase, phosphoglycerate kinase, and enolase were upregulated in *G. mosseae*-inoculated plants, suggesting a central role for glycolytic enzymes in As metabolism (Bona et al. 2010). Further proteomic analysis showed that As treatment affected the expression of 14 spots (12 upregulated and 2 downregulated) in *Pteris vittata* roots, while modulated 3 spots (1 upregulated and 2 downregulated) in *G. mosseae*-colonized roots, indicating a protective effect of AM symbiosis toward As stress (Bona et al. 2011). Lingua et al. (2012) found that both heavy metals (Cu and Zn) and AMF *Glomus intraradices* affected leaf protein expression of *Populus alba*. Likely, AMF are putatively able to alleviate other stresses on AM symbiosis through modulation of the expression of tolerance-related proteins.

11.3 AMF and Soil Quality

11.3.1 AMF and Soil Structure

Aggregates, especially water-stable aggregates, are one of the most important indicators for healthy soil structure. AMF provide the direct links between roots and soil and can potentially contribute to soil aggregation (Rothwell 1984) and soil structure stability by the combined action of extraradical hyphae and GRSP (Bedini et al. 2009). The role of AM fungal mycelium in aggregate formation or stabilization can be summarized as three mechanisms (Rillig and Mummey 2006): (1) biochemical mechanism: fungal mycelium products, including GRSP, mucilages, polysaccharides, and other extracellular compounds, hydrophobins, and related proteins; (2) biological mechanisms: mycelium-influenced microbiota and the soil food web; and (3) biophysical mechanisms: enmeshment, alignment, and altered water relations. GRSP has been widely studied, and it is thought to act as a rather stable hydrophobic glue that might decrease macroaggregate disruption during wetting and drying events via retarding water movement into the pores within the aggregate structure. Using *M. sativa* plants inoculated with or without different AMF isolates, Bedini et al. (2009) showed that: (1) AM plants all produced increased GRSP, (2) aggregate stability was significantly increased by AM inoculation; (3) GRSP concentration and soil aggregate stability were positively correlated with mycorrhizal root volume; (4) soil aggregate stability was positively correlated with total hyphal length and hyphal density; and (5) different AMF isolates differently affected GRSP concentration and stability of soil aggregates. In addition to AM mycelium, AMF can also potentially influence soil aggregation at other different scales, i.e., plant communities or plant roots (individual host) (Rillig and Mummey 2006). Bearden and Petersen (2000) demonstrated that the effect of AMF on stabilization of soil aggregates was associated with both AM hyphae and the stimulation of root growth by AMF. Plant species, root length, and percent cover also contribute strongly to water-stable aggregation (Rillig et al. 2002). AMF also influence soil microbial communities, which exert a strong influence on the formation and stabilization of microaggregates in a more direct manner (Rillig and Mummey 2006). Considering the function diversity of AMF species and AM fungal community, the role of AMF diversity (richness or community composition) in soil aggregation needs to be examined. It's necessary to select the most efficient associations between AMF isolates/community and host plants to be used for soil quality improvement and ecosystem restoration programs.

11.3.2 AMF and Other Soil Microorganisms

Biodiversity of microorganisms is an important indicator of soil fertility and health. In addition to roots, AMF may interact with numerous other soil organisms especially the microorganisms from rhizosphere and hyphosphere (Miransari 2014). AMF can directly influence bacterial communities via the deposition of mycelium products that act as substrates for bacterial growth and modify rhizodeposition products that result in alteration of the composition of the bacterial community (Rillig and Mummey 2006). On the other hand, soil microorganisms (mycorrhiza helper bacteria) modulate mycorrhizal symbiosis and assist mycorrhiza formation and/or interact positively with the functioning of the symbiosis (Frey-Klett et al. 2007). These mycorrhiza-associated bacteria, together with the fungal symbiont, contribute mutually to nutrient uptake and protection of the plants against abiotic stress and phytopathogens (Frey-Klett et al. 2007). To some extent, mycorrhizas can be defined as tripartite associations, including plants, AMF, and bacteria (Bonfante and Anca 2009). It's been confirmed that AMF and plant growth-promoting rhizobacteria (PGPR) can interact synergistically to stimulate plant growth through various mechanisms including improved nutrient acquisition and inhibition of soil pathogens (Artursson et al. 2006). These interactions may be of crucial importance for plant growth and hence ecosystem productivity.

11.4 AMF and Ecosystem Processes

11.4.1 AMF and Ecosystem Functioning

Biodiversity is essential for regulating and modulating ecosystem processes and functioning and maintaining ecosystem stability. AMF isolates are functionally diverse and the composition of AMF communities has a large impact on plant performance, plant community structure, plant succession, and ecosystem functioning

(Van der Heijden and Scheublin 2007; Janos 1980; Gange 1993). AMF are known to participate in most terrestrial ecosystem processes, such as primary productivity, energy flow, decomposition of organic matter, nutrient cycling, and resistance and resilience to perturbations, and therefore, they contribute greatly to ecosystem functions such as plant biodiversity, productivity, and variability.

11.4.2 AMF and Plant Community and Diversity

AMF play a key role in determining plant community composition through changing nutrients resource allocation among different plant species. Previous results have shown that the composition of AMF communities alters coexistence and resource (N and P) distribution between co-occurring plants (Van der Heijden et al. 2003). So, the competitive ability of a plant for soil resources therefore not only depends on its own ability to capture nutrients but also is influenced by AMF colonizing its roots. AMF can exert beneficial effect on the coexistence of warm-season grasses with more competitive cool-season grasses (Klabi et al. 2014). AMF species that co-occur as natural communities have the potential to determine plant community structure through their differential effects on growth of diverse plants (Van der Heijden et al. 1998a). Further study shows that plant biodiversity, nutrient capture, and productivity in macrocosms increase significantly with increasing AMF species richness (Van der Heijden et al. 1998b). The plant biodiversity and productivity were found lowest in those plots without AMF or with only a few AMF species, but highest when eight or fourteen AMF species were present. It's can be concluded that the diversity of AMF are considered a key factor to maintaining plant biodiversity.

11.4.3 AMF and Nutrient Cycling

AMF acquire up to 20% of plant-fixed carbon as a return for their benefits to plants (Parniske 2008); thus, they can fix large quantities of carbon and influence carbon fluxes between the biosphere and the atmosphere. AM symbiosis determines the flow of huge quantities of carbon worldwide—an estimate of 5 billion tons of carbon annually (Bago et al. 2000). In soil, it's estimated that the amount of organic carbon derived directly from AMF ranges from 54 to 900 kg ha^{-1} (Zhu and Miller 2003). Compared with that of orchid mycorrhizal and ectomycorrhizal plants, AM plant species showed comparatively high relative growth rate, high foliar N and P, and fast litter decomposition (Corradi and Charest 2011). Undoubtedly, AMF contribute greatly to ecosystem carbon cycling, influencing soil organic matter and the diversity of soil decomposers.

Numerous studies have shown that AMF play significant roles in activation of non-available nutrients and subsequent uptake of them by host plants, via hyphal exudates such as organic acids and diverse enzymes. In addition to their well-known role in plant P nutrient and P cycling (Hamel 2004), AMF can take up inorganic N outside the roots, and translocated from the extraradical to the intraradical mycelium as arginine (Govindarajulu et al. 2005). AMF can both enhance decomposition of and increase N capture from complex organic material (grass leaves) in soil (Hodge et al. 2001). Further studies have confirmed that AMF can obtain substantial amounts of N from decomposing organic materials (bone meal) (Hodge and Fitter 2010) and that the uptake from organic N could be important in AM symbiosis for both partners (Leigh et al. 2009). The large biomass and high N demand of AMF means that they represent a global N pool and play a substantial role in the N cycling.

11.5 AMF and Restoration of Degraded Ecosystem

11.5.1 Restoration of Mine Areas

Mining activities generally lead to loss of topsoil and vegetation and decreased soil biodiversity and yield extremely harsh surroundings, including severely infertile soils, large quantities of wastes, and high concentrations of pollutants, which in turn limit natural revegetation and soil restoration. Reestablishing native or pioneer plants is considered one of key steps in revegetation strategies. Numerous greenhouse and field (in situ) experiments have been conducted using native and/or pioneer plant/AMF species on mining-destroyed soil, tailings, or spoils, most of which confirm that AMF benefit plant survival, growth, and nutrition and improve soil structure and hence the reestablishment of vegetation (Table 11.2).

Native plants and AMF often develop good tolerance to stress especially pollutants and adapt well to local habitats because of long-term natural selection. The origin of AMF strains influences the mycorrhiza development of plants on industrial wastes (Orlowska et al. 2005a). A survey on Cu mine tailing indicates that revegetation and remediation of heavy metal-contaminated sites might be facilitated by selection of tolerant plant species, especially those dominant species in such surrounding, and isolation of tolerant AMF may also be warranted (Chen et al. 2005). The selection of suitable plants for restoration should focus on native species, and the introduced plants need to be accompanied by their mycorrhizal symbionts to ensure their proper establishment (Ryszka and Turnau 2007). Some results confirm that native AMF derived from polluted soil are more effective than those exotic ones (Taheri and Bever 2010; Weissenhorn et al. 1995; Toler et al. 2005; Biro et al. 2009), but some exceptions also show opposite results (Lagrange et al. 2011; Shetty et al. 1994). In addition, symbiotic associations can be successfully established between exogenous AMF and native plants, indicating use of local plant species in combination with exogenous AMF for ecological restoration (Guo et al. 2013a). Apparently, both origin and characteristics of plants and fungal partners should be included for selection of suitable combinations.

Table 11.2 The role of AMF in restoration of mine areas

Mine disturbed site/substrate	Host plants	AMF species	AMF effects	Experiment type	References
Field in coal mine area	*Populus alba*, *Fraxinus chinensis*	*Glomus mosseae*, *G. etunicatum*	Increased chest perimeter and plant height	Field	Du et al. (2008)
Field in coal mine area	*Amorpha fruticosa*	*G. mosseae*	Increased plant biomass, GRSP content, higher phosphatase activity, and microbial population in rhizosphere soil	Field	Li et al. (2013a)
Coal waste piles	*Fraxinus chinensis*, *Robinia pseudoacacia*	*G. mosseae*, *G. versiforme*	Increased plant growth and infectivity	Field	Hu et al. (2006)
Coal mine spoil	*Zea mays*	*G. etunicatum*, *G. versiforme*	Increased plant dry weight and N, P, and K uptake	Greenhouse	Zhao et al. (2013)
Coal mine spoil	*Zea mays*	*G. aggregatum*, *Rhizophagus intraradices*, *Funneliformis mosseae*	Increased plant dry weight and N, P, and K uptake	Greenhouse	Guo et al. (2014)
Coal mine spoil	*Glycine max*	*Acaulospora mellea*, *G. mosseae*, *G. caledonium*	Increased plant P uptake, height, biomass, yield	Greenhouse	Wang and Miao (2006)
Coal mine wastes	*Trifolium repens*, *Lolium perenne*, *Zea mays*	*G. versiforme*, *G. mosseae*	Increased plant growth	Greenhouse	Wang et al. (2009)
Coal mine spoil banks	*Zea mays*	*R. intraradices*	Increased plant P and K uptake and dry weight	Greenhouse	Bi et al. (2013)
Drastically disturbed semiarid surface coal mine soil	Art*emisia tridentata* ssp. *wyomingensis*	Native AMF	Improved soil structure, increased organic matter, decreased soluble salts and Na	Field	Frost et al. (2001)
Coal mine substrates	*Medicago sativa*	*G. mosseae*	Increased survival and dry weights	Greenhouse	Wu et al. (2009)

(continued)

Table 11.2 (continued)

Mine disturbed site/substrate	Host plants	AMF species	AMF effects	Experiment type	References
Coal mine spoil	*Fragaria vesca*	*G. macrocarpus* var. *geosporus*	Increased plant growth	Greenhouse	Daft et al. (1975)
Coal mine spoil	*Zea mays*	*G. mosseae*	Increased biomass and nutrient uptake	Greenhouse	Bi et al. (2002)
Coal spoil heaps	*Fraxinus chinensis*	*G. mosseae*, *G. etunicatum*	Increased survival, plant cover, growth, and height	Field	Bi et al. (2007)
Coal mine tailing	*Andropogon virginicus*, *Plantago lanceolata*	A mixture of indigenous AMF	Increased plant growth	Greenhouse	Taheri and Bever (2010)
Lignite mine spoils	*Alnus cordata*	*G. mosseae*, *G. fasciculatum*	Increased survival and biomass	Field	Lumini et al. (1994)
Lignite overburden	*Bouteloua curtipendula*, *Sorghastrum nutans*, *Panicum coloratum*	*G. fasciculatum*, *Gigaspora margarita*	Greater survival percentages, basal diameters, and aboveground biomass; increased N and P uptake	Greenhouse and field	Call and Davies (1988)
Anthracite spoil	*Acer rubrum*	*G. macrocarpus* var. *geosporus*, *Gi. gigantea*	Increased plant P uptake	Greenhouse	Daft and Hacskaylo (1977)
Fly ash, coal mine spoil	*Acer pseudoplatanus*, *Alnus glutinosa*, *Salix purpurea*	*G. mosseae*, a mixture of four AMF	Increased plant growth and P uptake	Greenhouse	Enkhtuya et al. (2005)
Fly ash	*Dendrocalamus strictus*	A mixture of indigenous AMF	Increased plant growth and N, P, and K uptake	Nursery	Babu and Reddy (2011)
Zn mine spoil	*Andropogon gerardii*, *Festuca arundinacea*	A mixture of AMF	Increased plant growth	Greenhouse	Shetty et al. (1994)

Zn mine tailings	*Andropogon gerardii*, *Festuca arundinacea*	A mixture of *G. ambisporum*, *G. constrictum*, *Scu. pellucida*, *G. mosseae*, *and G. etunicatum*	Increased plant growth	Greenhouse	Hetrick et al. (1994)
Zn mine tailings	*L. perenne*, *Festuca ovina*, *F. rubra*, *Poa pratensis*	*G. claroideum*	Improved development of grasses during early stages	Field	Ryszka and Turnau (2007)
Zn mine wastes	*Festuca rubra*, *Plantago lanceolata*	*G. claroideum*, *G. intraradices*, *G. geosporum*, *G. etunicatum*	Increased plant survival	Greenhouse	Orlowska et al. (2005b)
Heavy metal-polluted soil near a Zn smelter	*Glycine max*	A mixture of AMF	Increased plant dry weight and foliar P and N contents, reduced Zn, Cd, and Mn concentrations in leaves	Greenhouse	Heggo et al. (1990)
Zn-contaminated soil	*Thymus polytrichus*	A mixture of *G. mosseae*, *G. caledonium*, *G. claroideum*, and *G. intraradices*	Increased plant growth and P uptake in contaminated, low-P soil	Greenhouse	Whitfield et al. (2004)
Zn-polluted soil	*Solanum nigrum*	*G. claroideum*, *G. intraradices*	Increased Zn uptake	Greenhouse	Marques et al. (2007)
Cu mine tailings	*Coreopsis drummondii*, *Pteris vittata*, *Lolium perenne*, *Trifolium repens*	*G. mosseae*	Increased plant dry matter yield except for *L. perenne*, improved P nutrition and decreased shoot Cu, As, and Cd concentrations	Greenhouse	Chen et al. (2007b)

(continued)

Table 11.2 (continued)

Mine disturbed site/substrate	Host plants	AMF species	AMF effects	Experiment type	References
Cu mine tailings	*Zea mays*	*G. aggregatum*, *G. etunicatum*, *G. intraradices*, *G. mosseae*, *G. versiforme*	Decreased Cu and Zn uptake in shoots	Greenhouse	Guo et al. (2013d)
Cu-polluted soil	*Helianthus annuus*	*Glomus* spp.	Increased plant growth and Cu uptake	Greenhouse	Castañón-Silva et al. (2013)
Cu-polluted soil	*Medicago sativa*	*Glomus* spp.	Increased shoot dry weight	Greenhouse	Novoa M et al. (2010)
Cu mine tailings	*Trifolium repens*	*G. mosseae*, *G. versiforme*	Increased plant biomass, decreased Cu uptake	Greenhouse	Xiao et al. (2006)
Fe mine tailings	*Agropyron cristatum*, *Elymus dahuricus*	*G. mosseae*, *G. versiforme*	Increased plant N, P, and K uptake and dry weight, decreased heavy metal uptake	Greenhouse	Guo et al. (2013a)
Fe ore mine	*Acacia pyrifolia*	*Glomus* sp.	70% increases in dry matter production	Greenhouse	Jasper et al. (1988)
Taconite mine tailings	*Panicum virgatum*	*G. etunicatum*	Increased plant growth	Greenhouse and field	Johnson (1998)
Coarse taconite Fe ore tailing	*Andropogon gerardii*, *Schizachyrium scoparium*, *Elymus canadensis*, *Bromus kalmii*, *Lespedeza capitata*	*Glomus intraradices*, *Glomus claroideum*	Increased plant cover and AM infectivity	Field	Noyd et al. (1996)
Pb/Zn mine tailings	*Leucaena leucocephala*	*Glomus* spp.	Improved plant growth and N, P, and K uptake and N_2-fixing capacity; decreased mobility of Pb and Zn in soil	Greenhouse	Ma et al. (2006)

Pb/Zn mine tailings	*Leucaena leucocephala*	A mixture of *G. mosseae* and *G. intraradices*	Increased N, P, and K uptake and N_2-fixing capacity	Greenhouse	Ma et al. (2006)
Pb/Zn mine tailings	*Chrysopogon zizanioides*	*G. intraradices*, *G. mosseae*	Increased plant growth and N and P uptake	Field	Wu et al. (2011)
Acidic Pb/Zn mine tailings	*Prosopis juliflora*	*G. intraradices*, a mixture of *G. intraradices and G. deserticola*, native AMF	Increased dry biomass and root length	Greenhouse	Solis-Dominguez et al. (2011)
Soil near a Pb smelter	*Zea mays*	*G. intraradices*	Increased plant growth and decreased Pb uptake	Greenhouse	Sudova and Vosatka (2007)
Ultramafic soil with high Ni and low P	*Costularia comosa*	*G. etunicatum*, a mixture of indigenous AMF	Increased plant growth and decreased Ni uptake in root	Greenhouse	Lagrange et al. (2011)
Ni-enriched ultramafic soil	*Berkheya coddii*	Native AMF	Increased shoot biomass and Ni content	Greenhouse	Turnau and Mesjasz-Przybylowicz (2003)
Ni-enriched ultramafic soil	*Berkheya coddii*	*R. intraradices*	Increased plants growth, survival, and nutrient and Ni uptake	Greenhouse	Orlowska et al. (2011)
As mining contaminated soil	*Pteris vittata*	A mixture of *G. intraradices*, *G. geosporum* and *G. mosseae*, indigenous AMF, *G. mosseae*	Increased plant growth and As uptake	Greenhouse	Leung et al. (2010a,b)
Soil near As mine tailings	*Zea mays*	*G. mosseae*	Increased plant growth and decreased As uptake in shoot	Greenhouse	Xia et al. (2007)
As-contaminated soil near a closed silver mine	*Helianthus annuus*	*G. aggregatum*	Increased plant growth and root As but decreased shoot As uptake	Greenhouse	Ultra et al. (2007)

(continued)

Table 11.2 (continued)

Mine disturbed site/substrate	Host plants	AMF species	AMF effects	Experiment type	References
Soil near an As sulfide mine	*Zea mays*	*G. caledonium*	Increased plant growth and decreased As uptake	Greenhouse	Bai et al. (2008)
Soil near an As sulfide mine	*Nicotiana tabacum*	A mixture of indigenous AMF, *A. mellea*, *G. versiforme*, *G. caledonium*	Increased plant growth and decreased As uptake	Greenhouse	Hua et al. (2009)
As mine waste	*Plantago lanceolata*	*R. intraradices*, *Funneliformis geosporum*, *Rhizophagus clarus*, *Glomus* sp.	Higher shoot and root biomass and lower concentrations of As in roots	Greenhouse	Orlowska et al. (2012)
As mining-polluted soil	*Trifolium repens*, *Lolium perenne*	*G. mosseae*	Increased plant P nutrition, decreased root to shoot As translocation and shoot As concentrations	Greenhouse	Dong et al. (2008)
As-contaminated soil	*Pityrogramma calomelanos*, *Tagetes erecta*, *Melastoma malabathricum*	*G. mosseae*, *G. intraradices*, *G. etunicatum*	Decreased As accumulation in *P. calomelanos* and *T. erecta*, increased growth and As accumulation in *M. malabathricum*	Greenhouse	Jankong and Visoottiviseth (2008)
Arsenical sulfidic gold mine tailings	*Eucalyptus cladocalyx*	A mixture of *Glomus* sp., *Scu. auriglobа*, and *A. levis*	Increased plant biomass and nutrient uptake	Greenhouse	Madejon et al. (2012)
Rare earth mine tailings	*Zea mays*, *Sorghum bicolor*	*G. versiforme*	Increased plant growth and N, P, and K uptake	Greenhouse	Guo et al. (2013b)

Rare earth mine tailings	*Glycine max*	*G. versiforme*	Increased plant dry weight and P and K uptake; decreased rare earth elements concentrations in plants	Greenhouse	Guo et al. (2013c)
Rare earth mine tailings	*Lolium perenne*, *Cynodon dactylon*	*Diversispora spurcum*, *G. aggregatum*, *G. constrictum*	Increased plant dry weight, decreased leaf MDA and proline contents	Greenhouse	Chen et al. (2012)
U tailing	*M. truncatula*, *L. perenne*	*G. intraradices*	Increased U in roots	Greenhouse	Chen et al. (2008a)
U mining-impacted soil	*Pteris vittata*	*G. mosseae*, *G. caledonium*, *G. intraradices*	Depressed plant growth, increased root U concentrations	Greenhouse	Chen et al. (2006)
Soil mixed with Midnite Mine spoil	*Pseudoroegneria spicata* × *Elymus lanceolatus* ssp. *lanceolatus*	*G. mosseae*, *G. intraradices*, *Entrophospora* sp.	Increased plant growth or no effect	Greenhouse	Thorne et al. (1998)
Soil received Cd-contaminated sediment	*Sedum alfredii*, *Lolium perenne*	*G. caledonium*, *G. mosseae*	Increased plant biomass and Cd uptake, increased soil acid phosphatase activities, and available P concentrations, decreased phytoavailable Cd concentrations	Greenhouse	Hu et al. (2013a)
Soil received Cd-contaminated sediment	*Mimosa pudica*	*G. caledonium*, *G. versiforme*	Increased shoot biomass, decreased tissue Cd concentrations and soil phytoavailable Cd concentration	Greenhouse	Hu et al. (2014)

(continued)

Table 11.2 (continued)

Mine disturbed site/substrate	Host plants	AMF species	AMF effects	Experiment type	References
Soil contaminated with Cu, Zn, Pb, and Cd near a Cu refinery	*Elsholtzia splendens*	*G. caledonium*, indigenous AMF mixture	Increased plant dry weight and P uptake	Greenhouse and field	Wang et al. (2005)
Soil contaminated with Cd and Zn	*Lactuca sativa*	*G. fasciculatum*, *G. etunicatum*, *G. mosseae*	Cd and Zn decreased in shoots, and increased in roots	Greenhouse	Dehn and Schuepp (1990)
Soil near a former smelter (contaminated with Pb, Zn, Cd)	*Medicago truncatula*	*G. intraradices*	Increased plant biomass and decreased Cd uptake	Greenhouse	Redon et al. (2008)
Soil from a Cd-, Zn-, and Pb-polluted area affected by mining and smelting activities	*Thlaspi praecox*	Indigenous AMF	Increased plant P, K, S, Ca, Cu, and Ni uptake	Greenhouse	Vogel-Mikus et al. (2006)
Combined polluted soil with Cu, Cd, Pb, and Zn	*Salix viminalis*, *Populus* × *generosa*	*G. intraradices*	Decreased shoot/root ratio of Cu and Cd	Field	Bissonnette et al. (2010)
Abandoned mine tailing (Pb, Zn, Cd)	*Zea mays*	*G. mosseae*, *Glomus* sp.	Increased plant growth and decreased metal uptake	Greenhouse	Liang et al. (2009)
Soil from an abandoned paddy field near a Cu refinery	*Zea mays*	A mixture of indigenous AMF	Increased Cu, Zn, Pb, and Cd uptake in shoots and roots	Greenhouse	Wang et al. (2007)
Soil polluted with wastewater of the electroplating (Cu, Pb, Zn)	*Bidens pilosa* var. *radiata*, *Passiflora foetida* var. *hispida*	*G. macrocarpum*	Increased plant growth and metal uptake	Greenhouse and field	Tseng et al. (2009)
Soil polluted with heavy metal-rich industrial wastes	Native grasses	*G. clarum*, *G. intraradices*	Plants established well	Greenhouse and field	Turnau et al. (2008)

Combined heavy metal-contaminated soil	*Phacelia tanacetifolia*, *Sinapis alba*, *Trifolium pratense*, *Helianthus annuus*	*G. intraradices*	Increased plant growth	Greenhouse	Neagoe et al. (2013)
Soil contaminated by wastewater from local smelting factories	*Sesbania rostrata*, *Sesbania cannabina*, *Medicago sativa*	*G. mosseae*	Increased plant growth and N and P uptake, decreased metal uptake	Greenhouse	Lin et al. (2007)
Al ore tailing	*Zea mays*, *Glycine max*	*G. mosseae*	Improved plant nutrition, biomass, and yield	Field	Zhang et al. (1996)
Bauxite spoil	*Cedrela fissilis*, *Anadenanthera peregrina*	*Acaulospora scrobiculata*, *Gigaspora margarita*, *G. etunicatum*	Increased plant growth and P uptake	Greenhouse	Tótola and Borges (2000)
Soil contaminated with Al and Mn	*Vigna unguiculata*	*Scutellospora reticulata*, *G. pansihalos*	Decreased Al and Mn in soil	Greenhouse	Alori and Fawole (2012)
Alkaline gold mine tailing	*Dodonaea viscosa*, *Andropogon eucomus*, *Imperata cylindrica*	*G. intraradices*, indigenous AMF	Increased plant biomass and survival	Greenhouse	Orłowska et al. (2010)
Sulfidic gold mine tailings	*Bothriochloa macra*	A mixture of *Glomus* sp., *Scutellospora auriglob a*, and *Acaulospora levis*, indigenous AMF	Increased plant biomass and nutrient uptake	Greenhouse	Madejon et al. (2010)
Mined sand dune	*Tocoyena selloana*	*A. longula*, *Gigaspora albida*	Increased plant growth	Greenhouse and field	de Souza et al. (2010)
Sand mine tailings	*Acacia* spp.	*Glomus* sp., *Scutellospora calospora*	Increased plant dry weight and P uptake	Greenhouse	Jasper et al. (1989)

(continued)

Table 11.2 (continued)

Mine disturbed site/substrate	Host plants	AMF species	AMF effects	Experiment type	References
Processed oil shale and disturbed native soil	*Atriplex canescens*, *Artemisia tridentata* Nutt. ssp. *wyomingensis*, *Chrysothamnus nauseosus* (Pall.) Britton var. *nauseosus*, *Sarcobatus vermiculatus* (Hook.) Torr. var. *vermiculatus*	*G. fasciculatum*, *G. mosseae*	Increased shoot biomass and P contents	Greenhouse	Call and McKell (1985)
Strip mine spoil	*Lathyrus sylvestris*, *Coronilla varia*, *Lotus corniculatus*	Indigenous AMF	Increased plant yields and survival	Field	Lambert and Cole (1980)
Limestone mine spoil	*Acacia ampliceps*, *A. eriopoda*, F. Muell. ex Benth., *Albizia lebbeck*, *Azadirachta indica*, *A. juss*, *Colophospermum mopane*	*G. mosseae*	Increased plant biomass	Net house	Rao and Tak (2002)
Serpentine soil (low Ca/Mg ratio, deficiencies of nutrients, high heavy metals, and low water-holding capacity)	*Knautia arvensis*	*G. irregulare*, *Glomus* sp.	Increased plant growth and P uptake	Greenhouse	Doubkova et al. (2012)

Mine soil and tailings generally have low fertility because of disturbance and erosion. The rate of natural revegetation processes is usually slow due to deficient nutrients and the lack of organic matter, especially at the initial stages. AMF exhibit outstanding ability to improve P nutrition, but they are not so effective on other nutrients. Organic or inorganic fertilizer amendments are usually applied in AMF-assisted revegetation programs (de Souza et al. 2010; Hetrick et al. 1994; Jasper et al. 1989; Johnson 1998; Leung et al. 2010a; Noyd et al. 1996). However, excess available nutrients especially P are detrimental to mycorrhizal functioning, while organic amendments with slowly released nutrients may show positive effects on AMF (Gryndler et al. 2006), and are strongly recommended for revegetation (Noyd et al. 1996). Unlike inorganic fertilizers, organic amendments not only provide essential nutrients but also ameliorate metal toxicity and improve soil structure and the activities of soil microbes. There have been reports that cattle manure (Wang et al. 2012, 2013), composted yard waste (Noyd et al. 1996), biosolids (Madejon et al. 2012; Madejon et al. 2010), and sewage sludge (Thorne et al. 1998) could be reliable methods for AM restoration of polluted substrates.

As discussed above, soil beneficial microorganisms may interact positively with the AMF to exert better effects on plant growth and soil quality (Frey-Klett et al. 2007). Both plant growth-promoting bacteria and AMF can be used to facilitate the growth of plants in heavy metal-contaminated soils (Gamalero et al. 2009; Artursson et al. 2006). Plant growth-promoting bacteria can interact synergistically with AMF to improve plant growth and tolerance through various mechanisms (Gamalero et al. 2009; Artursson et al. 2006), indicating a potential in the revegetation of mine areas (Lumini et al. 1994; Wu et al. 2009). Among them, *Rhizobium* is the most widely used microorganism, because it has a capacity of fixing atmospheric N_2 within root nodules of leguminous plants. In mine areas, N deficiency is the most common problem for revegetation (Bradshaw and Wong 2002). Leguminous plants often develop tripartite symbioses (rhizobium-AMF-legume) and may respond positively to both rhizobial inoculants and AMF simultaneously, which enables them to survive in stressful conditions. The positive synergistic interactions among members of the tripartite symbiotic association generally result in improved N, P and K uptake and better plant growth, providing a promising approach for revegetation of coal mine substrates (Wu et al. 2009). In addition, dual inoculation with an indigenous Cd-adapted strain of *Glomus mosseae* and a Cd-adapted bacterium (*Brevibacillus* sp.) on *Trifolium repens* grown in Cd-polluted soil led to an improved symbiotic structures (nodules and AM colonization), increased plant growth, and nutrient acquisition (N and P) (Vivas et al. 2003b). Similar conclusions have been drawn by treating *T. repens* plants with AMF and *Brevibacillus* strains with Pb (Vivas et al. 2003a), Zn ((Vivas et al. 2006a), and Ni (Vivas et al. 2006b). The positive synergistic interactions on the growth of plants grown in polluted soil have been reported between AMF and other fungi such as *Fusarium concolor*; *Trichoderma koningii* (Arriagada et al. 2007), *Trichoderma harzianum* (Arriagada et al. 2009), and *Trichoderma pseudokoningii* (Nazir and Bareen 2011); phosphate-solubilizing fungus *Aspergillus tubingensis* (Babu and Reddy 2011); and even soil animals such as earthworm (Ma et al. 2006; Hua et al. 2010). These findings highly support the promising potential of dual inoculation of AMF and other soil organisms during revegetation.

Toxic metals are generally high in metal ore mining area or in soil contaminated by smelter or refinery. Under such conditions, AMF generally reduce the uptake of toxic metals by plants, or decrease them translocate from roots to shoots, thus protecting plants against toxicity (Chen et al. 2007b; Bissonnette et al. 2010; Ultra et al. 2007). As summarized in Table 11.2, AMF used commonly show little host specificity. The most widely studied species, such as *G. mosseae* and *G. intraradices*, have been shown to exhibit growth-promoting effects on most native or nonnative plants. This apparently facilitates the standardized production and commercialization of AMF inoculants. On the other hand, AMF are obligate biotrophs, so they typically gain by interaction with host plants. AMF often exhibit host-specific growth responses in various conditions. In the use in restoration of mine areas, AMF effects on plant growth and metal accumulation vary with plant type and fungal species. For plants with different heavy metal tolerance, different AMF species also show different mycorrhizal effects. When grown in a multi-metal-contaminated soil, maize growth showed no significant response to both AMF inoculants, while mycorrhizal *Elsholtzia splendens* plants grew better than nonmycorrhizal ones, and the concentrations of Cu, Zn, Pb, and Cd in plants also varied with plant species and AMF (Wang et al. 2005, 2007).

Hyperaccumulating plants represent a range of members with excellent ability to grow on metalliferous sites and to accumulate extraordinarily high amounts of metals in their aerial parts, far exceeding the levels found in the majority of species, without suffering phytotoxicity, and thus have a great potential in phytoremediation and revegetation of mine sites. However, most hyperaccumulators have a slow growth rate and small biomass, which greatly limits their application in ecosystem restoration. There have been reports that AMF colonize a number of hyperaccumulators, such as *Berkheya coddii* accumulating Ni (Turnau and Mesjasz-Przybylowicz 2003), *Solanum nigrum* accumulating Cd and Zn (Liu et al. 2015; Marques et al. 2006), *Pteris vittata* accumulating As (Al Agely et al. 2005; Leung et al. 2006; Liu et al. 2005), and *Sedum alfredii* accumulating Zn and Cd (Hu et al. 2013a; Wu et al. 2007). Most of these results show that AMF exert plant growth-promoting effects. Additionally, mycorrhizal effects also vary with AMF origin and their tolerance. Inoculation with *G. versiforme* significantly increased the growth and Cd concentrations of *S. nigrum* in 25 and 50 mg Cd kg^{-1} soils, but decreased Cd concentrations of the plants in 100 mg Cd kg^{-1} soil (Liu et al. 2015). When grown in soil with 1000 mg/kg Zn, Zn concentration in leaves of *S. nigrum* was increased by *G. claroideum* and *G. intraradices*, but not affected by *G. mosseae*, while all the AMF species exerted no significant influence on plant growth (Marques et al. 2006). The relationships between hyperaccumulators and AMF still need more efforts.

Unfortunately, most other hyperaccumulators belong to Brassicaceae, which are generally believed to be nonmycorrhizal (Regvar et al. 2003). Therefore, the chance is low to develop a symbiotic association between AMF and Brassicaceae hyperaccumulators used for restoration. *Thlaspi caerulescens* is among the most well-known hyperaccumulators (Assunção et al. 2003), which is able to grow in serpentine soils rich in Zn, Co, Pb, Cr, Cd, and Ni and accumulate high levels of Zn and Cd. However, most hyperaccumulating species belonging to the genus *Thlaspi* including *T. caerulescens* are difficultly colonized by AMF (Regvar et al. 2003). Although AMF can colonize the metal hyperaccumulating *T. praecox*, AMF inoculation showed no significant promoting effects and even negative effect on root biomass (Vogel-Mikus et al. 2006). So AMF can't be included in restoration using Brassicaceae plants.

11.5.2 Restoration of Desertified Areas

Over 40 million km^2, or 35% of the total land surface, can be described as drylands suffering from permanent, seasonal, or periodic significant moisture deficiency (Jeffries et al. 2002). Human disturbance activities, such as global warming, deforestation, and overgrazing, can cause or accelerate desertification processes, resulting in loss or disturbance of the vegetation cover, increased soil erosion or salinity, decreased available nutrients and organic matter, loss of microbial propagules, etc. Desertification in arid and semiarid lands may lead to decreased livestock and grain yields and possibly threaten ecological and food security. Therefore, restoration and revegetation of desertified ecosystems deserve high attention.

As summarized in Table 11.3, the role of AMF in restoration of desertified soil/areas has been widely investigated. Most results have shown that AMF exhibit various positive effects: enhanced drought tolerance, increased plant survival, cover, growth and nutrient uptake, improved soil physical-chemical and biological properties. Native and allochthonous AMF all exert positive plant growth-promoting effects, sometimes varying with plant species. *G. coronatum* native in the field site was more effective than the exotic *G. intraradices* on the growth and nutrition of *Anthyllis cytisoides* (Requena et al. 1997). A field experiment conducted in a semiarid area showed that the mixture of native AMF was more effective than the non-native *Glomus claroideum* with respect to increasing shoot biomass and foliar N, P, and K contents of *Olea europaea* subsp. *sylvestris*, *Retama sphaerocarpa*, and *Rhamnus lycioides* while exhibited equally effective for *Pistacia lentiscus*; mycorrhizal propagules were highest in soil from the center of the canopy of *P. lentiscus* and *R. lycioides* inoculated with native AMF, while the opposite results were obtained for plants of *O. europaea* and *R. sphaerocarpa* (Caravaca et al. 2005a). In another experiment, the most effective AMF-plant combinations were observed on *G. claroideum*-*R. sphaerocarpa* and mixed native AMF-*R. lycioides*, respectively, but in most cases, the mixture of native AMF was equally as or even more effective than the allochthonous *G. claroideum* regarding increases in plant growth (Caravaca et al. 2003c). Overall, native AMF mixture contains a consortium of fungal species that may develop better adaptability to local habitats, and are likely to make a collective contribution, and consequently functions better than sole or allochthonous species. However, both AMF and plants suitable for restoration need careful selection.

Table 11.3 The role of AMF in restoration of degraded/disturbed areas

Degraded or disturbed site/soil	Host plants	AMF species	AMF effects	Experiment type	References
Desertified semiarid Mediterranean area	*Anthyllis cytisoides*, *Spartium junceum*, *Robinia pseudoacacia*, *Medicago arborea*, *Acacia caven*, *Prosopis chilensis*	*Glomus mosseae*, *G. fasciculatum*, *Glomus* sp., *Scutellospora* sp.	Increased outplanting performance, plant survival, and biomass development	Greenhouse and field	Herrera et al. (1993)
Field in a desertified Mediterranean	*Pistacia lentiscus*, *Rhamnus lycioides*, *Olea europaea* subsp. *sylvestris*, *Retama sphaerocarpa*	*G. intraradices*	Increased plant N, P, and K uptake	Field	Palenzuela et al. (2002)
Field in a semiarid Mediterranean area	*Olea europaea* ssp. *sylvestris*, *Retama sphaerocarpa*, *Rhamnus lycioides*	*G. claroideum*, a mixture of native AMF	Increased shoot biomass and N, P, and K contents, antioxidant enzyme activities	Field	Alguacil et al. (2003)
Field in a semiarid Mediterranean area	*Retama sphaerocarpa*	*G. intraradices*, *G. deserticola*, *G. mosseae*	Increased plant growth and improved soil properties	Field	del Mar Alguacil et al. (2004)
Field in a semiarid Mediterranean area	*Olea europaea* subsp. *sylvestris*, *Retama sphaerocarpa*	A mixture of native AMF, *G. claroideum*	Increased plant growth and improved soil properties	Field	Alguacil et al. (2005)
Field in a degraded semiarid Mediterranean area	*Dorycnium pentaphyllum*	*G. intraradices*, *G. deserticola*, *G. mosseae*	Increased plant growth and improved soil properties	Field	Alguacil et al. (2008)
Soil from semiarid Mediterranean area	*Medicago arborea*	*G. mosseae*, *G. deserticola*, *G. fasciculatum*	Increased plant growth	Greenhouse	Valdenegro et al. (2001)
Field in a desertified Mediterranean area	*Anthyllis cytisoides*, *Lavandula multifida*	*G. intraradices*, indigenous AMF	Increased plant growth and improved soil properties	Field	Requena et al. (2001)

Field in a degraded semiarid Mediterranean area	*Pistacia lentiscus*	*G. intraradices*	Increased plant growth and foliar P uptake, improved soil properties	Field	Caravaca et al. (2002b)
Field in semiarid Mediterranean area	*Olea europaea* subsp. *sylvestris*	*G. intraradices*	Increased plant height and basal diameter, increased soil aggregate stability	Field	Caravaca et al. (2002a)
Field in semiarid Mediterranean area	*Olea europaea* subsp. *sylvestris*, *Rhamnus lycioides*	*G. intraradices*	Increased soil water-soluble carbon content and enzyme activities, aggregate stability	Field	Caravaca et al. (2002c)
Field in semiarid Mediterranean area	*Olea europaea* subsp. *sylvestris*, *Rhamnus lycioides*	*G. intraradices*	Increased shoot biomass and foliar N, P, and K contents	Field	Caravaca et al. (2003d)
Field in semiarid Mediterranean area	*Olea europaea* subsp. *sylvestris*, *Retama sphaerocarpa*, *Rhamnus lycioides*	*G. claroideum*, a mixture of native AMF	Increased shoot biomass and N, P, and K contents, increased antioxidant enzyme activities in leaves	Field	Alguacil et al. (2003)
Simulated drought stress	*Myrtus communis*, *Phillyrea angustifolia*	*G. intraradices*, a mixture of *G. intraradices*, *G. deserticola*, and *G. mosseae*	Increased shoot biomass and foliar N and P contents	Greenhouse	Caravaca et al. (2005b)
Soil from a semiarid Mediterranean area	*Juniperus oxycedrus*	*G. intraradices*, a mixture of *G. intraradices*,*G. deserticola*, and *G. mosseae*	Increased shoot dry weight	Nursery	Caravaca et al. (2006)
Field in semiarid Mediterranean area	*Olea europaea* subsp. *sylvestris*, *Retama sphaerocarpa*, *Rhamnus lycioides*, *Pistacia lentiscus*	*G. claroideum*, a mixture of native AMF	Increased shoot biomass and foliar N, P, and K contents	Field	Caravaca et al. (2005a)

(continued)

Table 11.3 (continued)

Degraded or disturbed site/soil	Host plants	AMF species	AMF effects	Experiment type	References
Field in semiarid Mediterranean area	*Olea europaea* subsp. *sylvestris*, *Pistacia lentiscus*, *Retama sphaerocarpa*, *Rhamnus lycioides*	*G. claroideum*, a mixture of native AMF	Increased shoot biomass and foliar N, P, and K contents	Field	Caravaca et al. (2003c)
Field in semiarid Mediterranean area	*Retama sphaerocarpa*	*G. intraradices*	Increased plant growth and N and P uptake, improved soil physical-chemical and biological properties	Field	Caravaca et al. (2003b)
Field in a semiarid, degraded Mediterranean area	*Retama sphaerocarpa*	*G. intraradices*, *G. deserticola*, *G. mosseae*	Increased plant growth and N and P uptake	Field	Caravaca et al. (2005c)
Field in a desertified Mediterranean area	*Dorycnium pentaphyllum*	*G. intraradices*, *G. deserticola*, *G. mosseae*	Increased plant growth and N, P, and K uptake	Field	Caravaca et al. (2004b)
Soil from a desertified Mediterranean area	*Anthyllis cytisoides*, *Spartium junceum*	*G. fasciculatum*	Increased plant growth, survival and P uptake	Greenhouse	Salamanca et al. (1992)
Soil from a semiarid Mediterranean area (nutrient deficient)	*Anthyllis cytisoides*	*G. coronatum*, *G. intraradices*	Increased plant growth and nutrient uptake	Greenhouse	Requena et al. (1997)
Degraded soils from desertification-threatened areas	*Lavandula spica*	*G. mosseae*	Increased plant growth and N, P, and K uptake	Greenhouse	Azcón and Barea (1997)
Semiarid disturbed soil	*Rosmarinus officinalis*	*G. intraradices*	Increased plant growth, survival, and cover	Greenhouse	Estaún et al. (1997)
Soil from the Gurbantünggüt Desert	*Plantago minuta*	*G. mosseae*, *G. etunicatum*	Increased plant growth and N, P, and K uptake	Greenhouse	Shi et al. (2015)

A semiarid Indian wasteland soil (erosion)	*Cassia siamea*	*G. fasciculatum*, *G. macrocarpum*	Increased plant growth and nutrient uptake, plant survival in field	Greenhouse and field	Giri et al. (2005)
Simulated saline soil	*Asteriscus maritimus*	*Rhizophagus intraradices*, *Claroideoglomus etunicatum*, *Septoglomus constrictum*	Increased shoot dry weight, efficiency of photosystem II, stomatal conductance, and accumulation of glutathione, increased survival	Greenhouse	Estrada et al. (2013)
Bare saline-alkaline soil	*Leymus chinensis*	*G. mosseae*, *G. geosporum*	Increased plant growth and survival, N, P, K, and Ca uptake, root/shoot ratio	Greenhouse and field	Zhang et al. (2011b)
Simulated saline soil	*Atriplex nummularia*	A mixture of six AMF	Increased plant growth and nutrient uptake	Greenhouse	Asghari et al. (2005)
Soil irrigation with saline water	*Lycopersicon esculentum*	*G. mosseae*	Increased plant biomass and yield, plant nutrient uptake, lower Na concentrations in shoot	Greenhouse	Al-Karaki (2006)
Simulated saline soil	*Lycopersicon esculentum*	*G. fasciculatum*	Increased plant growth	Greenhouse	Pond et al. (1984)
Simulated saline soil	*Olea europaea*	*G. mosseae*, *G. intraradices*, *G. claroideum*	Increased plant survival, growth, and nutrient acquisition	Nursery	Porras-Soriano et al. (2009)
Soil irrigated with seawater	*Vigna radiata*	*G. clarum*	Increased plant dry weights	Greenhouse	Rabie (2005)
Simulated saline soil	*Acacia nilotica*	*G. fasciculatum*	Increased plant biomass and nutrient uptake, K/Na ratio in tissues	Greenhouse	Giri et al. (2007)

(continued)

Table 11.3 (continued)

Degraded or disturbed site/soil	Host plants	AMF species	AMF effects	Experiment type	References
Simulated saline soil	*Kosteletzkya virginica*	*G. aggregatum*, *G. mosseae*	Increased plant growth and soil enzyme activity	Greenhouse	Zhang et al. (2011a)
Simulated saline-sodic soil	*Parthenium argentatum*	*G. intraradices*	Increased plant growth, decreased Na concentration in shoot	Greenhouse	Pfetffer and Bloss (1988)
Soil from Al-Kassab saline area	*Triticum aestivum*	*G. mosseae*, *G. deserticola*, *Gigaspora gregaria*	Increased plant growth and nutrient uptake, decreased Na uptake	Greenhouse	Abdel-Fattah and Asrar (2012)
Simulated saline soil	Native xeroriparian plants	*G. intraradices*	Increased plant survival and growth for most species	Greenhouse	Beauchamp et al. (2009)
Alkaline anthropogenic-stressed sediments	*Conyza bilbaoana*	*G. intraradices* *G. mosseae*	Increased plant growth	Greenhouse	Oliveira et al. (2005)
Naturally acid or alkaline soil	*Zea mays*	*G. etunicatum*, *G. diaphanum*, *G. intraradices*	Increased plant growth	Greenhouse	Clark and Zeto (1996)
Degraded spring grassland	*Erodium oxyrrhynchum*, *Hyalea pulchella*, *Trigonella arcuata*, *Schismus arabicus*	*G. mosseae*, *G. etunicatum*, *G. intraradices*	Increased plant cover and shoot dry weights	Field	Zhang et al. (2012)
Disturbed tallgrass prairie	A range of native grasses	A mixture of AMF	Increased native plant cover	Field	Smith et al. (1998)
Red soil (acid)	*Vigna radiata*	*G. caledonium*, *G. manihotis*	Increased plant growth	Greenhouse and field	Wu et al. (2002)
Naturally acid soil	*Panicum virgatum*	*G. clarum*, *G. diaphanum*	Increased plant biomass	Greenhouse	Clark et al. (1999)

Abandoned quarries (acid, low fertility)	*Rhynchelytrum repens*	*G. diaphanum*, *G. geosporum*, *G. mosseae*	Increased plant survival and biomass	Greenhouse	Chen et al. (2008b)
Fragile degraded tropical lands (acid, low P)	*Brachiaria decumbens*	A mixture of *Acaulospora laevis*, *Entrophospora colombiana*, *A. myriocarpa*, and *Scu. fulgida*	Increased plant cover, biomass, and uptake of nutrients	Field	Cuenca et al. (1998)
Old fields (low P)	*Acacia acuminata*, *Eucalyptus loxophleba* subsp. *loxophleba*, *Hakea preissii*	A mixture of *Scu. calospora*, *G. intraradices*, *G. mosseae*	Increased plant biomass and P uptake	Greenhouse	Standish et al. (2007)
Soil with drought stress	*Acacia nilotica*, *Leucaena leucocephala*	*G. etunicatum*, *G. mosseae*, *G. occultum*	Increased plant dry weight and P uptake	Greenhouse	Michelsen and Rosendahl (1990)
Dolomitic soil (low water-holding capacity)	*Thymus granatensis*, *T. mastichina*	Indigenous AMF	Increased plant growth	Greenhouse	Navarro-Fernández et al. (2011)
Marginal wasteland soil (received water erosion, low fertility)	*Acacia nilotica* var. *cupriciformis*	Indigenous AMF consortium	Increased plant dry weight and P uptake	Nursery	Sharma et al. (1996)
Simulated eroded soil	*Gliricidia sepium*, *Leucaena leucocephala*	*G. deserticola*	Increased growth of *Gliricidia sepium*	Greenhouse	Fagbola et al. (2001)
Abandoned agricultural field	*Sporobolus wrightii*	*G. mosseae*, *G. eburneum*, *G. microaggregatum*, *A. delicata*, *G. spurcum*, *G. macrocarpum*	Increased plant survival, basal diameter, and tiller and panicle production	Field	Richter and Stutz (2002)
Volcanic tephra field	*Leymus arenarius*, *Deschampsia beringensis*	*G. mosseae*, *G. intraradices*	Increased plant survival and growth, improved soil aggregation	Greenhouse and field	Enkhtuya et al. (2003)

Mediterranean ecosystems are characterized with scarce and irregular rainfall and a very long dry period in summer, usually lasting for several months, which make them easily suffer from degradation and even desertification. AMF associated with plant communities from semiarid SE Spain, a typical semiarid Mediterranean area, have been extensively surveyed, and almost 200 cultures of AMF ecotypes from diverse morphotypes within 11 genera, most within *Glomus*, were identified and cultured (Barea et al. 2011). Numerous field studies have been conducted in a semiarid Mediterranean area located in Spain (see Table 11.3) and confirmed that AMF play a substantial role in the establishment of vegetation. In these experiments, native shrub species, such as *A. cytisoides*, *O. europaea* subsp. *sylvestris*, *P. lentiscus*, *R. sphaerocarpa*, and *R. lycioides*, were the most common target plants, which have been shown to benefit from AMF (Caravaca et al. 2002c, 2003a, c, 2005a). In another 4-year trial, AMF effects in revegetation strategies were investigated using woody legumes including two native shrubs (*A. cytisoides* and *Spartium junceum*) and four nonnative tree legumes (*Robinia pseudoacacia*, *Acacia caven*, *Prosopis chilensis*, and *Medicago arborea*) (Herrera et al. 1993). The results showed that only the native shrubs were able to establish and develop under the local arid conditions, and inoculation with selected rhizobia and AMF improved outplanting performance, plant survival, and biomass production. A revegetation strategy using drought-tolerant *A. cytisoides* in combination with AMF may be feasible. In desertification-threatened Mediterranean areas, the growth of *Lavandula spica*, a small woody leguminous shrub, is generally being limited by an overall lack of nutrient supply but can be greatly improved by AMF inoculation (Azcón and Barea 1997). It's believed that lavender plants must be mycorrhizal in order to thrive in the degraded soils (deficient in nutrients) from desertification-threatened areas in typical Mediterranean ecosystems.

A series of field experiments have been conducted to test the interaction on plant performance and soil quality between AMF and compost, using *O. europaea* var. *sylvestris* (Caravaca et al. 2002a, c, 2003d; Palenzuela et al. 2002), *P. lentiscus* (Caravaca et al. 2002b, 2003d; Palenzuela et al. 2002), *R. sphaerocarpa* (Caravaca et al. 2003b; Palenzuela et al. 2002), and *R. lycioides* (Caravaca et al. 2002c, 2003e; Palenzuela et al. 2002). The results showed that water-soluble carbon, water-soluble carbohydrates, microbial biomass carbon content and enzyme activities, and rhizosphere aggregate stability were all increased in the mycorrhizosphere of the target plants. Similar results were also obtained using other organic amendments, such as sewage sludge (del Mar Alguacil et al. 2004), *A. niger*-treated dry olive cake (Alguacil et al. 2008; Caravaca et al. 2004b, 2006), and sugar beet residues (Medina et al. 2004; Caravaca et al. 2004a).

Native leguminous shrubs are mostly recommended to use for revegetation strategies, because of their strong drought tolerance, N_2-fixing capacity (forming symbionts with N_2-fixing rhizobia), and high mycorrhizal dependency (Salamanca et al. 1992; Valdenegro et al. 2001; Herrera et al. 1993; Requena et al. 1997, 2001). The results from two long-term experiments in a desertified Mediterranean ecosystem showed that dual inoculation with a mixture of indigenous AMF and rhizobial N_2-fixing bacteria not only enhanced the establishment of key plant species

(*A. cytisoides* and *Lavandula multifida*) but also increased soil fertility and quality (increased N content and higher levels of OM and soil aggregation) (Requena et al. 2001). In addition, PGPR also can be used in AMF-assisted revegetation. PGPR have been found to significantly increase growth of AMF-colonized *Medicago arborea* plants associated with *Rhizobium*, indicating selection of appropriate microbial combinations may be critical and decisive for the successful establishment of plants (Valdenegro et al. 2001). However, it's necessary to take into account the specific functional compatibility relationships between different microbial components (Requena et al. 1997).

11.5.3 Restoration of Salt-Affected Soils

Over 800 million hectares of land throughout the world is estimated to be salt affected, accounting for approximately 7% of the world total land area. This land covers a range of soils defined as saline, saline-sodic, and sodic (Ghassemi et al. 1995). High concentrations of Na^+ and Cl^- ions occur in salt-affected soil, which negatively affect vegetation and soil organisms directly or indirectly by extreme soil pH, low nutrient availability, and poor soil structure. AMF occur widely in various salty soils (Wang et al. 2004; García and Mendoza 2007; García and Mendoza 2008). Furthermore, AMF have been shown to alleviate salt stress to plants and to enhance plant growth and crop production in salty soils (Evelin et al. 2009; Miransari 2010; Porcel et al. 2012).

As shown in Table 11.3, numerous experiments have confirmed the positive effects of AMF on plants grown under salt stress; however, most of them were conducted under greenhouse conditions, and only one was carried out at field scale (Zhang et al. 2011b). *Leymus chinensis* inoculated with or without AMF (*Glomus mosseae* and *G. geosporum*) were grown in either pots filled with soil from bare saline-alkaline land, or transplanted seedlings into field plots, to determine the influence of AMF on the reestablishment of this dominant grass species in bare degraded land. The results showed that AMF increased plant biomass, root/shoot ratio, and the uptake of N, P, K^+, and Ca^{2+}, but decreased Mg^{2+}, Na^+, and Cl^- uptake under saline-alkaline stress, resulting in higher K/Na, Ca/Na, P/Na, and P/Cl ratios in AM plants. AMF also significantly increased the survival of seedlings outplanted to a bare saline-alkaline land. These results indicate AMF could improve the reestablishment of vegetation in saline-alkaline soil and drive the vegetation restoration to a community dominated by original species. Symbiotic effectiveness of four AMF strains (three native fungal isolates from a saline soil and one allochthonous from collection, *Rh. intraradices*) in *A. maritimus* under increasing salinity stress was compared, and the results showed that at the highest level of salinity, plants inoculated with native AMF had higher shoot dry weight, efficiency of photosystem II, stomatal conductance, and glutathione accumulation than those inoculated with allochthonous AMF (Estrada et al. 2013). Moreover, only 30% of plants inoculated with the collection AMF survived, while plant survival reached 100% for those inoculated with the three native AMF. Apparently, in addition to the mechanism underlying AMF tolerance to salt stress, more future efforts should made on selection of effective AMF species/strains and field-scale restoration cases.

11.5.4 Restoration of Degraded Grasslands

Climate change and anthropogenic disturbances such as overgrazing may result in grassland degradation. For example, a large area of grasslands in Inner Mongolia, China, has suffered from various extents of degradation because of excessive cutting and overgrazing (Ellis 1992). Although "degradation" seems less serious than "desertification," degraded grasslands have similar ecological imbalances to desertified lands, such as low plant productivity and soil fertility, poor soil structure, and reduced microbial activity.

AMF have been found to occur widely in various grasslands (van der Heijden 2004; Hempel et al. 2007; Jacobson 1997), including overgrazed or degraded steppe (Su and Guo 2007; Tian et al. 2009a, b), indicating their potential roles in maintenance of healthy grasslands and in restoration of degraded grasslands. AMF inoculation effects on four plant species (two grasses and two forbs) were tested using 1-year-old grassland microcosms, and the results showed that the AMF-inoculated seedlings grew larger and obtained more P, indicating AMF can act as a symbiotic support system for seedling establishment and reduce recruitment limitation in grassland (van der Heijden 2004). The restoration of prairie communities in North America has attracted considerable attention in recent years. The effect of AMF on the establishment of an early successional tallgrass prairie was investigated in field plots of a disturbed area in Minnesota, and the results showed that percent cover of native grasses was significantly increased by AMF, suggesting that AMF inoculation promotes the development of early successional tallgrass prairie communities (Smith et al. 1998). The effects of AMF on restoration of degraded grassland were determined using 3 years of field inoculation experiments in a Central Asian desert (Zhang et al. 2012). The results indicate that inoculation with AMF significantly improved the biomass, density, and cover of ephemerals, such as *Erodium oxyrrhynchum*, *Hyalea pulchella*, *Trigonella arcuata*, and *Schismus arabicus*, while did not affect nonmycorrhizal species *Alyssum linifolium* and *Ceratocarpus arenarius*. From the above results, AMF can speed up the regeneration process and restoration strategies of degraded grasslands.

11.6 Future Prospects

In addition to the abovementioned aspects, AMF have been shown to exhibit positive roles in the restoration of other fragile or disturbed habitats, such as acid soil (Clark et al. 1999; Wu et al. 2002), eroded soil (Burri et al. 2013; Sharma et al. 1996; Fagbola et al. 2001), abandoned soil (Richter and Stutz 2002), anthropogenic sediment (Oliveira et al. 2005), coastal sand dune (Gemma and Koske 1997), and even volcanic tephra field (Enkhtuya et al. 2003). Microbes can survive in various extreme environments (Rothschild and Mancinelli 2001). We undoubtedly believe that AMF can survive in any adverse environments, where plants can tolerate and grow, and can make substantial contributions to survival and growth of the plants suffering from various abiotic and biotic stresses.

However, AMF are just a promising helper but not a panacea for ecosystem restoration. AMF and plants can't be used in restoration of highly polluted soils where plants can't survive. Since AMF consume a proportion of photosynthate from their host plants, it is not appropriate to assume that AMF effects on plant growth always are positive, but sometimes neutral or even negative (Johnson et al. 1997). Stahl et al. (1988) found that the native AMF produced only limited amounts of mycorrhizal infection and no significant effect on the establishment, growth, or survival of sagebrush plants on a surface mine reclamation site. Commercial or native AMF inocula were used in a coastal sage scrub restoration project in Southern California (Aprahamian et al. 2016), but neither inoculum type significantly altered shoot biomass of native shrubs and forbs nor density of adult and seedling shrubs, indicating AMF inoculation did not improve restoration success in a coastal sage scrub ecosystem. Therefore, benefits and costs of AMF used for ecosystem restoration should be comprehensively evaluated.

Environmental stress is generally detrimental to AMF, but there may be a considerable hidden diversity of AMF existing in degraded, disturbed, fragile, and even extreme ecosystems. Using 454 sequencing of the large subunit rDNA region, the results revealed stochastic local reassembly and high disturbance tolerance within AMF communities (Lekberg et al. 2012). AMF species diversity survey deserves continuous attention. Because indigenous AMF might have developed excellent adaptation to adversity, the tolerant species/strains need to be investigated and selected especially from extreme environments. Non-tolerant or allochthonous AMF species may become tolerant after assimilation in polluted substrate. In addition, the tolerance of AMF species with different origins and the underlying mechanisms should be clearly elucidated.

Selection of suitable plant species is the key element for revegetation. More research needs to be conducted on the selection of combinations between AMF and hyperaccumulator plants, excluder plants, and transgenic plants, besides well-known native plants and leguminous plants. Both hyperaccumulator and excluder plants can normally survive in soil polluted with high levels of toxic metals, but hyperaccumulators accumulate very high contents of heavy metals especially in their aerial parts, while excluders restrict uptake of heavy metals into their plants especially in aerial parts. Symbiosis between AMF and hyperaccumulators or excluders may be promising for restoration of heavy metal-polluted sites; however, the following should be taken into account:

1. Most hyperaccumulators belonging to the Brassicaceae family are nonmycorrhizal and unable to form efficient symbiotic association with AMF.
2. Most hyperaccumulators can only accumulate one or two heavy metals and can't be used in multi-metal-polluted soil.

3. AMF may not perform efficiently in highly polluted soils.
4. AMF generally protect plants via decreasing heavy metal uptake by symbiotic associations; thus, AMF may be more suitable for excluders than hyperaccumulators in restoration process.

Most plants are unable to hyperaccumulate or exclude toxic pollutants and to grow normally in polluted sites. Transgenic plants modified with heavy metal-related genes acquire stronger tolerance than common nontransgenic plants and consequently grow better in polluted soils. It's an alternative approach, instead of native plants, to use transgenic plants in AMF-assisted restoration.

The application of other restoration approaches in combination with AMF needs sufficient investigations. The effects of organic amendments are overall positive, but attention should be paid on the pollutants in them, such as heavy metals and pesticides in livestock and poultry manure and sewage sludge. Benefits and costs and advantages and disadvantages of PGPR and chemical fertilizers all need to be carefully evaluated when they are applied together with AMF.

Most previous studies were conducted using pot culture in greenhouse conditions, which are much far different from the real sites that need restoration. More large-scale experiments including field cases and on-site trials should be conducted to confirm the results obtained from inside greenhouses and laboratories. Soil conditions, climate, water and irrigation, fertilization, disease control, and other factors affecting the field application of AMF in restoration all need to be taken into account comprehensively to enhance AMF effects.

Acknowledgments This work was sponsored by the National Natural Science Foundation of China (41471395, 41171369), the Plan for Scientific Innovation Talent of Henan Province (154100510010), and the Innovation Team Foundation of Henan University of Science and Technology (2015TTD002).

References

Abdel-Fattah GM, Asrar AWA (2012) Arbuscular mycorrhizal fungal application to improve growth and tolerance of wheat (*Triticum aestivum* L.) plants grown in saline soil. Acta Physiol Plant 34(1):267–277

Al Agely A, Sylvia DM et al (2005) Mycorrhizae increase arsenic uptake by the hyperaccumulator Chinese brake fern (*Pteris vittata* L.) J Environ Qual 34(6):2181–2186

Alguacil M, Hernández JA, Caravaca F et al (2003) Antioxidant enzyme activities in shoots from three mycorrhizal shrub species afforested in a degraded semi-arid soil. Physiol Plant 118(4):562–570

Alguacil M, Caravaca F, Roldán A (2005) Changes in rhizosphere microbial activity mediated by native or allochthonous AM fungi in the reafforestation of a Mediterranean degraded environment. Biol Fertil Soils 41(1):59–68

Alguacil MM, Caravaca F, Azcón R et al (2008) Changes in biological activity of a degraded Mediterranean soil after using microbially-treated dry olive cake as a biosolid amendment and arbuscular mycorrhizal fungi. Eur J Soil Biol 44(3):347–354

Al-Karaki GN (2006) Nursery inoculation of tomato with arbuscular mycorrhizal fungi and subsequent performance under irrigation with saline water. Sci Hortic 109(1):1–7

Alori E, Fawole O (2012) Phytoremediation of soils contaminated with aluminium and manganese by two arbuscular mycorrhizal fungi. J Agric Sci 4(8):246–252

Aloui A, Recorbet G, Gollotte A et al (2009) On the mechanisms of cadmium stress alleviation in *Medicago truncatula* by arbuscular mycorrhizal symbiosis: a root proteomic study. Proteomics 9(2):420–433

Aloui A, Recorbet G, Robert F et al (2011) Arbuscular mycorrhizal symbiosis elicits shoot proteome changes that are modified during cadmium stress alleviation in *Medicago truncatula*. BMC Plant Biol 11:75

Aprahamian AM, Lulow ME, Major MR et al (2016) Arbuscular mycorrhizal inoculation in coastal sage scrub restoration. Botany 94(6):493–499

Aroca R, Vernieri P, Ruiz-Lozano JM (2008) Mycorrhizal and non-mycorrhizal Lactuca sativa plants exhibit contrasting responses to exogenous ABA during drought stress and recovery. J Exp Bot 59(8):2029–2041

Arriagada CA, Herrera MA, Ocampo JA (2007) Beneficial effect of saprobe and arbuscular mycorrhizal fungi on growth of *Eucalyptus globulus* co-cultured with *Glycine max* in soil contaminated with heavy metals. J Environ Manag 84(1):93–99

Arriagada C, Aranda E, Sampedro I et al (2009) Contribution of the saprobic fungi *Trametes versicolor* and *Trichoderma harzianum* and the arbuscular mycorrhizal fungi *Glomus deserticola* and *G. claroideum* to arsenic tolerance of *Eucalyptus globulus*. Bioresour Technol 100(24):6250–6257

Artursson V, Finlay RD, Jansson JK (2006) Interactions between arbuscular mycorrhizal fungi and bacteria and their potential for stimulating plant growth. Environ Microbiol 8(1):1–10

Asghari HR, Marschner P, Smith SE et al (2005) Growth response of Atriplex nummularia to inoculation with arbuscular mycorrhizal fungi at different salinity levels. Plant Soil 273(1–2):245–256

Assunção AG, Schat H, Aarts MG (2003) Thlaspi caerulescens, an attractive model species to study heavy metal hyperaccumulation in plants. New Phytol 159(2):351–360

Azcón R, Barea J (1997) Mycorrhizal dependency of a representative plant species in Mediterranean shrublands (*Lavandula spica* L.) as a key factor to its use for revegetation strategies in desertification-threatened areas. Appl Soil Ecol 7(1):83–92

Azcón R, Perálvarez MC, Biró B et al (2009) Antioxidant activities and metal acquisition in mycorrhizal plants growing in a heavy-metal multicontaminated soil amended with treated lignocellulosic agrowaste. Appl Soil Ecol 41(2):168–177

Babu AG, Reddy MS (2011) Dual inoculation of arbuscular mycorrhizal and phosphate solubilizing fungi contributes in sustainable maintenance of plant health in fly ash ponds. Water Air Soil Pollut 219(1–4):3–10

Bago B, Pfeffer PE, Shachar-Hill Y (2000) Carbon metabolism and transport in arbuscular mycorrhizas. Plant Physiol 124(3):949–958

Bai J, Lin X, Yin R et al (2008) The influence of arbuscular mycorrhizal fungi on as and P uptake by maize (*Zea mays* L.) from As-contaminated soils. Appl Soil Ecol 38(2):137–145

Barea J, Palenzuela J, Cornejo P et al (2011) Ecological and functional roles of mycorrhizas in semi-arid ecosystems of Southeast Spain. J Arid Environ 75(12):1292–1301

Bearden BN, Petersen L (2000) Influence of arbuscular mycorrhizal fungi on soil structure and aggregate stability of a vertisol. Plant Soil 218(1–2):173–183

Beauchamp VB, Walz C, Shafroth PB (2009) Salinity tolerance and mycorrhizal responsiveness of native xeroriparian plants in semi-arid western USA. Appl Soil Ecol 43(2):175–184

Bedini S, Pellegrino E, Avio L et al (2009) Changes in soil aggregation and glomalin-related soil protein content as affected by the arbuscular mycorrhizal fungal species *Glomus mosseae* and *Glomus intraradices*. Soil Biol Biochem 41(7):1491–1496

Benedetto A, Magurno F, Bonfante P et al (2005) Expression profiles of a phosphate transporter gene (*GmosPT*) from the endomycorrhizal fungus *Glomus mosseae*. Mycorrhiza 15(8):620–627

Bi YL, Hu ZQ, Si JT et al (2002) Effects of arbuscular mycorrhizal fungi on nutrient uptake of maize in reclaimed soil. J China Univ Min Technol 31(3):252–257

Bi YL, Wu WY, Liu YP (2007) Application of arbuscular mycorrhizas in land reclamation of coal spoil heaps. Acta Ecol Sin 27(9):3738–3743

Bi N, Guo W, Guo JY et al (2013) Effect of arbuscular mycorrhizal fungi (*Rhizophagus intraradices*) on the maize growth in the three types of coal mine spoilt soil. J Saf Environ 16(2):194–199

Biro I, Nemeth T, Takács T (2009) Changes of parameters of infectivity and efficiency of different glomus mosseae arbuscular mycorrhizal fungi strains in cadmium-loaded soils. Commun Soil Sci Plant Anal 40(1–6):227–239

Bissonnette L, St-Arnaud M, Labrecque M (2010) Phytoextraction of heavy metals by two Salicaceae clones in symbiosis with arbuscular mycorrhizal fungi during the second year of a field trial. Plant Soil 332(1–2):55–67

Bona E, Cattaneo C, Cesaro P et al (2010) Proteomic analysis of *Pteris vittata* fronds: two arbuscular mycorrhizal fungi differentially modulate protein expression under arsenic contamination. Proteomics 10(21):3811–3834

Bona E, Marsano F, Massa N et al (2011) Proteomic analysis as a tool for investigating arsenic stress in *Pteris vittata* roots colonized or not by arbuscular mycorrhizal symbiosis. J Proteom 74(8):1338–1350

Bonfante P, Anca IA (2009) Plants, mycorrhizal fungi, and bacteria: a network of interactions. Annu Rev Microbiol 63:363–383

Bradshaw AD, Wong MH (2002) The development of nitrogen capital. In: The restoration and management of derelict land-modern approaches. World Scientific, Singapore, pp 128–137

Breuillin F, Schramm J, Hajirezaei M et al (2010) Phosphate systemically inhibits development of arbuscular mycorrhiza in *Petunia* hybrida and represses genes involved in mycorrhizal functioning. Plant J 64(6):1002–1017

Burleigh SH, Kristensen BK, Bechmann IE (2003) A plasma membrane zinc transporter from *Medicago truncatula* is up-regulated in roots by Zn fertilization, yet down-regulated by arbuscular mycorrhizal colonization. Plant Mol Biol 52(5):1077–1088

Burri K, Gromke C, Graf F (2013) Mycorrhizal fungi protect the soil from wind erosion: a wind tunnel study. Land Degrad Dev 24(4):385–392

Call CA, Davies F (1988) Effects of vesicular-arbuscular mycorrhizae on survival and growth of perennial grasses in lignite overburden in Texas. Agric Ecosyst Environ 24(4):395–405

Call CA, McKell C (1985) Endomycorrhizae enhance growth of shrub species in processed oil shale and disturbed native soil. J Range Manag 38(3):258–261

Caravaca F, Barea J, Figueroa D et al (2002a) Assessing the effectiveness of mycorrhizal inoculation and soil compost addition for enhancing reafforestation with *Olea europaea* subsp. *sylvestris* through changes in soil biological and physical parameters. Appl Soil Ecol 20(2):107–118

Caravaca F, Barea J, Roldán A (2002b) Synergistic influence of an arbuscular mycorrhizal fungus and organic amendment on *Pistacia lentiscus* L. seedlings afforested in a degraded semiarid soil. Soil Biol Biochem 34(8):1139–1145

Caravaca F, Hernandez T, Garcıa C et al (2002c) Improvement of rhizosphere aggregate stability of afforested semiarid plant species subjected to mycorrhizal inoculation and compost addition. Geoderma 108(1):133–144

Caravaca F, Alguacil M, Figueroa D et al (2003a) Re-establishment of *Retama sphaerocarpa* as a target species for reclamation of soil physical and biological properties in a semi-arid Mediterranean area. For Ecol Manag 182(1):49–58

Caravaca F, Barea J, Palenzuela J et al (2003b) Establishment of shrub species in a degraded semiarid site after inoculation with native or allochthonous arbuscular mycorrhizal fungi. Appl Soil Ecol 22(2):103–111

Caravaca F, Figueroa D, Azcón-Aguilar C et al (2003c) Medium-term effects of mycorrhizal inoculation and composted municipal waste addition on the establishment of two Mediterranean shrub species under semiarid field conditions. Agric Ecosyst Environ 97(1):95–105

Caravaca F, Figueroa D, Rolda A et al (2003d) Alteration in rhizosphere soil properties of afforested *Rhamnus lycioides* seedlings in short-term response to mycorrhizal inoculation with *Glomus intraradices* and organic amendment. Environ Manag 31(3):0412–0420

Caravaca F, Alguacil M, Azcón R et al (2004a) Comparing the effectiveness of mycorrhizal inoculation and amendment with sugar beet, rock phosphate and *Aspergillus niger* to enhance field performance of the leguminous shrub *Dorycnium pentaphyllum* L. Appl Soil Ecol 25(2):169–180

Caravaca F, Alguacil M, Vassileva M et al (2004b) AM fungi inoculation and addition of microbially-treated dry olive cake-enhanced afforestation of a desertified Mediterranean site. Land Degrad Dev 15(2):153–161

Caravaca F, Alguacil M, Barea J et al (2005a) Survival of inocula and native AM fungi species associated with shrubs in a degraded Mediterranean ecosystem. Soil Biol Biochem 37(2):227–233

Caravaca F, Alguacil M, Hernández JA et al (2005b) Involvement of antioxidant enzyme and nitrate reductase activities during water stress and recovery of mycorrhizal *Myrtus communis* and *Phillyrea angustifolia* plants. Plant Sci 169(1):191–197

Caravaca F, Alguacil MM, Diaz G et al (2005c) Nutrient acquisition and nitrate reductase activity of mycorrhizal *Retama sphaerocarpa* L. seedlings afforested in an amended semiarid soil under two water regimes. Soil Use Manag 21(1):10–16

Caravaca F, Alguacil M, Azcón R et al (2006) Formation of stable aggregates in rhizosphere soil of *Juniperus oxycedrus*: effect of AM fungi and organic amendments. Appl Soil Ecol 33(1):30–38

Castañón-Silva PA, Venegas-Urrutia MA, Lobos-Valenzuela MG et al (2013) Influence of arbuscular mycorrhizal *Glomus* spp. on growth and accumulation of copper in sunflower *Helianthus annuus* L. Agrociencia 47(4):309–317

Chen BD, Tang XY, Zhu YG et al (2005) Metal concentrations and mycorrhizal status of plants colonizing copper mine tailings: potential for revegetation. Sci China Ser C Life Sci 48:156–164

Chen BD, Zhu YG, Smith FA (2006) Effects of arbuscular mycorrhizal inoculation on uranium and arsenic accumulation by Chinese brake fern (*Pteris vittata* L.) from a uranium mining-impacted soil. Chemosphere 62(9):1464–1473

Chen A, Hu J, Sun S et al (2007a) Conservation and divergence of both phosphate-and mycorrhiza-regulated physiological responses and expression patterns of phosphate transporters in solanaceous species. New Phytol 173(4):817–831

Chen B, Zhu YG, Duan J et al (2007b) Effects of the arbuscular mycorrhizal fungus *Glomus mosseae* on growth and metal uptake by four plant species in copper mine tailings. Environ Pollut 147(2):374–380

Chen B, Roos P, Zhu YG et al (2008a) Arbuscular mycorrhizas contribute to phytostabilization of uranium in uranium mining tailings. J Environ Radioact 99(5):801–810

Chen Y, Yuan J, Yang Z et al (2008b) Associations between arbuscular mycorrhizal fungi and *Rhynchelytrum repens* in abandoned quarries in southern China. Plant Soil 304:257–266

Chen ZY, Cao YZ, Peng AP et al (2012) Impacts of arbuscular mycorrhizal fungi and water conditions on plant growth in rare earth tailings. J Agro-Environ Sci 31(11):2101–2107

Chen X, Wu F, Li H et al (2013) Phosphate transporters expression in rice (*Oryza sativa* L.) associated with arbuscular mycorrhizal fungi (AMF) colonization under different levels of arsenate stress. Environ Exp Bot 87:92–99

Cicatelli A, Lingua G, Todeschini V et al (2010) Arbuscular mycorrhizal fungi restore normal growth in a white poplar clone grown on heavy metal-contaminated soil, and this is associated with upregulation of foliar metallothionein and polyamine biosynthetic gene expression. Ann Bot 106(5):791–802

Cicatelli A, Lingua G, Todeschini V et al (2012) Arbuscular mycorrhizal fungi modulate the leaf transcriptome of a *Populus alba* L. clone grown on a zinc and copper-contaminated soil. Environ Exp Bot 75:25–35

Clark R, Zeto S (1996) Growth and root colonization of mycorrhizal maize grown on acid and alkaline soil. Soil Biol Biochem 28(10):1505–1511

Clark R, Zeto S, Zobel R (1999) Arbuscular mycorrhizal fungal isolate effectiveness on growth and root colonization of *Panicum virgatum* in acidic soil. Soil Biol Biochem 31(13):1757–1763

Corradi N, Charest C (2011) Some like it toxic. Mol Ecol 20(16):3289–3290

Cuenca G, De Andrade Z, Escalante G (1998) Arbuscular mycorrhizae in the rehabilitation of fragile degraded tropical lands. Biol Fertil Soils 26(2):107–111

Daft M, Hacskaylo E (1977) Growth of endomycorrhizal and nonmycorrhizal red maple seedlings in sand and anthracite spoil. For Sci 23(2):207–216

Daft M, Hacskaylo E, Nicolson T (1975) Arbuscular mycorrhizas in plants colonising coal spoils in-Scotland and Pennsylvania. In: Endomycorrhizas; Proceedings of a Symposium

de Souza RG, Goto BT, da Silva DKA et al (2010) The role of arbuscular mycorrhizal fungi and cattle manure in the establishment of *Tocoyena selloana* Schum. in mined dune areas. Eur J Soil Biol 46(3):237–242

Dehn B, Schuepp H (1990) Influence of VA mycorrhizae on the uptake and distribution of heavy-metals in plants. Agric Ecosyst Environ 29(1–4):79–83

del Mar AM, Caravaca F, Diaz G et al (2004) Establishment of *Retama sphaerocarpa* L. seedlings on a degraded semiarid soil as influenced by mycorrhizal inoculation and sewage-sludge amendment. J Plant Nutr Soil Sci 167(5):637–644

Dong Y, Zhu YG, Smith FA et al (2008) Arbuscular mycorrhiza enhanced arsenic resistance of both white clover (*Trifolium repens* Linn.) and ryegrass (*Lolium perenne* L.) plants in an arsenic-contaminated soil. Environ Pollut 155(1):174–181

Doubkova P, Suda J, Sudova R (2012) The symbiosis with arbuscular mycorrhizal fungi contributes to plant tolerance to serpentine edaphic stress. Soil Biol Biochem 44(1):56–64

Du SZ, Bi YL, Wu WY et al (2008) Ecological effects of arbuscular mycorrhizal fungi on environmental phytoremediation in coal mine areas. Trans CSAE 24(4):113–116

Ellis J (1992) Grasslands and grassland sciences in northern China. National Academies, Washington, DC

Enkhtuya B, Óskarsson Ú, Dodd JC et al (2003) Inoculation of grass and tree seedlings used for reclaiming eroded areas in Iceland with mycorrhizal fungi. Folia Geobot 38(2):209–222

Enkhtuya B, Pöschl M, Vosátka M (2005) Native grass facilitates mycorrhizal colonisation and P uptake of tree seedlings in two anthropogenic substrates. Water Air Soil Pollut 166(1–4):217–236

Estaún V, Savé R, Biel C (1997) AM inoculation as a biological tool to improve plant revegetation of a disturbed soil with *Rosmarinus officinalis* under semi-arid conditions. Appl Soil Ecol 6(3):223–229

Estrada B, Aroca R, Azcón-Aguilar C et al (2013) Importance of native arbuscular mycorrhizal inoculation in the halophyte *Asteriscus maritimus* for successful establishment and growth under saline conditions. Plant Soil 370(1–2):175–185

Evelin H, Kapoor R, Giri B (2009) Arbuscular mycorrhizal fungi in alleviation of salt stress: a review. Ann Bot 104(7):1263–1280

Fagbola O, Osonubi O, Mulongoy K et al (2001) Effects of drought stress and arbuscular mycorrhiza on the growth of *Gliricidia sepium* (Jacq). Walp, and *Leucaena leucocephala* (Lam.) de Wit. in simulated eroded soil conditions. Mycorrhiza 11(5):215–223

Feng G, Zhang F, Li X et al (2002) Uptake of nitrogen from indigenous soil pool by cotton plant inoculated with arbuscular mycorrhizal fungi. Commun Soil Sci Plant Anal 33(19–20):3825–3836

Feng G, Song Y, Li X et al (2003) Contribution of arbuscular mycorrhizal fungi to utilization of organic sources of phosphorus by red clover in a calcareous soil. Appl Soil Ecol 22(2):139–148

Frey-Klett P, Ja G, Tarkka M (2007) The mycorrhiza helper bacteria revisited. New Phytol 176(1):22–36

Frost S, Stahl P, Williams SE (2001) Long-term reestablishment of arbuscular mycorrhizal fungi in a drastically disturbed semiarid surface mine soil. Arid Land Res Manag 15(1):3–12

Gamalero E, Lingua G, Berta G et al (2009) Beneficial role of plant growth promoting bacteria and arbuscular mycorrhizal fungi on plant responses to heavy metal stress. Can J Microbiol 55(5):501–514
Gange AC (1993) Translocation of mycorrhizal fungi by earthworms during early succession. Soil Biol Biochem 25(8):1021–1026
García IV, Mendoza RE (2007) Arbuscular mycorrhizal fungi and plant symbiosis in a saline-sodic soil. Mycorrhiza 17(3):167–174
García IV, Mendoza RE (2008) Relationships among soil properties, plant nutrition and arbuscular mycorrhizal fungi – plant symbioses in a temperate grassland along hydrologic, saline and sodic gradients. FEMS Microbiol Ecol 63(3):359–371
Gemma J, Koske R (1997) Arbuscular mycorrhizae in sand dune plants of the North Atlantic coast of the US: field and greenhouse inoculation and presence of mycorrhizae in planting stock. J Environ Manag 50(3):251–264
Ghassemi F, Jakeman AJ, Nix HA (1995) Salinisation of land and water resources: human causes, extent, management and case studies. CAB International, Wallingford
Giri B, Kapoor R, Mukerji K (2005) Effect of the arbuscular mycorrhizae *Glomus fasciculatum* and *G. macrocarpum* on the growth and nutrient content of *Cassia siamea* in a semi-arid Indian wasteland soil. New For 29(1):63–73
Giri B, Kapoor R, Mukerji KG (2007) Improved tolerance of *Acacia nilotica* to salt stress by arbuscular mycorrhiza, *Glomus fasciculatum* may be partly related to elevated K/Na ratios in root and shoot tissues. Microb Ecol 54(4):753–760
Glassop D, Smith SE, Smith FW (2005) Cereal phosphate transporters associated with the mycorrhizal pathway of phosphate uptake into roots. Planta 222(4):688–698
González-Guerrero M, Azcon-Aguilar C, Mooney M et al (2005) Characterization of a *Glomus intraradices* gene encoding a putative Zn transporter of the cation diffusion facilitator family. Fungal Genet Biol 42(2):130–140
González-Guerrero M, Cano C, Azcón-Aguilar C et al (2007) *GintMT1* encodes a functional metallothionein in glomus intraradices that responds to oxidative stress. Mycorrhiza 17(4):327–335
González-Guerrero M, Benabdellah K, Valderas A et al (2010a) *GintABC1* encodes a putative ABC transporter of the MRP subfamily induced by Cu, Cd, and oxidative stress in *Glomus intraradices*. Mycorrhiza 20(2):137–146
González-Guerrero M, Oger E, Benabdellah K et al (2010b) Characterization of a CuZn superoxide dismutase gene in the arbuscular mycorrhizal fungus *Glomus intraradices*. Curr Genet 56(3):265–274
Gonzalez-Chavez M, Carrillo-Gonzalez R, Wright S et al (2004) The role of glomalin, a protein produced by arbuscular mycorrhizal fungi, in sequestering potentially toxic elements. Environ Pollut 130(3):317–323
Gonzalez-Chavez MCA, del Pilar Ortega-Larrocea M, Carrillo-Gonzalez R et al (2011) Arsenate induces the expression of fungal genes involved in as transport in arbuscular mycorrhiza. Fungal Biol 115(12):1197–1209
Govindarajulu M, Pfeffer PE, Jin H et al (2005) Nitrogen transfer in the arbuscular mycorrhizal symbiosis. Nature 435(7043):819–823
Gryndler M, Larsen J, Hršelová H et al (2006) Organic and mineral fertilization, respectively, increase and decrease the development of external mycelium of arbuscular mycorrhizal fungi in a long-term field experiment. Mycorrhiza News 16(3):159–166
Guether M, Neuhäuser B, Balestrini R et al (2009) A mycorrhizal-specific ammonium transporter from *Lotus japonicus* acquires nitrogen released by arbuscular mycorrhizal fungi. Plant Physiol 150(1):73–83
Guo W, Zhao R, Yang H et al (2013a) Using native plants to evaluate the effect of arbuscular mycorrhizal fungi on revegetation of iron tailings in grasslands. Biol Fertil Soils 49(6):617–626

Guo W, Zhao R, Zhao W et al (2013b) Effects of arbuscular mycorrhizal fungi on maize (*Zea mays* L.) and sorghum (*Sorghum bicolor* L. Moench) grown in rare earth elements of mine tailings. Appl Soil Ecol 72:85–92

Guo W, Zhao RX, Zhao WJ et al (2013c) Effects of arbuscular mycorrhizal fungi on the growth and rare earth elements uptake of soybean grown in rare earth mine tailings. Environ Sci 34(5):1915–1921

Guo W, Zhao WJ, Zhao RX et al (2013d) Role of arbuscular mycorrhizal fungi in promoting the vegetation restoration of the grassland eco-environment in copper mine tailings. J Saf Environ 13(4):54–59

Guo W, Zhao R, Fu R et al (2014) Contribution of arbuscular mycorrhizal fungi to the development of maize (*Zea mays* L.) grown in three types of coal mine spoils. Environ Sci Pollut Res 21(5):3592–3603

Halary S, Daubois L, Terrat Y et al (2013) Mating type gene homologues and putative sex pheromone-sensing pathway in arbuscular mycorrhizal fungi, a presumably asexual plant root symbiont. PLoS One 8(11):e80729

Hamel C (2004) Impact of arbuscular mycorrhizal fungi on N and P cycling in the root zone. Can J Soil Sci 84(4):383–395

Hammer EC, Rillig MC (2011) The influence of different stresses on glomalin levels in an arbuscular mycorrhizal fungus-salinity increases glomalin content. PLoS One 6(12):e28426

Harrison MJ, van Buuren ML (1995) A phosphate transporter from the mycorrhizal fungus *Glomus versiforme*. Nature 378:626–629

Harrison MJ, Dewbre GR, Liu J (2002) A phosphate transporter from *Medicago truncatula* involved in the acquisition of phosphate released by arbuscular mycorrhizal fungi. Plant Cell Online 14(10):2413–2429

Heggo A, Angle JS, Chaney RL (1990) Effects of vesicular arbuscular mycorrhizal fungi on heavy-metal uptake by soybeans. Soil Biol Biochem 22(6):865–869

Hempel S, Renker C, Buscot F (2007) Differences in the species composition of arbuscular mycorrhizal fungi in spore, root and soil communities in a grassland ecosystem. Environ Microbiol 9(8):1930–1938

Herrera MA, Salamanca CP, Barea JM (1993) Inoculation of woody legumes with selected arbuscular mycorrhizal fungi and rhizobia to recover desertified Mediterranean ecosystems. Appl Environ Microbiol 59(1):129–133

Hetrick B, Wilson G, Figge D (1994) The influence of mycorrhizal symbiosis and fertilizer amendments on establishment of vegetation in heavy metal mine spoil. Environ Pollut 86(2):171–179

Hodge A, Fitter AH (2010) Substantial nitrogen acquisition by arbuscular mycorrhizal fungi from organic material has implications for N cycling. Proc Natl Acad Sci India Sect B (Biol Sci) 107(31):13754–13759

Hodge A, Campbell CD, Fitter AH (2001) An arbuscular mycorrhizal fungus accelerates decomposition and acquires nitrogen directly from organic material. Nature 413(6853):297–299

Hu ZQ, Liu J, Cai B et al (2006) Application study on mycorrhizas in revegetating coal waste piles in Dawukou coal processing plant. Energy Environmental Protection 21(1):14–16

Hu J, Wu S, Wu F et al (2013) Arbuscular mycorrhizal fungi enhance both absorption and stabilization of Cd by Alfred stonecrop (*Sedum alfredii* Hance) and perennial ryegrass (*Lolium perenne* L.) in a Cd-contaminated acidic soil. Chemosphere 93(7):1359–1365

Hu J, Wang H, Wu F et al (2014) Arbuscular mycorrhizal fungi influence the accumulation and partitioning of Cd and P in bashfulgrass (*Mimosa pudica* L.) grown on a moderately Cd-contaminated soil. Appl Soil Ecol 73:51–57

Hua J, Lin X, Yin R et al (2009) Effects of arbuscular mycorrhizal fungi inoculation on arsenic accumulation by tobacco (*Nicotiana tabacum* L.) J Environ Sci China 21(9):1214–1220

Hua JF, Lin XG, Bai JF et al (2010) Effects of arbuscular mycorrhizal fungi and earthworm on nematode communities and arsenic uptake by maize in arsenic-contaminated soils. Pedosphere 20(2):163–173

Jacobson KM (1997) Moisture and substrate stability determine VA-mycorrhizal fungal community distribution and structure in an arid grassland. J Arid Environ 35(1):59–75
Jankong P, Visoottiviseth P (2008) Effects of arbuscular mycorrhizal inoculation on plants growing on arsenic contaminated soil. Chemosphere 72(7):1092–1097
Janos DP (1980) Mycorrhizae influence tropical succession. Biotropica (12):56–64
Jasper DA, Robson AD, Abbott LK (1988) Revegetation in an iron ore mine – nutrient requirements for plant growth and the potential role of vesicular arbuscular (VA) mycorrhizal fungi. Aust J Soil Res 26(3):497–507
Jasper DA, Abbott LK, Robson AD (1989) Acacias respond to additions of phosphorus and to inoculation with VA mycorrhizal fungi in soils stockpiled during mineral sand mining. Plant Soil 115(1):99–108
Jeffries P, Craven-Griffiths A, Barea JM et al (2002) Application of arbuscular mycorrhizal fungi in the revegetation of desertified Mediterranean ecosystems. In: Mycorrhizal technology in agriculture. Springer, Heidelberg, pp 151–174
Johnson N (1998) Responses of Salsola kali and Panicum virgatum to mycorrhizal fungi, phosphorus and soil organic matter: implications for reclamation. J Appl Ecol 35:86–94
Johnson NC, Graham JH, Smith FA (1997) Functioning of mycorrhizal associations along the mutualism-parasitism continuum. New Phytol 135(4):575–585
Karandashov V, Bucher M (2005) Symbiotic phosphate transport in arbuscular mycorrhizas. Trends Plant Sci 10(1):22–29
Kivlin SN, Hawkes CV, Treseder KK (2011) Global diversity and distribution of arbuscular mycorrhizal fungi. Soil Biol Biochem 43(11):2294–2303
Klabi R, Hamel C, Schellenberg MP et al (2014) Interaction between legume and arbuscular mycorrhizal fungi identity alters the competitive ability of warm-season grass species in a grassland community. Soil Biol Biochem 70:176–182
Kobae Y, Tamura Y, Takai S et al (2010) Localized expression of arbuscular mycorrhiza-inducible ammonium transporters in soybean. Plant Cell Physiol 51(9):1411–1415
Lagrange A, Ducousso M, Jourand P et al (2011) New insights into the mycorrhizal status of Cyperaceae from ultramafic soils in New Caledonia. Can J Microbiol 57(1):21–28
Lambert DH, Cole H (1980) Effects of mycorrhizae on establishment and performance of forage species in mine spoil. Agron J 72(2):257–260
Lanfranco L, Bolchi A, Ros EC et al (2002) Differential expression of a metallothionein gene during the presymbiotic versus the symbiotic phase of an arbuscular mycorrhizal fungus. Plant Physiol 130(1):58–67
Lanfranco L, Novero M, Bonfante P (2005) The mycorrhizal fungus *Gigaspora margarita* possesses a CuZn superoxide dismutase that is up-regulated during symbiosis with legume hosts. Plant Physiol 137(4):1319–1330
Leigh J, Hodge A, Fitter AH (2009) Arbuscular mycorrhizal fungi can transfer substantial amounts of nitrogen to their host plant from organic material. New Phytol 181(1):199–207
Lekberg Y, Schnoor T, Kjøller R et al (2012) 454-sequencing reveals stochastic local reassembly and high disturbance tolerance within arbuscular mycorrhizal fungal communities. J Ecol 100(1):151–160
Leung HM, Ye ZH, Wong MH (2006) Interactions of mycorrhizal fungi with *Pteris vittata* (As hyperaccumulator) in As-contaminated soils. Environ Pollut 139(1):1–8
Leung HM, Wu FY, Cheung KC et al (2010a) The effect of arbuscular mycorrhizal fungi and phosphate amendment on arsenic uptake, accumulation and growth of *Pteris vittata* in As-contaminated soil. Int J Phytoremediation 12(4):384–403
Leung HM, Wu FY, Cheung KC et al (2010b) Synergistic effects of arbuscular mycorrhizal fungi and phosphate rock on heavy metal uptake and accumulation by an arsenic hyperaccumulator. J Hazard Mater 181(1–3):497–507
Li XL, George E, Marschner H (1991) Extension of the phosphorus depletion zone in VA-mycorrhizal white clover in a calcareous soil. Plant Soil 136(1):41–48

Li XL, Zhang JL, George E et al (1997a) Phosphorus acquisition from compacted soil by hyphae of a mycorrhizal fungus associated with red clover (*Trifolium pratense*). Can J Bot 75(5):723–729

Li XL, George E, Marschner H (1997b) Phosphorus acquisition of VA mycorrhizal hyphae from compacted soil in clover. Can J Bot 75(5):723–729

Li SP, Bi YL, Kong WP et al (2013a) Effects of the arbuscular mycorrhizal fungi on environmental phytoremediation in coal mine areas. Environ Sci 34(11):4455–4459

Li T, Hu YJ, Hao ZP et al (2013b) First cloning and characterization of two functional aquaporin genes from an arbuscular mycorrhizal fungus *Glomus intraradices*. New Phytol 197(2):617–630

Liang C, Li T, Xiao Y et al (2009) Effects of inoculation with arbuscular mycorrhizal fungi on maize grown in multi-metal contaminated soils. Int J Phytorem 11(8):692–703

Lin AJ, Zhang XH, Wong MH et al (2007) Increase of multi-metal tolerance of three leguminous plants by arbuscular mycorrhizal fungi colonization. Environ Geochem Health 29(6):473–481

Lingua G, Bona E, Todeschini V et al (2012) Effects of heavy metals and arbuscular mycorrhiza on the leaf proteome of a selected poplar clone: a time course analysis. PLoS One 7(6):e38662

Liu Y, Zhu YG, Chen BD et al (2005) Influence of the arbuscular mycorrhizal fungus *Glomus mosseae* on uptake of arsenate by the As hyperaccumulator fern *Pteris vittata* L. Mycorrhiza 15(3):187–192

Liu H, Yuan M, Tan S et al (2015) Enhancement of arbuscular mycorrhizal fungus (*Glomus versiforme*) on the growth and Cd uptake by Cd-hyperaccumulator *Solanum nigrum*. Appl Soil Ecol 89:44–49

López-Pedrosa A, González-Guerrero M, Valderas A et al (2006) *GintAMT1* encodes a functional high-affinity ammonium transporter that is expressed in the extraradical mycelium of *Glomus intraradices*. Fungal Genet Biol 43(2):102–110

Loth-Pereda V, Orsini E, Courty P-E et al (2011) Structure and expression profile of the phosphate *Pht1* transporter gene family in mycorrhizal *Populus trichocarpa*. Plant Physiol 156(4):2141–2154

Lumini E, Bosco M, Puppi G et al (1994) Field performance of *Alnus cordata* Loisel (Italian alder) inoculated with Frankia and VA-mycorrhizal strains in mine-spoil afforestation plots. Soil Biol Biochem 26(5):659–661

Ma Y, Dickinson N, Wong M (2006) Beneficial effects of earthworms and arbuscular mycorrhizal fungi on establishment of leguminous trees on Pb/Zn mine tailings. Soil Biol Biochem 38(6):1403–1412

Madejon E, Doronila AI, Sanchez-Palacios JT et al (2010) Arbuscular mycorrhizal fungi (AMF) and biosolids enhance the growth of a native australian grass on sulphidic gold mine tailings. Restor Ecol 18:175–183

Madejon E, Doronila AI, Madejon P et al (2012) Biosolids, mycorrhizal fungi and eucalypts for phytostabilization of arsenical sulphidic mine tailings. Agrofor Syst 84(3):389–399

Madiba OF (2014) Alleviation of soil stresses by arbuscular mycorrhizal fungi. In: Mycorrhizal fungi: use in sustainable agriculture and land restoration. Springer, New York, pp 215–224

Maeda D, Ashida K, Iguchi K et al (2006) Knockdown of an arbuscular mycorrhiza-inducible phosphate transporter gene of *Lotus japonicus* suppresses mutualistic symbiosis. Plant Cell Physiol 47(7):807–817

Maldonado-Mendoza IE, Dewbre GR, Harrison MJ (2001) A phosphate transporter gene from the extra-radical mycelium of an arbuscular mycorrhizal fungus *Glomus intraradices* is regulated in response to phosphate in the environment. Mol Plant-Microbe Interact 14(10):1140–1148

Marques APGC, Oliveira RS, Rangel AOSS et al (2006) Zinc accumulation in *Solanum nigrum* is enhanced by different arbuscular mycorrhizal fungi. Chemosphere 65(7):1256–1263

Marques APGC, Oliveira RS, Samardjieva KA et al (2007) *Solanum nigrum* grown in contaminated soil: effect of arbuscular mycorrhizal fungi on zinc accumulation and histolocalisation. Environ Pollut 145(3):691–699

Medina A, Vassileva M, Caravaca F et al (2004) Improvement of soil characteristics and growth of Dorycnium pentaphyllum by amendment with agrowastes and inoculation with AM fungi and/or the yeast *Yarrowia lipolytica*. Chemosphere 56(5):449–456

Meier S, Borie F, Bolan N et al (2012) Phytoremediation of metal-polluted soils by arbuscular mycorrhizal fungi. Crit Rev Environ Sci Technol 42(7):741–775

Michelsen A, Rosendahl S (1990) The effect of VA mycorrhizal fungi, phosphorus and drought stress on the growth of *Acacia nilotica* and *Leucaena leucocephala* seedlings. Plant Soil 124(1):7–13

Miransari M (2010) Contribution of arbuscular mycorrhizal symbiosis to plant growth under different types of soil stress. Plant Biol 12(4):563–569

Miransari M (2013) Arbuscular mycorrhizal fungi and uptake of nutrients. In: Aroca R (ed) Symbiotic endophytes, soil biology, vol 37. Springer, Berlin, pp 253–270

Miransari M (2014) The interactions of soil microbes, arbuscular mycorrhizal fungi and N-fixing bacteria, rhizobium, under different conditions including stress. In: Use of microbes for the alleviation of soil stresses. Springer, New York, pp 1–21

Nagy R, Karandashov V, Chague V et al (2005) The characterization of novel mycorrhiza-specific phosphate transporters from *Lycopersicon esculentum* and *Solanum tuberosum* uncovers functional redundancy in symbiotic phosphate transport in solanaceous species. Plant J 42(2):236–250

Navarro-Fernández CM, Aroca R, Barea JM (2011) Influence of arbuscular mycorrhizal fungi and water regime on the development of endemic *Thymus* species in dolomitic soils. Appl Soil Ecol 48(1):31–37

Nazir A, Bareen F (2011) Synergistic effect of *Glomus fasciculatum* and *Trichoderma pseudokoningii* on *Helianthus annuus* to decontaminate tannery sludge from toxic metals. Afr J Biotechnol 10(22):4612–4618

Neagoe A, Iordache V, Bergmann H et al (2013) Patterns of effects of arbuscular mycorrhizal fungi on plants grown in contaminated soil. J Plant Nutr Soil Sci 176(2):273–286

Novoa MD, Palma SS, Gaete OH (2010) Effect of arbuscular mycorrhizal fungi *Glomus* spp. inoculation on alfalfa growth in soils with copper. Chil J Agric Res 70(2):259–265

Noyd RK, Pfleger F, Norland MR (1996) Field responses to added organic matter, arbuscular mycorrhizal fungi, and fertilizer in reclamation of taconite iron ore tailing. Plant Soil 179(1):89–97

Oliveira R, Vosátka M, Dodd J et al (2005) Studies on the diversity of arbuscular mycorrhizal fungi and the efficacy of two native isolates in a highly alkaline anthropogenic sediment. Mycorrhiza 16(1):23–31

Orlowska E, Godzik B, Turnau K (2012) Effect of different arbuscular mycorrhizal fungal isolates on growth and arsenic accumulation in *Plantago lanceolata* L. Environ Pollut 168:121–130

Orlowska E, Jurkiewicz A, Anielska T et al (2005a) Influence of different arbuscular mycorrhiza fungal (AMF) strains on heavy metal uptake by *Plantago lanceolat*a (Plantaginaceae). Pol Bot Stud 19:65–72

Orlowska E, Ryszka P, Jurkiewicz A et al (2005b) Effectiveness of arbuscular mycorrhizal fungal (AMF) strains in colonisation of plants involved in phytostabilisation of zinc wastes. Geoderma 129(1–2):92–98

Orłowska E, Orłowski D, Mesjasz-Przybyłowicz J et al (2010) Role of mycorrhizal colonization in plant establishment on an alkaline gold mine tailing. Int J Phytorem 13(2):185–205

Orlowska E, Przybylowicz W, Orlowski D et al (2011) The effect of mycorrhiza on the growth and elemental composition of Ni-hyperaccumulating plant *Berkheya coddii* Roessler. Environ Pollut 159(12):3730–3738

Palenzuela J, Azcon-Aguilar C, Figueroa D et al (2002) Effects of mycorrhizal inoculation of shrubs from Mediterranean ecosystems and composted residue application on transplant performance and mycorrhizal developments in a desertified soil. Biol Fertil Soils 36(2):170–175

Pallara G, Todeschini V, Lingua G et al (2013) Transcript analysis of stress defence genes in a white poplar clone inoculated with the arbuscular mycorrhizal fungus *Glomus mosseae* and grown on a polluted soil. Plant Physiol Biochem 63:131–139

Parniske M (2008) Arbuscular mycorrhiza: the mother of plant root endosymbioses. Nat Rev Microbiol 6(10):763–775

Paszkowski U, Kroken S, Roux C et al (2002) Rice phosphate transporters include an evolutionarily divergent gene specifically activated in arbuscular mycorrhizal symbiosis. Proc Natl Acad Sci U S A 99(20):13324–13329

Pérez-Tienda J, Testillano PS, Balestrini R et al (2011) *GintAMT2*, a new member of the ammonium transporter family in the arbuscular mycorrhizal fungus *Glomus intraradices*. Fungal Genet Biol 48(11):1044–1055

Pfetffer CM, Bloss HE (1988) Growth and nutrition of guayule (Parthenium argentatum) in a saline soil as influenced by vesicular–arbuscular mycorrhiza and phosphorus fertilization. New Phytol 108(3):315–321

Pond EC, Menge JA, Jarrell WM (1984) Improved growth of tomato in salinized soil by vesicular-arbuscular mycorrhizal fungi collected from saline soils. Mycologia 76(1):74–84

Porcel R, Azcón R, Ruiz-Lozano JM (2004) Evaluation of the role of genes encoding for Δ 1-pyrroline-5-carboxylate synthetase (P5CS) during drought stress in arbuscular mycorrhizal *Glycine max* and *Lactuca sativa* plants. Physiol Mol Plant Pathol 65(4):211–221

Porcel R, Aroca R, Azcon R et al (2006) PIP aquaporin gene expression in arbuscular mycorrhizal *Glycine max* and *Lactuca sativa* plants in relation to drought stress tolerance. Plant Mol Biol 60(3):389–404

Porcel R, Aroca R, Ruiz-Lozano JM (2012) Salinity stress alleviation using arbuscular mycorrhizal fungi. A review. Agron Sustain Dev 32(1):181–200

Porras-Soriano A, Soriano-Martín ML, Porras-Piedra A et al (2009) Arbuscular mycorrhizal fungi increased growth, nutrient uptake and tolerance to salinity in olive trees under nursery conditions. J Plant Physiol 166(13):1350–1359

Pozo MJ, Azcón-Aguilar C (2007) Unraveling mycorrhiza-induced resistance. Curr Opin Plant Biol 10(4):393–398

Rabie G (2005) Influence of arbuscular mycorrhizal fungi and kinetin on the response of mungbean plants to irrigation with seawater. Mycorrhiza 15(3):225–230

Rao A, Tak R (2002) Growth of different tree species and their nutrient uptake in limestone mine spoil as influenced by arbuscular mycorrhizal (AM)-fungi in Indian arid zone. J Arid Environ 51(1):113–119

Rausch C, Daram P, Brunner S et al (2001) A phosphate transporter expressed in arbuscule-containing cells in potato. Nature 414(6862):462–470

Redon PO, Beguiristain T, Leyval C (2008) Influence of *Glomus intraradices* on Cd partitioning in a pot experiment with *Medicago truncatula* in four contaminated soils. Soil Biol Biochem 40(10):2710–2712

Regvar M, Vogel K, Irgel N et al (2003) Colonization of pennycresses (*Thlaspi* spp.) of the Brassicaceae by arbuscular mycorrhizal fungi. J Plant Physiol 160(6):615–626

Repetto O, Bestel-Corre G, Dumas-Gaudot E et al (2003) Targeted proteomics to identify cadmium-induced protein modifications in *Glomus mosseae*-inoculated pea roots. New Phytol 157(3):555–567

Requena N, Jimenez I, Toro M et al (1997) Interactions between plant-growth-promoting rhizobacteria (PGPR), arbuscular mycorrhizal fungi and *Rhizobium* spp. in the rhizosphere of *Anthyllis cytisoides*, a model legume for revegetation in mediterranean semi-arid ecosystems. New Phytol 136(4):667–677

Requena N, Perez-Solis E, Azcón-Aguilar C et al (2001) Management of indigenous plant-microbe symbioses aids restoration of desertified ecosystems. Appl Environ Microbiol 67(2):495–498

Richter BS, Stutz JC (2002) Mycorrhizal inoculation of big sacaton: implications for grassland restoration of abandoned agricultural fields. Restor Ecol 10(4):607–616

Rillig MC, Mummey DL (2006) Mycorrhizas and soil structure. New Phytol 171(1):41–53

Rillig MC, Wright SF, Eviner VT (2002) The role of arbuscular mycorrhizal fungi and glomalin in soil aggregation: comparing effects of five plant species. Plant Soil 238(2):325–333

Rosewarne GM, Barker SJ, Smith SE et al (1999) A Lycopersicon esculentum phosphate transporter (*LePT1*) involved in phosphorus uptake from a vesicular–arbuscular mycorrhizal fungus. New Phytol 144(3):507–516
Rothschild LJ, Mancinelli RL (2001) Life in extreme environments. Nature 409(6823):1092–1101
Rothwell FM (1984) Aggregation of surface mine soil by interaction between VAM fungi and lignin degradation products of lespedeza. Plant Soil 80(1):99–104
Ruiz-Lozano JM (2003) Arbuscular mycorrhizal symbiosis and alleviation of osmotic stress. New perspectives for molecular studies. Mycorrhiza 13(6):309–317
Ruiz-Lozano JM, Porcel R, Aroca R (2006) Does the enhanced tolerance of arbuscular mycorrhizal plants to water deficit involve modulation of drought-induced plant genes? New Phytol 171(4):693–698
Ryszka P, Turnau K (2007) Arbuscular mycorrhiza of introduced and native grasses colonizing zinc wastes: implications for restoration practices. Plant Soil 298(1–2):219–229
Salamanca CP, Herrera MA, Barea JM (1992) Mycorrhizal inoculation of micropropagated woody legumes used in revegetation programmes for desertified Mediterranean ecosystems. Agronomie 12:869–872
Sharma MP, Gaur A, Bhatia NP et al (1996) Growth responses and dependence of *Acacia nilotica* var. *cupriciformis* on the indigenous arbuscular mycorrhizal consortium of a marginal wasteland soil. Mycorrhiza 6(5):441–446
Shetty KG, Hetrick BAD, Figge DAH et al (1994) Effects of mycorrhizae and other soil microbes on revegetation of heavy-metal contaminated mine spoil. Environ Pollut 86(2):181–188
Shi Z, Mickan B, Feng G et al (2015) Arbuscular mycorrhizal fungi improved plant growth and nutrient acquisition of desert ephemeral *Plantago minuta* under variable soil water conditions. J Arid Land 7(3):414–420
Smith SE, Read DJ (2008) Mycorrhizal Symbiosis, 3rd edn. Academic, Amsterdam
Smith MR, Charvat I, Jacobson RL (1998) Arbuscular mycorrhizae promote establishment of prairie species in a tallgrass prairie restoration. Can J Bot 76(11):1947–1954
Smith SE, Smith FA, Jakobsen I (2003) Mycorrhizal fungi can dominate phosphate supply to plants irrespective of growth responses. Plant Physiol 133(1):16–20
Sokolski S, Dalpé Y, Piché Y (2011) Phosphate transporter genes as reliable gene markers for the identification and discrimination of arbuscular mycorrhizal fungi in the genus *Glomus*. Appl Environ Microbiol 77(5):1888–1891
Solis-Dominguez FA, Valentin-Vargas A, Chorover J et al (2011) Effect of arbuscular mycorrhizal fungi on plant biomass and the rhizosphere microbial community structure of mesquite grown in acidic lead/zinc mine tailings. Sci Total Environ 409(6):1009–1016
Stahl PD, Williams SE, Christensen M (1988) Efficacy of native vesicular-arbuscular mycorrhizal fungi after severe soil disturbance. New Phytol 110(3):347–354
Standish RJ, Stokes BA, Tibbett M et al (2007) Seedling response to phosphate addition and inoculation with arbuscular mycorrhizas and the implications for old-field restoration in Western Australia. Environ Exp Bot 61(1):58–65
Stommel M, Mann P, Franken P (2001) EST-library construction using spore RNA of the arbuscular mycorrhizal fungus *Gigaspora rosea*. Mycorrhiza 10(6):281–285
Su YY, Guo LD (2007) Arbuscular mycorrhizal fungi in non-grazed, restored and over-grazed grassland in the Inner Mongolia steppe. Mycorrhiza 17(8):689–693
Sudova R, Vosatka M (2007) Differences in the effects of three arbuscular mycorrhizal fungal strains on P and Pb accumulation by maize plants. Plant Soil 296(1–2):77–83
Suzuki H, Sakamoto K, Inubushi K (1999) Effects of arbuscular mycorrhizal colonization on ammonium-N uptake activity of the host plant. Jpn J Soil Sci Plant Nutr 70(1):59–63
Taheri WI, Bever JD (2010) Adaptation of plants and arbuscular mycorrhizal fungi to coal tailings in Indiana. Appl Soil Ecol 45(3):138–143
Tamura Y, Kobae Y, Mizuno T et al (2011) Identification and expression analysis of arbuscular mycorrhiza-inducible phosphate transporter genes of soybean. Biosci Biotechnol Biochem 76(2):309–313

Tan Z, Hu Y, Lin Z (2012) *PhPT4* is a mycorrhizal-phosphate transporter suppressed by lysophosphatidylcholine in petunia roots. Plant Mol Biol Report 30(6):1480–1487
Thorne M, Zamora B, Kennedy A (1998) Sewage sludge and mycorrhizal effects on secar bluebunch wheatgrass in mine spoil. J Environ Qual 27(5):1228–1233
Tian H, Gai J, Zhang J et al (2009a) Arbuscular mycorrhizal fungi associated with wild forage plants in typical steppe of eastern Inner Mongolia. Eur J Soil Biol 45(4):321–327
Tian H, Gai J, Zhang J et al (2009b) Arbuscular mycorrhizal fungi in degraded typical steppe of Inner Mongolia. Land Degrad Dev 20(1):41–54
Tótola MR, Borges AC (2000) Growth and nutritional status of Brazilian wood species *Cedrela fissilis* and *Anadenanthera peregrina* in bauxite spoil in response to arbuscular mycorrhizal inoculation and substrate amendment. Braz J Microbiol 31(4):257–265
Toler H, Morton J, Cumming J (2005) Growth and metal accumulation of mycorrhizal sorghum exposed to elevated copper and zinc. Water Air Soil Pollut 164(1–4):155–172
Treseder KK, Cross A (2006) Global distributions of arbuscular mycorrhizal fungi. Ecosystems 9(2):305–316
Tseng CC, Wang JY, Yang L (2009) Accumulation of copper, lead, and zinc by in situ plants inoculated with AM fungi in multicontaminated soil. Commun Soil Sci Plant Anal 40(21–22):3367–3386
Turnau K, Mesjasz-Przybylowicz J (2003) Arbuscular mycorrhiza of *Berkheya coddii* and other Ni-hyperaccumulating members of Asteraceae from ultramafic soils in South Africa. Mycorrhiza 13(4):185–190
Turnau K, Anielska T, Ryszka P et al (2008) Establishment of arbuscular mycorrhizal plants originating from xerothermic grasslands on heavy metal rich industrial wastes–new solution for waste revegetation. Plant Soil 305(1–2):267–280
Ultra VUY Jr, Tanaka S, Sakurai K et al (2007) Arbuscular mycorrhizal fungus (*Glomus aggregatum*) influences biotransformation of arsenic in the rhizosphere of sunflower (*Helianthus annuus* L.) Soil Sci Plant Nutr 53(4):499–508
Valdenegro M, Barea J, Azcón R (2001) Influence of arbuscular-mycorrhizal fungi, *Rhizobium meliloti* strains and PGPR inoculation on the growth of *Medicago arborea* used as model legume for re-vegetation and biological reactivation in a semi-arid mediterranean area. Plant Growth Regul 34(2):233–240
van der Heijden MG (2004) Arbuscular mycorrhizal fungi as support systems for seedling establishment in grassland. Ecol Lett 7(4):293–303
van der Heijden MG, Scheublin TR (2007) Functional traits in mycorrhizal ecology: their use for predicting the impact of arbuscular mycorrhizal fungal communities on plant growth and ecosystem functioning. New Phytol 174(2):244–250
van der Heijden MG, Boller T, Wiemken A et al (1998a) Different arbuscular mycorrhizal fungal species are potential determinants of plant community structure. Ecology 79(6):2082–2091
van der Heijden MG, Klironomos JN, Ursic M et al (1998b) Mycorrhizal fungal diversity determines plant biodiversity, ecosystem variability and productivity. Nature 396(6706):69–72
van der Heijden MG, Wiemken A, Sanders IR (2003) Different arbuscular mycorrhizal fungi alter coexistence and resource distribution between co-occurring plant. New Phytol 157(3):569–578
Varma A (1999) Hydrolytic enzymes from arbuscular mycorrhizae: the current status. In: Mycorrhiza. Springer, Berlin, pp 373–389
Vivas A, Azcon R, Biro B et al (2003a) Influence of bacterial strains isolated from lead-polluted soil and their interactions with arbuscular mycorrhizae on the growth of *Trifolium pratense* L. under lead toxicity. Can J Microbiol 49(10):577–588
Vivas A, Biro B, Campos E et al (2003b) Symbiotic efficiency of autochthonous arbuscular mycorrhizal fungus (*G. mosseae*) and *Brevibacillus* sp. isolated from cadmium polluted soil under increasing cadmium levels. Environ Pollut 126(2):179–189
Vivas A, Barea JM, Biro B et al (2006a) Effectiveness of autochthonous bacterium and mycorrhizal fungus on *Trifolium* growth, symbiotic development and soil enzymatic activities in Zn contaminated soil. J Appl Microbiol 100(3):587–598

Vivas A, Biro B, Nemeth T et al (2006b) Nickel-tolerant *Brevibacillus brevis* and arbuscular mycorrhizal fungus can reduce metal acquisition and nickel toxicity effects in plant growing in nickel supplemented soil. Soil Biol Biochem 38(9):2694–2704

Vodnik D, Grcman H, Macek I et al (2008) The contribution of glomalin-related soil protein to Pb and Zn sequestration in polluted soil. Sci Total Environ 392(1):130–136

Vogel-Mikus K, Pongrac P, Kump P et al (2006) Colonisation of a Zn, Cd and Pb hyperaccumulator *Thlaspi praecox* Wulfen with indigenous arbuscular mycorrhizal fungal mixture induces changes in heavy metal and nutrient uptake. Environ Pollut 139(2):362–371

Wang FY, Miao YF (2006) Effects of different arbuscular mycorrhizal fungi on the growth and yield of soybean in coal mine spoil. World J Agric Sci 2(4):383–389

Wang FY, Liu RJ, Lin XG et al (2004) Arbuscular mycorrhizal status of wild plants in saline-alkaline soils of the Yellow River Delta. Mycorrhiza 14(2):133–137

Wang FY, Lin XG, Yin R (2005) Heavy metal uptake by arbuscular mycorrhizas of *Elsholtzia splendens* and the potential for phytoremediation of contaminated soil. Plant Soil 269(1–2):225–232

Wang FY, Lin XG, Yin R (2007) Effect of arbuscular mycorrhizal fungal inoculation on heavy metal accumulation of maize grown in a naturally contaminated soil. Int J Phytorem 9(4):345–353

Wang LP, Zhang WW, Guo GX et al (2009) Selection experiments for the optimum combination of AMF-plant-substrate for the restoration of coal mines. Mining Sci Technol (China) 19(4):479–482

Wang FY, Wang L, Shi ZY et al (2012) Effects of AM inoculation and organic amendment, alone or in combination, on growth, P nutrition, and heavy-metal uptake of tobacco in Pb-Cd-contaminated soil. J Plant Growth Regul 31(4):549–559

Wang FY, Shi ZY, Xu XF et al (2013) Contribution of AM inoculation and cattle manure to lead and cadmium phytoremediation by tobacco plants. Environ Sci Process Impacts 15(4):794–801

Weissenhorn I, Leyval C, Belgy G et al (1995) Arbuscular mycorrhizal contribution to heavy-metal uptake by maize (*Zea mays* L.) in pot culture with contaminated soil. Mycorrhiza 5(4):245–251

Whitfield L, Richards AJ, Rimmer DL (2004) Effects of mycorrhizal colonisation on *Thymus polytrichus* from heavy-metal-contaminated sites in northern England. Mycorrhiza 14(1):47–54

Wu T, Hao W, Lin X et al (2002) Screening of arbuscular mycorrhizal fungi for the revegetation of eroded red soils in subtropical China. Plant Soil 239(2):225–235

Wu FY, Ye ZH, Wu SC et al (2007) Metal accumulation and arbuscular mycorrhizal status in metallicolous and nonmetallicolous populations of *Pteris vittata* L. and *Sedum alfredii* Hance. Planta 226(6):1363–1378

Wu F, Bi Y, Wong M (2009) Dual inoculation with an arbuscular mycorrhizal fungus and *Rhizobium* to facilitate the growth of alfalfa on coal mine substrates. J Plant Nutr 32(5):755–771

Wu SC, Wong CC, Shu WS et al (2011) Mycorrhizo-remediation of lead/zinc mine tailings using vetiver: a field study. Int J Phytorem 13(1):61–74

Xia YS, Chen BD, Christie P et al (2007) Arsenic uptake by arbuscular mycorrhizal maize (*Zea mays* L.) grown in an arsenic-contaminated soil with added phosphorus. J Environ Sci China 19(10):1245–1251

Xiao XY, Chen BD, Zhu YG (2006) The influences of arbuscular mycorrhizal fungi on growth and mineral nutrition of plants grown in copper mine tailing. Acta Sci Circums 26(2):312–317

Xie X, Huang W, Liu F et al (2013) Functional analysis of the novel mycorrhiza-specific phosphate transporter *AsPT1* and *PHT1* family from *Astragalus sinicus* during the arbuscular mycorrhizal symbiosis. New Phytol 198(3):836–852

Xu G, Chague V, Melamed-Bessudo C, Kapulnik Y et al (2007) Functional characterization of *LePT4*: a phosphate transporter in tomato with mycorrhiza-enhanced expression. J Exp Bot 58(10):2491–2501

Yao Q, Li XL, Feng G et al (2001) Mobilization of scarcely-soluble inorganic phosphate by external mycelium of arbuscular mycorrhizal fungus. Plant Soil 230:279–285
Zhang WM, Ma YQ, Meng N et al (1996) Field study on vesicular-arbuscular mycorrhizae used in mine reclamation. Mine Metall 5(3):17–21
Zhang H, Wu X, Li G et al (2011a) Interactions between arbuscular mycorrhizal fungi and phosphate-solubilizing fungus (*Mortierella* sp.) and their effects on *Kosteletzkya virginica* growth and enzyme activities of rhizosphere and bulk soils at different salinities. Biol Fertil Soils 47(5):543–554
Zhang YF, Wang P, Yang YF et al (2011b) Arbuscular mycorrhizal fungi improve reestablishment of *Leymus chinensis* in bare saline-alkaline soil: implication on vegetation restoration of extremely degraded land. J Arid Environ 75(9):773–778
Zhang T, Sun Y, Shi Z et al (2012) Arbuscular mycorrhizal fungi can accelerate the restoration of degraded spring grassland in Central Asia. Rangel Ecol Manag 65(4):426–432
Zhao RX, Guo W, Fu RY et al (2013) Effects of arbuscular mycorrhizal fungi on the vegetation restoration of different types of coal mine spoil banks. Environ Sci 34(11):4447–4454
Zhu YG, Miller MR (2003) Carbon cycling by arbuscular mycorrhizal fungi in soil–plant systems. Trends Plant Sci 8(9):407–409

Chapter 12
Arbuscular Mycorrhizal Fungi and Plant Growth on Serpentine Soils

Husna, Faisal Danu Tuheteru, and Asrianti Arif

Abstract Arbuscular mycorrhizal fungi (AMF) are obligate fungi (root symbionts) of the phylum of Glomeromycota that associated with 70–90% of land's plants. AMF are found in many types of soils and ecosystems. AMF can colonize plant roots on serpentine soils, and 11 AMF genera and Glomeraceae as dominant family are found. Diversity of AMF on serpentine soil is influenced by soil chemical properties (metal content, Ni and Mg/Ca ratio), plant species, and vegetation types as well as AMF types. Inoculation of AMF improved growth, biomass, and nutrient uptake (especially P) for sensitive plant and nickel accumulators. Ni uptake by inoculated plants is inconsistent, showing that AMF reduced Ni in sensitive plant tissues. Otherwise, AMF increased Ni uptake in hyperaccumulator plants. Effectiveness of AMF is determined by plant species and AMF. AMF colonization is essential for vegetation successional acceleration and revegetation success in nickel post-mining land. AMF are potential to be developed as a biological fertilizer to support revegetation of nickel post-mining land on serpentine soil.

Keywords Glomeraceae • Nickel hyperaccumulators • Revegetation • Sulawesi • Serpentine soil

12.1 Introduction

Serpentine soil covers less than 3% of the earth's surface and distributed in several regions in the world of California, Cuba, South Africa, South Europe, New Caledonia, Southeast Asia, and West Australia (Coleman and Jove 1992; Guillot and Hattori 2013). Lodging in Southeast Asia, serpentine soil spreads over the northern part of Borneo, Palawan, Mindanao, Sabah, and the majority of Sulawesi and Halmahera (Whitten et al. 1987; Proctor 2003; Ent et al. 2013, 2015). Serpentine soil has the following characteristics: high Mg/Ca ratio, heavy metal (Co, Cu, Cr, Mn, Ni) concentrations, and deficiencies of macronutrients (N, P, K) (Brooks 1987;

Husna (✉) • F.D. Tuheteru (✉) • A. Arif
Departement of Forestry, Faculty of Forestry and Environmental Science,
Halu Oleo University, Kendari, Southeast Sulawesi 93121, Indonesia
e-mail: husna.faad19@yahoo.com; faisaldanu_28@yahoo.com

Q.-S. Wu (ed.), *Arbuscular Mycorrhizas and Stress Tolerance of Plants*,
DOI 10.1007/978-981-10-4115-0_12

Kruckeberg 1984; Proctor 2003). This is a characteristic of the syndrome known as serpentine. In addition to the phenomenon of serpentine syndrome, nickel mining activities also contribute to the degradation of serpentine soil and vegetation removal (Review of O'Dell and Claassen 2009). Conditioning the land can be toxic as well as restrict the growth and development of plants and soil microbial activity.

Each species has a specific mechanism for adaptation, tolerance, and survival on serpentine soil. One of the mechanisms of plant adaptation on serpentine soil may be symbiosed with arbuscular mycorrhizal fungi (AMF). AMF are an obligate fungi (root symbionts) of the phylum Glomeromicota with 80% of land plants (Smith and Read 2008). Southworth et al. (2014) reported that soil fertility was low in serpentine soil and stress that strong serpentine habitat was thought to stimulate the formation of mycorrhizae. AMF play an important role in supporting the growth and development of sensitive plants grown in serpentine soils through improved water and nutrient status and heavy metal stress tolerance. The presence of AMF on serpentine soil plays an important role in adaptation, distribution, and succession (community dynamics) in various types of vegetation in conditions of serpentine (Hopkins 1987; Castelli and Casper 2003; Perrier et al. 2006; Husna et al. 2016a). AMF influencing on crop tolerance to heavy metal toxicity is seriously affected by many factors such as soils, plant species, isolates, and type of symbiont and metal concentrations (Amir et al. 2013).

Various studies related to AMF symbiosis with plants in serpentine soil have been carried out. AMF ecological studies in the rhizosphere of plants on serpentine soil have been reviewed by Southworth et al. (2014). However, studies had examined the roles of AMF in the growth and uptake of unavailable plant nutrient. Therefore, this article is to review the AMF ecology, inoculated plant growth, and researches on the AMF symbiosis with plant species and planting practices in Indonesia.

12.2 Ecology of AMF on Serpentine Soils

AMF symbiosis with plants in serpentine soil has been reported in both temperate and tropical regions (Southworth et al. 2014; Husna et al. 2015). AMF have been associated with 27 species of herbaceous plants colonized in serpentine grassland in California (Hopkins 1987). On serpentine soils in Italy, four plant species, namely, *Centaurea paniculata*, *Genista salzmannii*, *Helichrysum italicum*, and *Thymus striatus*, were colonized by native AMF (Lioi and Giovannetti 1989). There is also a symbiosis between AMF plant species in grassland of Pennsylvania dominated by *Andropogon gerardii*, *Schizachyrium scoparium*, *Sorghastrum nutans*, and *Sporobolus heterolepis* (Castelli and Casper 2003). Perrier et al. (2006) found a symbiosis between AMF and 10 endemic species of plants that grow on ultramafic lands in Koniambo Massif, New Caledonia. Cumming and Kelly (2007) also reported six AMF species in the rhizosphere of three vegetation types in Maryland, USA, site, while the abundance and sporulation of AMF were more in grassland site.

According to Lagrange et al. (2011), roots of nine types of Cyperaceae homegrown ultramaphic New Caledonia were colonized by AMF. AMF had also been found in serpentine revegetated lands of Southeast and South Sulawesi, Indonesia (Setiadi and Setiawan 2011; Husna et al. 2014, 2015, 2016a).

In addition to ultramaphic adaptive and tolerant plants, the presence of AMF also reported in symbiosis with plants hyperaccumulator of Ni. AMF colonization was found to be high on hyperaccumulator of Ni *Berkheya coddii* and three other types of family Asteraceae, namely, *Senecio anomalochrous*, *S. coronatus*, and B. *zeyherri* in South Africa (Turnau and Mesjasz-Przbylowicz 2003). There are variations in mycorrhizal colonization on plant hyperaccumulation on nickel site of *Phyllanthus favieri* in humid forest dominated by *Nothofagus*. The Koniambo Massif, New Caledonia, and plants that are not colonized tend to have high nickel content in leaves (Perrier et al. 2006). According to Amir et al. (2007), nine endemic species of New Caledonia were symbiosed with the AMF, while *Sebertia acuminata, Psychotria douarrei*, and *Phyallanthus favieri* plants represented lower root AMF colonization than the others. Amir et al. (2013) have again reported that AMF colonized all types of hyperaccumulators of nickel and AMF colonization was relatively low on conditions of strong hyperaccumulator while weaker than hyperacumulators. Rhizosphere of AMF of strong nickel hyperaccumulators of plant has a high tolerance to the high Ni and is able to colonize the roots (Amir et al. 2007).

AMF genera found in the root zone of plants in the serpentine soil were *Archaespora*, *Acaulopsora*, *Diversispora*, *Scutellospora*, *Pacispora*, *Glomus*, *Sclerocystis*, *Funneliformis*, *Rhizophagus*, *Claroideoglomus*, and *Paraglomus* (Table 12.1). Glomeraceae is the largest family. *F. mosseae*, *C. etunicatum*, and *R. fasciculatum* are types of dominant AMF and widespread (Hopkins 1987; Lioi and Giovannetti 1989; Castelli and Casper 2003; Gustafson and Casper 2004; Lagrange et al. 2011; Orlowska et al. 2011; Husna et al. 2015). In addition to the AMF, the ectomycorrhiza type has also been found to associate with plants in serpentine soil, in Japan (Kayama et al. 2006), Portugal (Gonçalves et al. 2009), New Caledonia (Perrier et al. 2006; Jourand et al. 2010a, b), and the USA (Gladish et al. 2010).

Various factors that affect colonization, spore abundance, and species richness AMF in serpentine soil include soil chemical properties (metal content, Ni and Mg/Ca ratio), plant species, and vegetation types as well as type of AMF. Ni content can lower sporulation and AMF species richness (Perrier et al. 2006; Cumming and Kelly 2007; Amir et al. 2007) and root colonization (Lagrange et al. 2011; Amir et al. 2013). Nonetheless, Doubková et al. (2011) found a low AMF colonization in rooting types of *Knautia sparsiflora* on serpentine than non-serpentine, and there is a positive correlation between colonization of AMF and soil pH, Ca, and K or Ca/Mg ratio. Colonization level and diversity of AMF are reported to be relatively similar between on and off serpentine soil (Schechter and Bruns 2008). Perrier et al. (2006) reported that there was a positive relationship between the availability of metals (Ni and Co) and abundance of black spore. Relationships of fungi and plants in serpentine grassland in Pensylvannia were heavily influenced by the performance of the plant or regulatory diversity (Castelli and Casper 2003). Moreover, there was a reciprocal relationship between plants and fungal-specific serpentine soil, where

Table 12.1 Researches of AMF ecology on serpentine soils

References	Hopkins (1987)	Lioi and Giovannetti (1989)	Castelli and Casper (2003)	Cumming and Kelly (2007)	Schechter and Bruns (2008)	Lagrange et al. (2011)	Ji et al. (2010)
Vegetation types	Serpentine grassland/ California	Italy	Serpentine grassland/ Pennsylvania	Serpentine grassland/ Maryland	*Collinsia sparsiflora*/ California	Cyperaceae/New Caledonia	Maryland
AMF morphology observed	Vesikel, hifa	Hifa, coils	–	–	–	Hifa, vesikel, arbuskula	–
Species richness	2	17	7	6	15	2	10
Recorded genera	*Rhizophagus*	*Glomus*, *Funneliformis*, *Rhizophagus*, *Sclerocystis*	*Gigaspora*, *Scutellospora*, *Glomus*, *Rhizophagus*, *Claroideoglomus*	*Acaulospora*, *Archaespora*, *Diversispora*, *Paraglomus*, *Cetraspora*	*Acaulopsora*, *Archaespora*, *Diversispora*, glomus, *Pacispora*, *Scutellospora*	*Claroideoglomus*	*Archaespora*, *Acaulospora*, *Entrophospora*, *Gigaspora*, *Scutellospora*, *Glomus*, *Funneliformis*, *Sclerocystis*, *Septoglomus*, *Claroideoglomus*
AMF Col (%)	5–100	0–47	–	20–43	44–57	8–57	–
Spore abundances	–	30–116 per 100 g	16–40/50 g	0–102/50 ml	–	–	224–506/50 ml
Focus studied	Vegetation type	Plant species	Plant species	Vegetation type, soil properties	Plant species	Plant species	Soil properties

Gigaspora gigantea increased plant biomass of *Schizachyrium* and *Andropogon*, but a decline in biomass of *Andropogon* inoculated with *Glomus etunicatum*.

AMF isolated from the roots of hyperaccumulator ultramaphic Ni in soil are more tolerant than nonhyperaccumulators isolated from ultramaphic Ni in soil (Amir et al. 2008). Amir et al. (2008) revealed that five isolates of *Glomus* spp. in ultramaphic soil were capable of germination under the condition of up to 30 μg/g Ni, but the spores of non-ultramaphic overall stunted at 15 μg/g Ni. According to Doubková et al. (2012), a high tolerance on populations of plants and fungi on serpentine was recorded.

12.3 Plant Growth Affected by AMF on Serpentine Soils

AMF presence on serpentine soil conditions can contribute to tolerance of crops to Ni, decreasing/increasing the uptake and accumulation of Ni. AMF is also suitable for phytoremediation activities with phytoextraction mechanism and phytostabilization (reducing nickel translocation root shoots) and phyto-mining of nickel, as well as local AMF isolated from ultramaphic soil has the potential to support the success of ecological restoration in degraded ecosystems.

Inoculation with indigenous AMF strains significantly enhanced the growth, biomass, and survival of *Berkheya coddii* than non-mycorrhizal plants (Orlowska et al. 2011). Colonization by *Glomus intraradices* increased the weight of dry matter sunflower on level 0 and 100 Ni mg/kg treatment, compared to non-mycorrhizal control (Ker and Christine 2009). *Glomus* sp. and a mixture of AMF improved plant growth and P uptake in *Knautia arvensis* with increasing intensity of drought (Doubkova et al. 2013). Under greenhouse, native AMF inoculation increased biomass of hyperaccumulator plant shoots in Ni. *Berkheya coddii* in South Africa had a positive correlation between AMF colonization and shoot height (Turnau and Mesjasz-Przybylowics 2003). Gustafson and Casper (2004) reported that *G. intraradices* promoted plant growth in *Andropogon gerardii* on serpentine soil and sand mixture.

Two types of ultramaphic endemic in New Caledonia, *Alphitonia neocaledonica*, and *Cloezia artensis* inoculated with *Glomus etunicatum* at the age of 12 months accumulated dry matter and high P (Amir et al. 2013). Boulet and Lambers (2005) reported that inoculation with AMF induced lower plant growth but increased levels of P and K in the shoot. The concentration of P, Ca, Zn, and Cu in mycorrhizal *Berkheya coddii* plants was higher than in non-mycorrhizal plants, while an increase in the levels of P in inoculated plant was ten times, compared to the control (Orlowska et al. 2008). AMF inoculation with *Glomus* sp. increased the biomass of shoots and roots in *Costularia comosa* by 124% and 246%, respectively (Lagrange et al. 2011). Inoculation with local AMF improved P uptake of the *Knautia arvensis* plant (Doubková et al. 2011). Doubková et al. (2012) reported that isolates of AMF on serpentine soil not only had high root colonization but also efficiently supported plant growth and P uptake in *Knautia arvensis*.

Ni uptake by inoculated plants is inconsistent. Inoculation with AMF increased Ni content types such as nickel hyperaccumulator in *Berkheya coddii* (Turnau and Mesjasz-Przybylowics 2003; Orlowska et al. 2011) and sunflower (Ker and Christine 2009). Inoculation with AMF reduced the Ni content at the top of *Phaseolus vulgaris* (Guo et al. 1996), shoots of *Trifolium repens* (Vivas et al. 2006), the root of *Costularia comosa* (Lagrange et al. 2011, 2013), *Knautia arvensis* (Doubková et al. 2011), *Alphitonia neocaledonica*, and *Cloezia artensis* (Amir et al. 2013) compared to the control. Ni content in leaves was not influenced by the presence of the AMF (Boulet and Lambers 2005). High Ni accumulation in plants is allegedly due to AMF which enhances the activity of glutamine synthesis (chelate nickel) in the root (Ker and Christine 2009).

12.4 Plant-AMF Symbiosis on Serpentine Soil in Indonesia

Indonesia is one of the regions in the world which has fairly large ultramaphic bedrock scattered on the island of Sulawesi and Halmahera (Whitten et al. 1987; Proctor 2003; Ent et al. 2013). Ultramaphic bedrock in Indonesia has productive potential for nickel mining operations. Nickel mining operations in Sulawesi have been conducted by PT. INCO (known as PT. Vale Indonesia) in South Sulawesi, Central and Southeast, as well as PT. ANTAM, Tbk in Pomalaa, Southeast Sulawesi, and ± 200an mining business license (IUP) has also been operating in Southeast Sulawesi. Nickel exploitation activities are generally carried out using opencast mining. This method usually ruins the landscape and leaving land nutrient deficiency and toxic heavy metals and reduces the biological activity, waterlogging, etc. (Review O'Dell and Claassen 2009). These lands should be restored with revegetation techniques. Selection of seed types and supplying seedling with soil microbes (Mycorrhizae and Rhizobium) are very important to support the success of the nickel post-mining site revegetation (Marpaung et al. 1994; Ambodo 2002; Mansur 2010; Husna et al. 2012).

Exploration, extraction, identification, and propagation of AMF on serpentine soils in Sulawesi began in the mid-1990s. Based on the identification of spore morphology, we discovered the eight AMF genera, including *Glomus*, *Septoglomus*, *Claroideoglomus*, *Rhizopagus*, *Acaulospora*, *Gigaspora*, *Racocetra*, and *Scutellospora*, in the rhizosphere of plant revegetation and vegetation succession on nickel post-mining lands (Review Husna et al. 2012; Husna et al. 2014, 2015; Husna et al. 2016a; Setiadi and Setiawan 2011). Husna et al. (2014, 2015) found seven kinds of AMF in revegetation serpentine lands in PT. Vale Indonesia, Southeast Sulawesi Pomalaa, namely, *Glomus canadense*, *G. boreale*, *Rhizopagus diaphanous*, *Septoglomus constrictum*, *Claroideoglomus etunicatum*, *Racocetra gregaria*, and *Scutellospora auriglobosa*. According to Husna et al. (2016a), AMF colonized 15 pioneer species nickel post-mining land in Pomalaa, Southeast Sulawesi. Plant types *Cynodon dactylon*, *Ipomoea* sp., and *Sarcotheca celebica* have the highest root mycorrhizal colonization ($\geq$70%). Information on AMF colonization in

Fig. 12.1 Growth performance of *P. mooniana* at 3 months of age inoculated with local AMF on serpentine soil media (Husna et al. 2016b)

rooting in hyperaccumulator of nickel (*S. celebica*) enriches knowledge of AMF symbiosis with hyperaccumulator types of nickel across the world. AMF colonization in all 15 types of pioneer speciesis also indicated that the AMF alleged role accelerating vegetation succession in post-mining land nickel.

On a nursery and greenhouse scale, inoculation with exotic (GMRT-17 and ET-17) and local isolates (INCO-12) effectively increased the growth and improvement of seed quality of *Paraserianthes falcataria*, *Acacia mangium*, and *Trichospermum buretti* (Marpaung et al. 1994). According to Husna (2010), seedling of *Pericopsis mooniana* inoculated with 5 g of AMF Mycofer inocula (mix *Glomus manihotis*, *Glomus etunicatum*, *Acaulospora tuberculata*, *Gigaspora margarita*) and the addition of 20 g of pulp sago had high growth, dry weight, nodule, and the uptake of K and Ca with an increase in each of the controls 51.4%, 41.3%, 19.7%, 8.8%, and 25.6% and reduced by 32% of Ni. Inoculation with AMF Mycofer gave a high boost of nodules and biomass of *Albizia saponaria* plant age of 3 months with an increase over the control by 76%, 447%, and 309% (Tuheteru et al. 2011). In the serpentine soil media, *P. mooniana* desperately need mycorrhizal (MIE> 75%) to support growth (Husna et al. 2016b). Local AMF isolated from the roots of *P. mooniana* of CA Lamedai (non-serpentine) and PT. Vale Indonesia (serpentine) better increased the biomass of *P. mooniana* at 3 months in the media serpentine soil, which increased 442–472% over the controls and 64% and 73% over the commercial Mycofer inoculum (Fig. 12.1).

On the field scale, *P. mooniana* inoculated with AMF (CA Lamedai and PT. Vale Indonesia) planted in post-mining land of nickel (Mg/Ca ratio of 3.55 and a concentration of 2.1 ppm Ni) 3 months after implantation has the tolerance for survival, growth, biomass, and accumulation of N, P, and K, which is higher than the

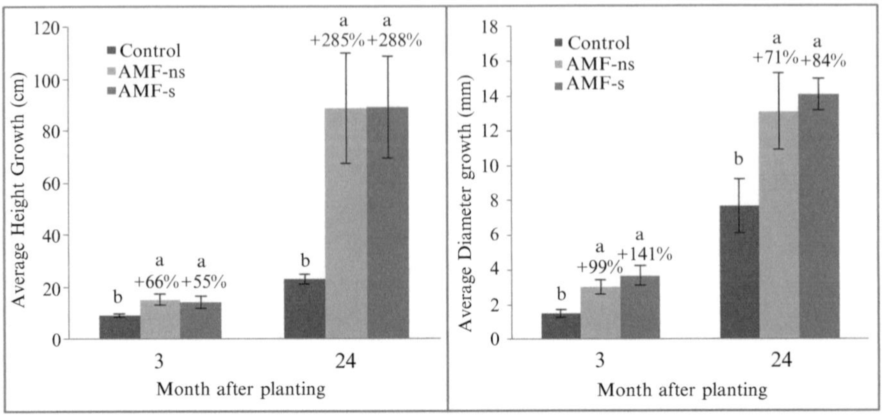

Fig. 12.2 Growth responses in height and diameter of inoculated *P. mooniana*. Thw. at 3 and 24 months after planting

control and low Ni content (Husna et al. 2015). At the age of 24 months after planting, inoculation with local AMF also increased the height and diameter growth of *P. mooniana* on the field over the control (Fig. 12.2).

Based on the important role of the AMF, PT. Vale Indonesia (PT.) Sorowako South Sulawesi has been producing AMF-inoculated seeds on nursery scale for revegetation purposes (Ambodo 2002; Mansur 2010).

12.5 Conclusion Remarks

AMF, assosiated with plant root on serpentine soil, have important roles to support growth and improved nutrient status on serpentine soil media. Furthermore, AMF are important components in ecological restoration for revegetation degraded land (nickel post-mining land). AMF diversity study has been done in various countries (the center of serpentine soil) using morphology and DNA molecular approach, details the response of non-mycorrhizal and mycorrhizal plants on serpentine soil conditions, and includes adaptation mechanisms of anatomy, physiology, molecular and signaling pathways, as well as the response of AMF spores (serpentine and non-serpentine isolates) to heavy metal stress, especially Ni.

Acknowledgments This research was supported by a competitive grant (No. 228c.1/UN29.20/PP/2016) and a fundamental grant (No. 229.a.1/UN.29.20/PPM/2016) of the Ministry of Research, Technology and Higher Education. The author would like to thank the leadership of the PT. Vale Indonesia (PT.) Pomalaa Region, Kolaka, and PT. Stargate Pacific Resources, North Konawe, Southeast Sulawesi.

References

Ambodo AP (2002) Mine reclamation – the PT Inco experience. Proceedings of the 26th Annual British Columbia Mine Reclamation Symposium in Dawson Creek, BC

Amir H, Parrier N, Riagault F et al (2007) Relationships between Ni-hyperaccumulation and mycorrhizal status of different endemic plant species from New Caledonian ultramafic soils. Plant Soil 293:23–35

Amir H, Jasper DA, Abbott LK (2008) Tolerance and induction of tolerance to Ni of arbuscular mycorrhizal fungi from New Caledonian ultramafic soils. Mycorrhiza 19:1–6

Amir H, Lagrange A, Hassaïne N et al (2013) Arbuscular mycorrhizal fungi from New Caledonian ultramafic soils improve tolerance to nickel of endemic plant species. Mycorrhiza 23:585–595

Boulet FM, Lambers H (2005) Characterisation of arbuscular mycorrhizal fungi colonisation in cluster roots of *Hakea verrucosa* F. M. (Proteaceae), and its effect on growth and nutrient acquisition in ultramafic soil. Plant Soil 269:357–367

Brooks RR (1987) Serpentine and its vegetation. A multidisciplinary approach. Dioscorides Press, Portland

Castelli JP, Casper BB (2003) Intraspecific AM fungal variation contributes to plant-fungal feedback in a serpentine grassland. Ecology 84:323–336

Coleman R, Jove C (1992) Geological origin of serpentinites. In: Baker AJM, Proctor J, Reeves RD (eds) The vegetation ofultramafic (serpentine) soils. Intercept Ltd, Andover, pp 1–18

Cumming JR, Kelly CN (2007) *Pinus virginiana* invasion influences soils and arbuscular mycorrhizae of a serpentine grassland. J Torrey Bot Soc 134(1):63–73

Doubková P, Suda J, Sudová R (2011) Arbuscular mycorrhizal symbiosis on serpentine soils: the effect of native fungal communities on different *Knautia arvensis* ecotypes. Plant Soil 345:325–338

Doubková P, Suda J, Sudová R (2012) The symbiosis with arbuscular mycorrhizal fungi contributes to plant tolerance to serpentine edaphic stress. Soil Biol Biochem 44:56–64

Doubkova P, Vlasakova E, Sudova R (2013) Arbuscular mycorrhizal symbiosis alleviates drought stress imposed on *Knautia arvensis* plants in serpentine soil. Plant Soil 370:149–161

Ent AVD, Baker AJM, Van Balgooy MMJ et al (2013) Ultramafic nickel laterites in Indonesia (Sulawesi, Halmahera): mining nickel hyperaccumulators and opportinities for phytomining. University Of Queensland. J Chem Explor 128:72–79

Ent AVD, Erskine P, Sumail S (2015) Ecology of nickel hyperaccumulator plants from ultramafic soils in Sabah (Malaysia). Chemoecology 25:243–259

Gladish S, Frank JL, Southworth D (2010) The serpentine syndrome belowground: ectomycorrhizas and hypogeous fungi associated with conifers. Can J For Res 40:1671–1679

Gonçalves SC, Martins-Loução MA, Freitas H (2009) Evidence of adaptive tolerance to nickel in isolates of Cenococcum geophilum from serpentine soils. Mycorrhiza 19:221–230

Guillot S, Hattori K (2013) Serpentinites: essential roles in geodynamics, arc volcanism, sustainable development, and the origin of life. Elements 9(2):95–98

Guo Y, George E, Marschner H (1996) Contribution of an arbuscular mycorrhizal fungus to the uptake of cadmium and nickel in bean and maize plants. Plant Soil 184:195–205

Gustafson DJ, Casper BB (2004) Nutrient addition affects AM fungal performance and expression of plant/fungal feedback in three serpentine grasses. Plant Soil 259:9–17

Hopkins NA (1987) Mycorrhizae in a California serpentine grassland community. Can J Bot 65:484–487

Husna (2010) Pertumbuhan bibit kayu kuku (*Pericopsis mooniana* THW) melalui aplikasi fungi mikoriza arbuskula (FMA) dan ampas sagu pada media tanah bekas tambang nikel [thesis]. Univeristas Halu Oleo, Kendari

Husna, Tuheteru FD, Arif A (2012) Post-mine land re-vegetation in Southeast Sulawesi biotechnology-based Mycorrhizal Fungi. Proceeding of International Conference on Perspectives of Tropical Forest Rehabilitation Better Forest Functions and Management. Faculty of Forestry UGM, Yogyakarta, pp 186–190

Husna, Budi SWR, Mansur I, Kusmana C, Kramadibrata K (2014) Arbuscular Mycorrhizal Fungi from Rhizosphere of *Pericopsis mooniana* (Thw.) Thw. in South-East Sulawesi. Berita Biologi 13(3):263–273

Husna, Budi SWR, Mansur I, Kusmana C (2015) Diversity of arbuscular mycorrhizal fungi in the growth habitat of kayu kuku (*Pericopsis mooniana* Thw.) in Southeast Sulawesi. Pak J Biol Sci 18(1):1–10

Husna, Tuheteru FD, Khalifah N (2016a)Symbiosis arbuscular mycorrhizal fungi with pioneer plants on nickel post mining land. Presented paper on national seminar of Silviculture IV. Faculty of Forestry, Mulawarman University, Balikpapan (Indonesia) 19–20 Juli 2016

Husna, Sri Wilarso Budi R, Mansur I, Kusmana C (2016b) Growth and nutrient status of kayu kuku (*Pericopsis mooniana* Thw.) with mycorrhiza in soil media of nickel post mining. Pak J Biol Sci 19:158–170

Ji BM, Bentivenga SP, Casper BB (2010) Evidence for ecological matching of whole AM fungal communities to the local plant–soil environment. Ecology 91:3037–3046

Jourand P, Ducousso M, Loulergue-Majorel C et al (2010a) Ultramafic soils from New Caledonia structure *Pisolithus albus* in ecotype. FEMS Microbiol Ecol 72:238–249

Jourand P, Ducousso M, Reid R et al (2010b) Nickel-tolerant ectomycorrhizal *Pisolithus albus* ultramafic ecotype isolated from nickel mines in New Caledonia strongly enhance growth of a host plant at toxic nickel concentrations. Tree Physiol 30:1311–1319

Kayama M, Choi D, Tobita H et al (2006) Comparison of growth characteristics and tolerance to serpentine soil of three ectomycorrhizal spruce seedlings in northern Japan. Trees 20:430–440

Ker K, Christine C (2009) Nickel remediation by AM-colonized Sunflower. Mycorrhiza 20:399–406

Kruckeberg AR (1984) California serpentines: flora, vegetation,geology, soils and management problems. University of California Press, Berkeley

Lagrange A, Ducousso M, Jourand P et al (2011) New insights into the mycorrhizal status of Cyperaceae from ultramafic soils in New Caledonia. Can J Microbiol 57:21–28

Lagrange A, L'Huillier L, Amir H (2013) Mycorrhizal status of Cyperaceae from new Caledonian ultramafic soils: effects of phosphorus availability on arbuscular mycorrhizal colonization of *Costularia comosa* under field conditions. Mycorrhiza 23:655–661

Lioi L, Giovannetti M (1989) Vesicular-arbuscular mycorrhizae and species of the Endogonaceae in an Italian serpentine soil. G Bot Ital 123:1–8

Mansur I (2010) Teknik Silvikultur untuk Reklamasi Lahan Bekas Tambang. SEAMEO BIOTROP, Bogor. (in Indonesia)

Marpaung P, Setiadi Y, Tobing B (1994) Revegetation development and progress in nickel mine sites at PT. International Nickel Indonesia. In: Simatupang M, Wahju BN (eds) Mineral development in Asia Pasific into the year 2000. Proceeding of the 4th Asia Pacific mining Conference, Jakarta 26–29 Oktober 1994. Asean Federation of Mining Association, Jakarta

O'Dell RE, Claassen VP (2009) Serpentine revegetation: a review. Soil and biota of serpentine: a world view 2009. Northeast Nat 16:253–271

Orlowska E, Mesjasz-Przybylowicz J, Przybylowicz W et al (2008) Nuclear microprobe studies of elemental distribution in mycorrhizal and non-mycorrhizal roots of ni-hyperaccumulator *Berkheya coddii*. X-Ray Spectrom 37:129–132

Orlowska E, Przybylowicz W, Orlowski D et al (2011) The effect of mycorrhiza on the growth and elemental composition of ni-hyperaccumulating plant *Berkheya coddii* roessler. Environ Pollut 159:3730–3738

Perrier N, Amir H, Colin F (2006) Occurrence of mycorrhizal symbioses in the metal-rich lateritic soils of the *Koniambo massif*, New Caledonia. Mycorrhiza 16:449–458

Proctor J (2003) Vegetation and soil and plant chemistry on ultramic rocks in the tropical far east. Perspect Plant Ecol Evol Syst 6(1,2):105–124

Schechter SP, Bruns TD (2008) Serpentine and non-serpentine ecotypes of *Collinsia sparsiflora* associate with distinct arbuscular mycorrhizal fungal assemblages. Mol Ecol 17:3198–3210

Setiadi Y, Setiawan A (2011) Study of arbuscular mycorrhizal fungi status at rehabilitation post-nickel mining area (Case study at PT INCO Tbk. Sorowako, South Sulawesi). J Silvikultur Tropika 3(1):88–95

Smith SE, Read DJ (2008) Mycorrhizal symbiosis, 3rd edn. Academic Press, USA

Southworth D, Tackaberry LE, Massicotte HB (2014) Mycorrhizal ecology on serpentine soils. Plant Ecol Divers 7(3):445–455

Tuheteru FD, Husna, Arif A (2011) Response of growth and dependency of *Albizia saponaria* (Lour.) Miq on local arbuscular mycorrhizae fungi from Southeast Sulawesi in post-nickel mining soil. Berita Biologi 5:605–611

Turnau K, Mesjasz-Przybylowicz J (2003) Arbuscular mycorrhiza of *Berkheya coddii* and other ni-hyperaccumulating members of *Asteraceae* from ultramafic soils in South Africa. Mycorrhiza 13:185–190

Vivas A, Biró B, Németh T et al (2006) Nickel-tolerant *Brevibacillus brevis* and arbuscular mycorrhizal fungus can reduce metal acquisition and nickel toxicity effects in plant growing in nickel supplemented soil. Soil Biol Biochem 38:2694–2704

Whitten AJ, Mustafa M, Henderson GS (1987) Ekologi Sulawesi. Gadjah Mada University Press, Yogyakarta

Chapter 13
Application of Arbuscular Mycorrhizal Fungi into Agriculture

Ibrahim Ortaş, Mazhar Rafique, and İbrahim A.M. Ahmed

Abstract In the natural ecosystem, rhizospheric soils have various biological organisms to favour the plant growth, nutrient absorption, stress tolerance, disease prevention, carbon capturing and many more. These organisms include mycorrhizal fungi, bacteria, actinomycetes, etc. which solubilize nutrients and assist the plants in uptaking by roots. Among them, arbuscular mycorrhizal (AM) fungi have key importance in natural ecosystem, but high rate of chemical fertilizer in agricultural fields is diminishing its importance. In this chapter, indigenous AM fungi efficiency is discussed with various doses of chemical fertilizer against number of cereal, cash, horticultural and fruit crops. Moreover, their effects on the plant growth, yield enhancement, fruit quality and soil quality are discussed. In the rhizosphere, AM fungi have main interaction with multipurpose bacteria such as phosphorus solubilizing bacteria, nitrogen fixers, plant growth-promoting rhizobacteria and stress tolerance bacteria. AM fungi contribute in building rhizospheric carbon stock, and, recently, addition of biochar in the soil for enhancing soil physicochemical properties and nutrient release has been studied with AM fungi. In order to manage the indigenous AM fungal spores, soil and crop management is important in association with carbon amendments for soils. One of the greatest challenges for the society is food insecurity, which should be changed into 'food security' by improving our knowledge and practicality to double the food production through sustainable farming approaches.

I. Ortaş (✉)
Department of Soil Science and Plant Nutrition, Faculty of Agriculture,
University of Çukurova, Adana, Turkey
e-mail: iortas@cu.edu.tr

M. Rafique
Department of Soil Science and Plant Nutrition, Faculty of Agriculture,
University of Çukurova, Adana, Turkey

Department of Plant Science, Faculty of Biological Sciences, Quaid-I-Azam University,
Islamabad, Pakistan

İ.A.M. Ahmed
Department of Soil Science and Plant Nutrition, Faculty of Agriculture,
University of Çukurova, Adana, Turkey

Department of Soil and Environment Sciences, University of Khartoum, Khartoum, Sudan

Q.-S. Wu (ed.), *Arbuscular Mycorrhizas and Stress Tolerance of Plants*,
DOI 10.1007/978-981-10-4115-0_13

Keywords Biochar • Mycorrhizal efficiency • Phosphorus solubilizing bacteria • Plant growth-promoting rhizobacteria

13.1 Introduction

Microorganisms (bacteria and fungi) in the rhizosphere infect roots and create nutrient-rich environment for the plant growth and food security (Rodriguez and Sanders 2015). Moreover, they repel the hazardous chemicals from the roots or alternatively dilute their effects in rhizosphere. Mycorrhizal fungi included in natural resources which develop symbiosis with 70–90% of the terrestrial plant roots (Parniske 2008) and its worldwide presence in forest and agroecosystems form 50% of microbial biomass in tropical ecosystem (Olsson et al. 1999). Among various important functions of the plant-fungal symbiosis, plant growth promotion by the phosphorus uptake is a key function (Ortas et al. 2001). Mycorrhizal fungi increase nutrient uptake for the plants, particularly immobile nutrients such as phosphorus (P), copper (Cu) and zinc (Zn) in soil which are not accesible to plant roots in normal condition due to slow immobility (Marschner 2012; Ortas 2003). These macro and micro nutrients are taken up by AM fungi in poorly fertile soils (Kayama and Yamanaka 2014). Moreover, water-use efficency for the plant has been reportedly improved by AM fungal application (Bowles et al. 2016). Additionaly, AM fungi assist the plants to tolerate various environemental stresses such as salinity, drought, heat and pollutants in rhizosphere (Aranda et al. 2013; Bowles et al. 2016; Chandrasekaran et al. 2016; Maya et al. 2014). Besides all these direct benefits to the plants, AM fungi have significant role in soil structure development and aggregate formation which improves soil quality and culminates to better plant health (Jansa et al. 2016; Zou et al. 2016).

Mycorrhizal application also reduces the quantitative use of chemical fertilizer input especially P fertilizer (Charron et al. 2001; Ortas 2012b). By keeping in mind the importance of mycorrhizae for plant growth promotion and soil qualaity improvement, it is necessary to manage the indigenously present mycorrhizal fungi in the field especially in nutrient-deficient soil. For sustainable agriculture, proper nutrient management resides an inevitable status where mycorrhizal fungi contribution cannot be neglected. In this regard, indegenious mycorrhizal fungal spores infect the plant roots and contribute in sustainable management of nutrients, water, soil properties and crop yield (Ortas 2003; Ortas and Coskan 2016). The effect of nutrient- and water-deficient conditions in field can be reduced in well-managed rhizosphere with increased microbial community and nutrient avaialability which benefit plant-soil quality (Barea et al. 2005).

Mycorrhizal fungi feed the plant by transfering P predominantly along with other immobile or slowly nutrients and balancing water uptake for ecological farming. As mycorrhizal fungi directly influence the plants' physiology and improve food quality, a recent demand in the organically produced crops has increased where mycorrhizal fungi feed the plants to balance nutrients and water requirements (Ortaş and Varma 2007). Mycorrhizal fungi can be an important and primarily useful biofertilizer source for organic agriculture whose spores are indegeniously present in the soil. For several reasons, especially for the future in terms of water deficiency, global climate change and soil and water pollution, management of indigenous mycorrhizae and selected microbial inoculum application is very important for food chain security, food safety and agricultural sustainability. Using the organic fertilizer for crop productivity, such as mycorrhizal fungi inoculant and other microbial inoculants, can ensure agricultural sustainability for a longer time period with reducing consumption of chemical fertilizer. Using these agricultural adaptation mechanisms is environmentally sound to ensure food security.

Proper use of mycorrhizal fungi in agriculture has not been classically adopted around the world particularly in Europe and North America, while in some of the developing countries, mycorrhizal fungi use has been developed due to expensive chemical input (Roy-Bolduc and Hijri 2012). To apply mycorrhizal fungi in commercial fields, large-scale production of mycorrhizal fungi (IJdo et al. 2011) and its coating on seed (Vosátka et al. 2012) could be the conceivable solution of main hurdles in vast area application. Furthermore, Roy-Bolduc and Hijri (2012) suggested two approaches of using mycorrhizal fungi in the field: inoculum selection and adopting cultural practices that enhance indigenous mycorrhizal fungi population. Cultural practices optimize the efficiency of indigenous mycorrhizae, but still it does not make the best and specific strain/plant combination for productivity which is a big challenge for researchers to achieve food security.

Selection of suitable plant as host for the mycorrhizal fungi production is another big question, although some of the researchers tried to explore host-specific strain by exploration of indigenous mycorrhizal fungi (Abdullahi et al. 2014; Ortas 2015; Ortas and Ustuner 2014a, b). The answer of this question depends on the type of mycorrhizal species, on the plant genotype and undoubtedly on the environment. In the field, targeted application of mycorrhizal fungi can be the large-scale option in terms of cost-effectiveness.

Until now, several studies have been conducted around the world on various crops in different soil conditions with mycorrhizal fungi. In greenhouse conditions, it has been extensively shown that mycorrhizal fungi can increase plant growth and nutrient uptake. However, there is still a lack of information concerned with increase in plant growth and nutrient uptake under field conditions.

Few leading experiments established knowledge of mycorrhizal fungi on main agricultural crops including cereals, cash crops, vegetables and fruits have been mentioned in this chapter. This chapter emphasizes on knowledge gap in mycorrhizal fungi application on a large scale for food security and quality along with environmental stress tolerance for various crops. Mycorrhizal fungi application in the field for various cash crops, cover crops, vegetables and fruits will be discussed in this chapter. Mycorrhizal fungi activities in association with various bacteria, termed as 'mycorrhizae helper bacteria', have also been discussed. Moreover, emerging technology of using mycorrhizal fungi with biochar as soil amendment will be subjected to discuss for its promising potential in agriculture.

13.2 Mycorrhization in Field Crops

13.2.1 Wheat

Wheat (*Triticum aestivum* L.) is one of the main crops fundamental to human civilization. It is improving world food security by providing 20% of the total dietary calories and proteins worldwide (Shiferaw et al. 2013). To feed the world population, proper and healthy growth of plants is inevitable in limited resources. Supply of macro- and micronutrients, particularly P by mycorrhizal fungi, is a way out to feed the plants and save the nine billion humans on earth. During field application, large amount of inoculum is required, which is the main limitation for mycorrhizal fungi use on large scale. To overcome this limitation, a study was conducted by using *Glomus intraradices* on wheat seed as coating material to test its feasible delivery system. The use of mycorrhizal fungi as seed coating material has the potential to significantly reduce required amount of inoculum, increasing efficiency and cost reduction (Oliveira et al. 2016). Root colonization was not different than the conventional method of mycorrhizal fungi application which ensured the uptake of P, K, S and Zn, whereas significant increase in the dry shoot was also observed. On the other hand, this methodology also ensures the reduction in chemical fertilizer consumption which adds up heavy metals in the soil.

Similarly, a meta-analysis of 38 field trials of wheat was conducted, and 333 observations for plant growth and nutritional parameters such as grain yield, root biomass, root colonization and nutrient uptake were considered. Results showed that aboveground biomass, grain yield, harvest index and nutrient concentration were increased, and positive correlation was observed between root colonization and grain yield (Pellegrino et al. 2015). Besides that, mycorrhizal fungi species specificity to the plant genotype is a crucial driver to achieve notable plant response. Meta-analysis study concluded that mycorrhizal fungi inoculation in wheat field could be an effective approach for economically sustainable cropping system. Addition of mycorrhizal fungi is strongly related with dose of P_2O_5 application. In an experiment, wheat yield was observed under three different P doses and compared with non-mycorrhizal treatments. Results showed that maximum yield was gained in limited supply of P_2O_5 fertilizer and both have inverse relation (Fig. 13.1). Pandey et al. (2005) found P uptake by wheat was near 10% more than without mycorrhiza, while another cereal, rye, was taking up phosphorous 64% more in the presence of mycorrhizas.

In addition to giving a direct benefit to the plants, mycorrhizal fungi alter bacterial community to assist plant in growth promotion and stress tolerance. In a study conducted on wheat crop, *G. intraradices* increased the gram-negative and gram-positive bacteria by 37 and 56%. Integrated approach of using mycorrhizal fungi to boost bacterial community can assist plant against various stress conditions such as temperature, water, nutrient and pollutants (Ingrid et al. 2016).

At molecular and genetic level, few studies have been conducted to evaluate the influence of mycorrhizal fungi on nutrient uptake. Valuable information is available

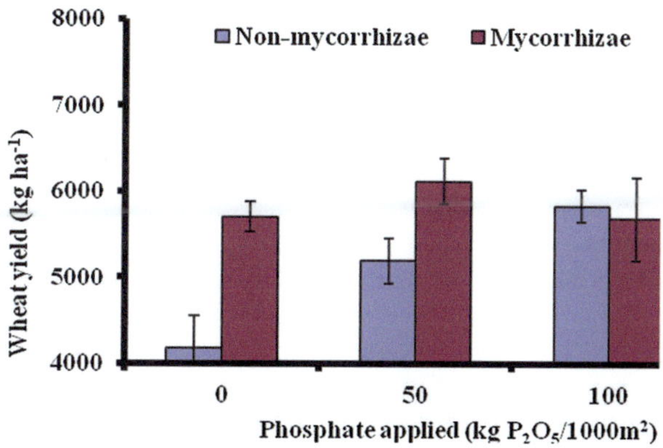

Fig. 13.1 Wheat yield in mycorrhizal and non-mycorrhizal treated treatments under different P_2O_5 levels (Ortas unpublished results)

for the upregulation of phosphate (Pi) and nitrogen (N) transporter genes in plant roots infected by mycorrhizal fungi, but detailed mechanism still needed exploration to answer (Duan et al. 2015). Four species of mycorrhizal fungi, *Acaulospora delicata*, *G. etunicatum*, *G. mosseae* and *G. intraradices*, were used to evaluate gene upregulation in the roots of winter wheat, which increased mycorrhiza-inducible Pi transporter genes, while on addition of N in soil, expression of mycorrhiza-inducible Pi transporter was increased.

13.2.2 Maize

Maize (*Zea mays* L.) is another important crop around the world after wheat. To meet the demand of maize consumption, nutrient availability, environmental stress in terms of drought, heat fluctuation and contaminants are main shortcomings. The use of mycorrhizal fungi in stress tolerance emerged as a promising approach to cope these limitations. To evaluate mycorrhizal fungi assistance in stress tolerance against P- and Zn-deficient soils, a study was conducted in calcareous soils from the Central Anatolia Region of Turkey. Maize plants were grown in *G. mosseae*- and *G. etunicatum*-inoculated soil up to 7 weeks with three rates of P_2O_5 application (0, 25, 125 mg P kg^{-1} soil) and two rates of Zn (0 and 5 mg Zn kg^{-1} soil). Maize plant is highly mycorrhizal dependent, and under low application of nutrients, mycorrhizal fungi managed uptake of P and Zn assisting plants. When the soil was inoculated with mycorrhizal inoculation, the increasing effects of P and Zn fertilization on plant growth remained less pronounced (Ortas 2012a).

Sterile and non-sterile soils have different impacts on the mycorrhizal fungi inoculation on plant growth and maize dependency. To evaluate *G. caledonium*, *G. etunicatum* and *G. mosseae* effect on maize growth and nutrient uptake, a study was established with different doses of P_2O_5 application. Results showed that inoculation in sterilized condition with low P_2O_5 input increased the plant growth and P uptake. Moreover, mycorrhizal dependency of maize plants grown in sterilized soil and low P_2O_5 dose was observed higher than the non-sterilized soils. Specificity of *G. caledonium* with maize plant was observed higher (Ortas 2003).

Similarly, a 2-year field trial was conducted to evaluate growth promotion and salinity stress tolerance of maize and cotton by assistance of indigenous mycorrhizal fungi. Biomass production and P uptake were increased notably, while, physiologically, leaf proline accumulation and K^+/Na^+ ratio were promoted through selective preferential uptake of K^+ and Na^+. Although measuring the effects of indigenous mycorrhizal fungi has some methodological limitations, its effect on plant growth and stress tolerance cannot be compromised (Liu et al. 2016). Native mycorrhizal fungi isolated from the saline field improve maize growth by boosting antioxidant system in plants; moreover, in Mediterranean saline habitat, mycorrhizal fungi enhanced ion homeostasis system in maize which assist plant growth in saline condition.

The use of mycorrhizal fungi to overcome low temperature stress by the maize plants is evaluated in a study. Inoculation with *G. etunicatum* alters host water status and photosynthesis which assists the plants in low temperature stress elevation by providing nutrients and alleviating water loss in the maize leaves (Zhu et al. 2010). To evaluate the relation between mycorrhizal fungi and maize genotypes, a greenhouse experiment was conducted where six maize genotypes (Luce, Vero, Darva, Pegasso, P.3394 and P.32K61) and eight mycorrhizal fungal species (*G. mosseae*, *G. caledonium*, *G. etunicatum*, *G. clarium*, *G. macrocarpum*, *G. fasciculatum*, Dr. Kinkon (Japanese species) and *G. intraradices*) were used (Ortas and Akpinar 2011). Maize genotype growth was found strongly linked with mycorrhizal fungi species inoculated. The uptake of nutrients and biomass production is variable for maize genotypes depending on mycorrhizal fungal species inoculation, but all were positively correlated (Ortas 2003; Ortas and Akpinar 2011).

13.2.3 Cotton

Cotton (*Gossypium arboreum* L.) being a cash crop and primary source of fabric is important. Application of mycorrhizal fungi in the field has various limitations such as storage, propogation and bulk application. Besides that, in a study, G. *clarium* inoculum was applied in the field for soybean (*Glycine max* L.) and cotton crops with different doses of P_2O_5 fertilizer. The results of Ortas (2012b) indicate that under 3-year field conditions, mycorrhiza-inoculated cotton plant had higher P and Zn concentrations than their non-inoculated counterparts. However, for cotton plants phosphorus application had a significant ($P < .0.02$) effect on yield as well. Mycorrhizal fungi inoculation increased 20% root colonization which assisted in nutrient absorption and decreased chemical fertizlizer consumption (Cely et al. 2016). Similarly, to investigate the integerative effect of mycorrhizal fungi and phosphate-solubilizing bacteria, a pot experiment was conducted on cotton plant. *G. mosseae* and *A. laevis* were used with *Pseudomonas fluorescens* and different doses of superphosphate. The combination of *G. mosseae* and *Pseudomonas fluorescens* was verified promising in various physiological and growth parameters such as stomatal conductance, chlorophyll content, acidic and alkaline phosphatase activity, shoot and root dry biomass, root and shoot length and leaf area (Badda et al. 2015).

13.2.4 Soybean

Soybean (*Glycine max* L.) is an important leguminous oilseed crop rather than pulse which produces significant amount of proteins in a hectare than other crops. Mycorrhizal fungus and bacterial effectivity was evaluated in a study where two varieties of soybean were inoculated with *G. intraradices* and *Thiobacillus* sp. as biological amendments along with different chemical fertilizer inputs in a low available P soil. Seed yield, oil proteins and nutrient concentration were analysed. In co-inoculation of mycorrhizal fungi and bacteria with chemical fertilizer, seed yield and seed nutrient concentration (P and Mn) were improved. Under field conditions soybean plants showed high mycorrhizal inoculation efficiency, and also mycorrhizal inoculation increased P and Zn concentration (Ortas 2012b). Mycorrhizal fungi and chemical fertilizer in combination enhanced oil and protein content percentage (Mostafavian et al. 2008).

13.3 Mycorrhizal Application for Vegetables

Mycorrhizal inoculums' potential can be adversely affected by soil management practices such as fertilizer application, pesticide use, crop rotation, fallowing, tillage and topsoil removal. Soil moisture content and irrigation are also important factors for the success of seedling transplantation. Soil biological properties are important, and competition among microorganisms such as native mycorrhizal species, bacterial species, soilborne fungi and nematodes affects the plant health and nutrient uptake efficiency. Inoculated mycorrhizal seedlings reduce the cost of chemical fertilizer input without compromising on plant growth and yield. Moreover, inoculating the seedlings could also reduce the cost of inoculum production on large scale for field application. Mycorrhizal fungi and P_2O_5 fertilizer are correlated; under high rates of P_2O_5 application, mycorrhizal fungi effects are masked. Plant growth under stress conditions may be enhanced by the application of microbial inoculation including plant growth-promoting rhizobacteria (PGPR) and mycorrhizal fungi. These microbes can promote plant growth by regulating nutritional and hormonal balance, producing plant growth regulators, solubilizing nutrients and inducing resistance against plant pathogens. In addition to their interactions with plants, these microbes also show synergistic as well as antagonistic interactions with other microbes in the soil. These interactions may be vital for sustainable agriculture because they mainly depend on biological processes rather than on agrochemicals to maintain plant growth and development as well as proper soil health

Table 13.1 Effect of different host plant and mycorrhizal species on mycorrhizal propagation

Host plant	Mycorrhizal species	Shoot dry weight (g/pot)	% infection	Spore number in (10 g inoculums)
	Control	0.27	0	0
Onion	*G. mosseae*	1.35	68	150
	Acaulospora laevis	1.27	59	146
	G. intraradix	1.21	75	140
	Control	0.45	0	0
	G. mosseae	1.72	74	231
Leek	*Acaulospora laevis*	1.35	66	195
	G. intraradix	1.25	73	218

Ortas (unpublished data)

under stress conditions. Biological interactions between PGPR and mycorrhizal fungi are believed to cause a cumulative effect on all rhizosphere components, and these interactions are also affected by environmental factors such as soil type, nutrition, moisture and temperature (Nadeem et al. 2014).

13.3.1 Onion

Onion (*Allium cepa* L.) is the most widely grown vegetable and it is a biennial crop. It is strongly a mycorrhizal-dependent plant. Application of mycorrhizal fungi to the onion is supposed to be promising for root-disease control, stress amelioration, biocontrol activities and fruit quality enhancement for human nutrition. To achieve all these benefits, considering the plant genotype and mycorrhizal fungi combination is important followed by soil properties and inoculation method (Baum et al. 2015). Mycorrhization improves bulb formation and growth of onion seedlings. Accumulation of proteins, proline and soluble sugars on leaves becomes high in mycorrhizal fungi-treated plants. This interaction is beneficial for enhancing the tolerance of onion seedlings to environmental stresses (Bettoni et al. 2014). Similarly, to evaluate the mycorrhizal fungi efficiency under drought stress, *G. versiforme*, *G. intraradices* and *G. etunicatum* were applied on onion plants. Inoculation with *G. versiforme* and *G. etunicatum* improved P and K uptake (Aliasgharzad et al. 2009). Mycorrhizae species also have significant effect on plant growth and number of spore production (Table 13.1).

13.3.2 Pepper

Pepper from Piperaceae family is a widely consumed vegetable, and it has shallow root system which cannot absorb sufficient nutrient elements in deep soil. It makes pepper a strongly mycorrhizal-dependent plant for nutrient and water absorption

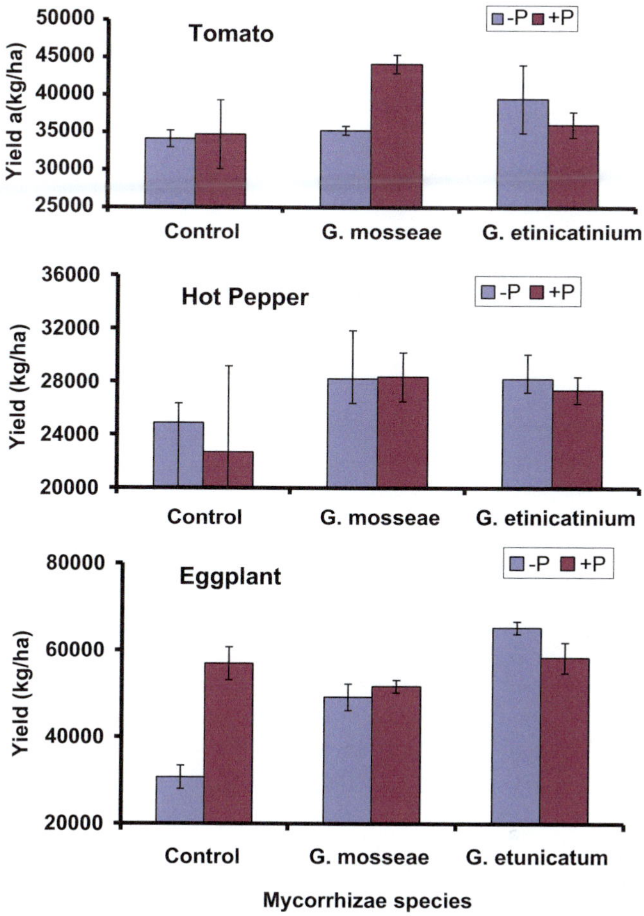

Fig. 13.2 Effect of different mycorrhizal inoculations on yield of tomato, pepper and eggplant under +/− P addition under field conditions

(Şimşek et al. 1998). A 3-year field experiment was conducted in Mediterranean climate to evaluate indigenous mycorrhizal performance in fumigated and non-fumigated soil conditions. Non-fumigated soil leads to high root growth, mycorrhizal root infection and better plant growth. Moreover, inoculation effectiveness was high in limited P_2O_5 application (Ortas 2012b). Similarly, in a field experiment, *G. mosseae* and *G. etunicatum* were applied on tomato, hot pepper and eggplant to observe yield enhancement in the absence and presence of P_2O_5 (Fig. 13.2).

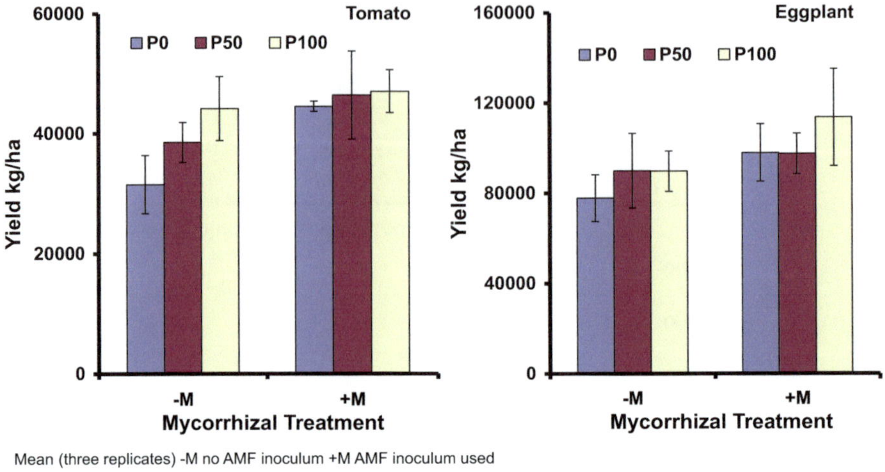

Fig. 13.3 Effect of mycorrhizae on tomato and eggplant yield with several P levels under field conditions

13.3.3 Tomato

Mycorrhizal inoculation in the plants reduces chemical P_2O_5 requirement in comparison to the non-inoculated plants. In tomatoes, reduced application of P_2O_5 and enhanced mycorrhizal infection may depict the inherited response on tomato growth in low to moderate soil P_2O_5 environment (Sylvia and Chellemi 2001). A 3-year long-term field experiment concluded that fumigated fields reduce indigenous mycorrhizal community which immediately affect the root infection and plat growth promotion (Ortas et al. 2003). Similarly, in another 3-year field experiment on tomato, *G. mosseae*, *G. clarum*, *G. etunicatum*, *G. intraradices*, *G. caledonium* and a cocktail of them were used to inoculate tomato plants. Plant biomass was significantly higher than the non-inoculated, and interesting results were reported that mycorrhizal fungi pre-inoculation (seed stage inoculation) had high mycorrhizal dependency and growth performance than reinoculation (seedling stage). Even the inoculated plants flower earlier than the non-inoculated and produce high yield. Specificity of *G. etunicatum* is reported to be high for tomato plant growth promotion than other mycorrhizal fungi (Ortas et al. 2013). In a field experiment, at three different doses of P_2O_5 and in the absence and presence of mycorrhizal fungi, tomato and eggplants were grown. Results showed that P_2O_5 had direct relation with tomato yield in the absence of mycorrhizal fungi, whereas, on mycorrhizal inoculation, fertilizer addition was non-significant for yield enhancement. Besides that, yield of mycorrhizal inoculated plant treatment without P_2O_5 application was higher than the non-mycorrhizal but highest dose of P_2O_5 (Fig. 13.3).

In another study, cherry tomato growth was evaluated by applying *Azospirillum brasilense* and *G. intraradices* together in comparison to the conventional inorganic fertilizer under shade conditions. After 15 and 30 days of transplantation, suspension

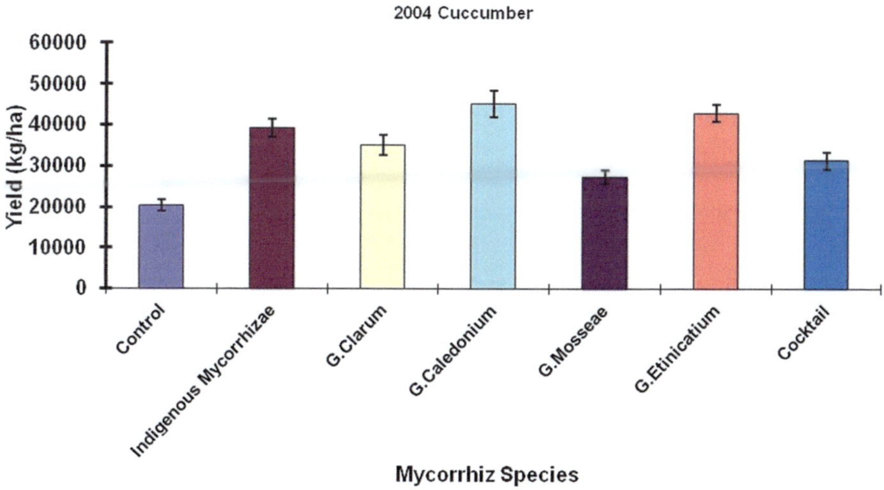

Fig. 13.4 Cucumber yield under different AM fungal inoculations

of the bacteria and mycorrhizal fungi was applied and notable data was observed. There were 6% increase in plant height, 10.5% dry biomass and 11% leaf area; 16% cherry tomato yield increase was recorded against the plants grown by conventional use of fertilizer (Lira-Saldivar et al. 2014).

13.3.4 Cucumber

Cucumber (*Cucumis sativus*) is also grown usually by seedlings, and mycorrhizal contribution for seedling survival and plant growth promotion is broadly tested. In this regard, two experiments were conducted where *G. mosseae* and *G. etunicatum* were used to check the seedling survival, root colonization and different plant growth promotion parameters in the absence and presence of P_2O_5 fertilizer. In another trial, these mycorrhizal fungi were inoculated with cucumber in addition to the indigenous mycorrhizae (*G. mosseae*, *G. etunicatum*, *G. clarum*, *G. caledonium*). Significant survival of seedlings, fruit yield and P and Zn shoot concenteration were observed in inoculated plants (Ortaş 2010) (Figs. 13.4 and 13.5).

Similarly, *G. fasciculatum* and *G. mosseae* were used as inoculants for cucmuber seedlings under sterilized and non-sterilized conditions in Mediterranean climate. Results depicted that using mycorrhizal fungi increased yield and nutrient concenteration such as P, Zn and Mn in the absence of fertilizer (Charron et al. 2001; Ortas 2008). In a 1-year field experiment, five mycorrhizal species were applied on cucumber, and their cocktail was also tried. Fruit was picked 12 times and average total yield was noted. Mycorrhizal inoculation was promising for yield enhancement than the control, whereas *G. etinicatinum* and *G. caledonium* were more specific and promising in cucumber yield (Fig. 13.6).

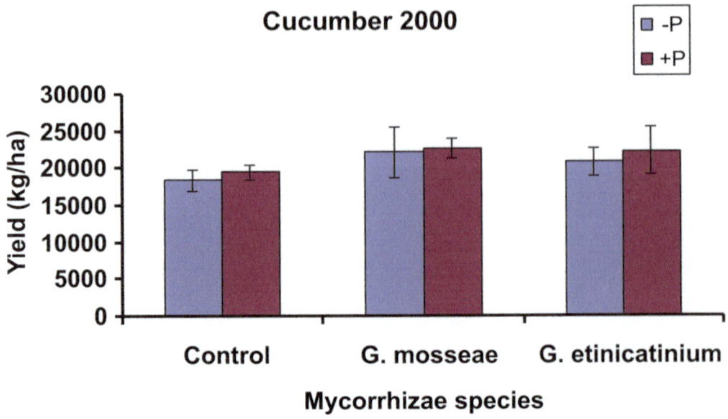

Fig. 13.5 Effect of AM fungi on cucumber yield in 2000 under field conditions

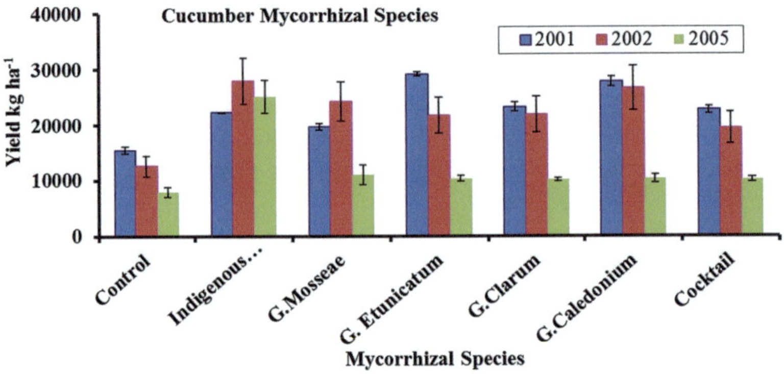

Fig. 13.6 Effect of several AM fungi on cucumber yield 3-year field experiment

13.3.5 Eggplant

Eggplant (*Solanum melongena* L.) is also a major horticultural crop produced from the seedlings. Many studies have been conducted to evaluate the suitability and specificity of the mycorrhizal fungi species for eggplant growth promotion and yield enhancement. Various *Glomus* species were tested in three greenhouse experiments for the three consecutive growing seasons to evaluate seedling production and growth before their transplantation in the fields. Treatments were *G. mosseae*, *G. clarum*, *G. caledonium*, *G. intraradices* and *G. etunicatum* and their mixture. Results showed that eggplant was significantly dependent on the mycorrhizal fungi inoculation and seedling stage was suitable for achieving maximum benefit from the mycorrhization in eggplant (Ortas et al. 2011).

13.4 Mycorrhizae for the Tree Plants

A number of woody plants are the source of fruits, and they are propagated using tissue culture technique for disease-free plants. This technique ensures the production of high-quality plant species, but their transplantation to the field often arises environmental stress problem. To make the plants strong against stress, mycorrhizal fungi inoculation could be the possibility as many trials have been conducted to explore the function of mycorrhizal fungi in stress tolerance and has been explained in the previous chapters.

13.4.1 *Citrus*

Citrus is an important fruit tree with strong dependency on mycorrhizal fungi. Many greenhouse and field trials have been conducted to evaluate mycorrhizal fungi influence on nutrient uptake and its growth. In a study, five mycorrhizal fungi species (*G. mosseae*, *G. mosseae*, *G. clarium*, *G. caledonium* and *G. etunicatum*) were tested for citrus specificity by using two host plants, i.e. clover and maize. *Citrus sinensis* L. were inoculated with respective mycorrhizal fungi in greenhouse, and various growth parameters were observed. Among all, *G. clarium* was more specific for growth improvement, nutrient uptake, plant biomass and leaf area in both experiments (Table 13.2). Even the height of *G. clarium*-inoculated plants was observed maximum along with root length and mycorrhizal infection. The sequence of mycorrhizal effectiveness in clover host was *G. clarium* > *G. mosseae* (1) > *G. mosseae* (2) > *G. caledonium* > *G. etunicatum*, whereas for the maize host, it was *G. clarium* > *G. caledonium* > *G. etunicatum* > *G. mosseae* (2) *G. mosseac* (1) (Ortas et al. 2002a).

In another study, sour orange (*Citrus aurantium* L.) was inoculated with *G. clarium* with 3 levels of P_2O_5 (0, 100 and 200 ppm) and 3 levels of Zn (0, 2.5 and 5 ppm) to evaluate mycorrhizal fungus influence on plant growth and nutrient uptake. Results showed that tenfold increase in the shoot and root dry weight was observed

Table 13.2 The effect of different arbuscular mycorrhizal inoculations on citrus shoot and root dry weight

Mycorrhizal species	Shoot dry weight (g plant^{-1})	Root dry weight (g plant^{-1})	Shoot/root dry weight ratio
Experiment I			
Control	0.90b ± 0.28	1.13b ± 0.35	0.79b
G. mosseae	1.04b ± 0.14	1.30b ± 0.17	1.01b
G. clarium	4.68a ± 0.96	2.60a ± 0.53	1.80a
G. caledonium	0.60b ± 0.06	0.63b ± 0.03	0.95b
G. etunicatum	0.63b ± 0.21	0.66b ± 0.12	0.96b

Ortas et al. (2002a)

than non-inoculated plants (Ortas et al. 2002b). Moreover, the suitability of five mycorrhizal fungi was tested for 12 months with the sour orange (*Citrus aurantium* L.), and their different combinations were also tried in two growth media separately, i.e. andesitic tuff + peat (1:1, V: V) and andesitic tuff + peat + soil (4 + 5 + 1, V: V: V). Results showed that performance of inoculants is strongly dependent on the availability of growth media type. As in first growth media, *G. mosseae* and *R. clarus* were promising, while in second, *G. mosseae* and indigenous mycorrhizae were the best inoculum which lead to healthy growth of citrus seedlings (Ortas and Ustuner 2014c).

Similarly, to testify the interaction of growth media and mycorrhizal fungi species in plant growth promotion, another study was conducted where growth of citrus was monitored for 10 months grown in four different growth media. These growth media were GM-A, andesitic tuff + peat (1:1, v/v); GM-B, andesitic tuff + compost (1:1, v/v); GM-C, andesitic tuff + peat + compost (2:1:1, v/v/v) and GM-D andesitic tuff + peat + soil (from the Balcali region) (2:1:1, v/v/v). A trend of the plant growth was quite interesting in the following descending order: GM-C > GM-A > GM-D > and GM-B. *G. clarium*, *G. margarita* and *G. mosseae* were promising mycorrhizal species for citrus growth promotion in these growth media (Ortas and Ustuner 2014a).

13.4.2 Apple

Apple (*Malus pumila* L.) is a cold climate crop which is strongly mycorrhizal dependent. To evaluate impact of mycorrhizal fungi on apple plant growth, different application methods were tested under different levels of P_2O_5 (0, 20, 40 and 80 mg P dm^{-3} of soil). Mycorrhizal fungi were applied as granular, quick root dip and irrigation which had valuable impact on mycorrhizal frequency by 83, 99 and 100%; moreover, mycorrhizal abundance was 36, 63 and 67% by respective application methodology. Physiologically, PSII efficiency of apple plant was enhanced, and significant increase in nutrient uptake (N, P, K, S, Cu, Fe, Mn, Mo, Ti) of shoot and root was observed. Besides that, concentration of Al, Ba, Li, Cd, Pb and V was much reduced on mycorrhizal application (Gastol et al. 2016).

Compost and mycorrhizal inoculation was tested for two rootstock varieties of Honeycrisp apple (M.26 EMLA and G.16) in a 9-year experiment. Mycorrhizal infection was measured in the first, fifth, seventh and eighth year of experiment; moreover, plant growth, yield and foliar nutrient analyses were done. After 7 years, concentration of foliar Zn was increased (17 mg kg^{-1}) in G.16 rootstock while foliar Cu (8 mg kg^{-1}) in the M.26 by mycorrhizal application. Besides that, compost had immediate effect on shoot growth, whereas after 2 years, mycorrhizal fungi also made impact on shoot growth in G.16 (Moran 2014).

13.4.3 Plum

Deciduous fruit trees (plum) are known to be salt sensitive, and they show the symptoms depending on the age and height of plant. Mycorrhizal fungi compensate the adverse effects of salinity by assisting plant in growth promotion. To evaluate the mycorrhizal fungi contribution in stress tolerance, *G. mosseae* and phosphate-solubilizing fungus (*Mortierella* sp. SM-1) were used as inoculant on rhizosphere of beach plum (*Prunus maritima*) with 1% (w/v) NaCl stress. In a saline environement, where available P is reduced, most of the enzymes required for nutrient availability such as urease phosphatase and proteinase are also adversely affected. Besides that, mycorrhizal fungus colonization also reduces along with dry weight of plant. When *G. mosseae* and phosphate-solubilizing fungus were allied collectively, adverse effects of the salinity were reduced and dry weight was increased. Phosphatase, urease and protease activities were also boomed in rhizosphere along with available P. Synergistic effects of mycorrhizal fungi and phosphate-solubilizing fungus improve micro-ecological environment which increase the soil fertility in rhizosphere and assist in stress tolerance of plants (Zai et al. 2015).

In another study, *Paraglomus occultum*, *G. etunicatum* and *G. mosseae* were used as inoculants for plum growth and nutrient uptake in saline stress condition. In non-inoculated set of experiment, plants were severely affected and their biomass was reduced, whereas in inoculated treatments, shoot dry biomass, root colonization and nutrient (N, Ca, Mg, Cu, Zn and Mn) acquisition were high. Mycorrhizal spores, root infection and spore density were high. *G. mosseae* and *G. etunicatum* significantly increased plant biomass by 48 and 43%, while leaf area was increased by 34 and 33%, respectively (Zai et al. 2014).

13.4.4 Grapevine

In a 2-year study, Crimson grapevines were grown by inoculation of *G. iranicum* through drip irrigation, and nutritional activity was observed along with physiological progress. Besides that, persistence of mycorrhizal fungi and continuity of their effects were evaluated for grapevine plants. Root colonization was satisfactory, and impact of colonization was notable as enhanced yield, improved water-use efficiency, photosynthetic performance, starch reservation and nutrient uptake, particularly P, K and Ca. In the second year, persistence of the mycorrhizal effects was continued, but efficieny was reduced. To achieve the persistent and efficient mycorrhizal effects for a long term, periodic monitoring of root colonization and reinoculation is mandatory (Nicolas et al. 2015).

Under greenhouse conditions mycorrhiza species strongly influenced grapevine seedling growth (Fig. 13.7). As can be seen in the picture and Table 13.3, different mycorrhiza species have different growth under different growth media. GM-A consisted of a mixture of soil, sand and peat at a ratio of 3:1:1 by volume, and GM-B

Fig. 13.7 Effects of different mycorrhizae species on micro-propagated seedlings of grapevines

Table 13.3 Effect of mycorrhizal species and growth media on shoot and root dry weight of grapevine seedlings

	GM-A		GM-B	
Mycorrhizal species	Shoot	Root	Shoot	Root
Control	0.28 ± 0.3 ab	0.11 ± 0.1 a	0.29 ± 0.2 c	0.06 ± 0.0 bc
G. mosseae	0.43 ± 0.1 a	0.13 ± 0.0 a	0.56 ± 0.3 a–c	0.16 ± 0.1 ab
G. caledonium	0.19 ± 0.2 b	0.06 ± 0.1 a	0.72 ± 0.4 a	0.17 ± 0.1 a
G. etunicatum	0.40 ± 0.2 a	0.13 ± 0.1 a	0.34 ± 0.2 bc	0.14 ± 0.1 ab
G. clarium	0.37 ± 0.2 ab	0.13 ± 0.1 a	0.28 ± 0.1 c	0.02 ± 0.0 c
G. margarita	0.31 ± 0.2 ab	0.12 ± 0.1 a	0.39 ± 0.3 bc	0.07 ± 0.1 a–c
G. macrocarpum	0.30 ± 0.0 ab	0.08 ± 0.0 a	0.61 ± 0.2 ab	0.13 ± 0.1 ab
G. fasciculatum	0.32 ± 0.2 ab	0.08 ± 0.0 a	0.61 ± 0.2 ab	0.12 ± 0.0 a–c
Mean	0.32 ± 0.11	0.10 ± 0.04	0.47 ± 0.19	0.11 ± 0.05

consisted of a mixture of sand, soil and organic matter at a ratio of 6:3:1 by volume. Compared to GM-A, seedlings grown in GM-B produced higher dry weight. Cocktail and indigenous mycorrhizae also significantly increased the seedling growth. In GM-A plants inoculated with *G. mosseae* and in GM-B plants inoculated with *G. caledonium* show high shoot biomass. Also, root colonization was very high compared to control treatments. These patterns may be explained by the stimulant effects of AM fungi on root growth due to improved nutrient acquisition of the host plant.

The results of Ozdemir et al. (2010) showed that *G. mosseae*-inoculated grapevine rootstocks increased shoot dry weight and also increased seedlings P and Zn concentration. Also, grapevine seedling growth is strongly mycorrhizal dependent. Similarly, a comprehensive survey study was conducted to evaluate the relation between the presence and abundance of mycorrhizal fungi in roots of vines, and mineral nutrition was also conducted for 2 years. Analysis of variance, principal component analysis and cluster analysis were used for the analysis of root infection and nutrient uptake data. Most frequently present mycorrhizal fungal genera were *Acaulospora*, *Gigaspora*, *Glomus* and *Scutellospora* which showed high percentage of root colonization. Besides that, strong inverse correlation was found between root colonization percentage and soil P. Vineyards with high mycorrhizal colonization showed a greater concentration of blade and petiole Mn, petiole Zn and blade Fe (Karagiannidis and Nikolaou 1999).

Root exploration largely determines the ability of plants to acquire mineral nutrients from soils. Therefore, root architecture, the three-dimensional configuration of the plant's root system in the soil, is of great importance for improving crop nutrient efficiency. Furthermore, the symbiotic associations between host plants, mycorrhizal fungi and bacteria are additional important strategies to enhance nutrient acquisition. Their interaction develops a synergistic mechanism which develops a win-win situation for all the factors and improve plant growth even in the stress condition. Root structuring is largely dependent on the surrounding environment such as soil, microbes, aeration, water and nutrient availability (Li et al. 2016).

13.5 Tripartite Relationship: Plant-Bacteria-Mycorrhizae

In natural ecosystem, mycorrhizal fungi are surrounded by a number of complex microbial communities, and possibly they modulate symbioses. When plant-bacteria-mycorrhizae work together in the rhizosphere for plant growth promotion, nutrient uptake and plant defence are termed as tripartite relationship. It is dependent on plant-soil feedback (PSF) system, and all the contributors are involved in multidirectional exchange of goods and services which induce changes in the aboveground and belowground interactions. Studies showed that that limitation of resources could be a key driver for local adaptation among organisms for the nutritional symbioses. There are different models to explain the way of interaction where optimal resource allocation (ORA) model emphasises in resources-limited condition leading to natural selection of various taxa by adjusting with biomass and energy allocation (Revillini et al. 2016) (Fig. 13.8).

To evaluate the plant-mycorrhizae-bacteria interaction, several studies have been conducted which showed quite interesting results. The mycorrhiza helper bacteria (MHB) assist the AM fungi up to 3.8-fold increase in root colonization (Duponnois and Plenchette 2003). Until now, the following bacterial genera have been reported for their association with mycorrhizal fungi such as gram-negative *Proteobacteria* (*Agrobacterium*, *Azospirillum*, *Azotobacter*, *Burkholderia*, *Bradyrhizobium*, *Enterobacter*, *Pseudomonas*, *Klebsiella* and *Rhizobium*), gram-positive *Firmicutes* (*Bacillus*, *Brevibacillus* and *Paenibacillus*) and gram-positive *Actinomycetes* (*Rhodococcus*, *Streptomyces* and *Arthrobacter*) (Frey-Klett et al. 2007). Similarly, *Thiobacillus* and mycorrhizal fungus were used in an experiment on maize crop with sulphur application. Different combinations of these treatments were tried on alkaline soil. Results showed that inoculation of mycorrhizal fungus, *Thiobacillus* and sulphur application decreased soil pH which made nutrients available to the plant roots as most of the nutrients are available at neutral pH while increasing the grain yield and oil contents in seed. Bacterial and sulphur (500 kg ha^{-1}) inoculation somehow decreased root colonization, but Fe contents were higher in the mycorrhizal fungal inoculated plants. Similarly, P contents were also increased along with grain yield (Ansori and Gholami 2015). Consequently, plant-mycorrhizae-bacteria association gets the benefit in terms of three core functions of bacteria, i.e. nutrient mobilization from soil minerals, fixation of atmospheric nitrogen and protection of plants against root pathogens.

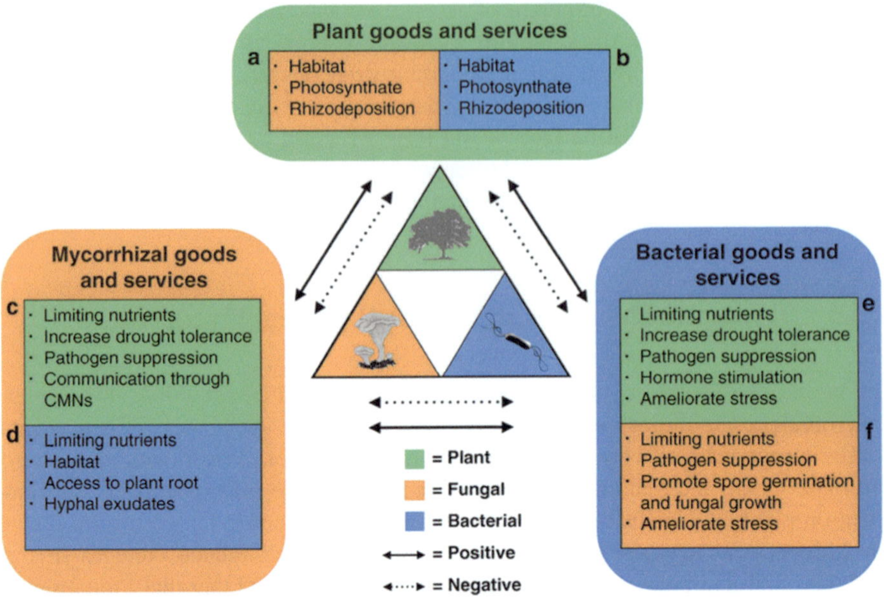

Fig. 13.8 Multidirectional exchanges of goods and services among plants, mycorrhizal fungi and rhizobacteria. Exchanges presented here include plant benefits to mycorrhizal fungi (**a**) and rhizobacteria (**b**), mycorrhizal fungi benefit to plants (**c**) and rhizobacteria (**d**) and rhizobacterial benefits to plants (**e**) and mycorrhizal fungi (**f**) (Adapted from Revillini et al. 2016)

13.6 Emerging Interaction of Mycorrhizal Fungi and Biochar

The use of biochar in the fields for altering soil physicochemical properties, aeration and environment for the microbiota living in soil is an emerging engineering technology for agricultural fields. To evaluate the impact of biochar potential as compensatory fertilizer with mycorrhizal fungi in a P-fixing soil, a study was conducted recently. Plant root-mycorrhizae-biochar interaction in P acquisition was evaluated for maize in the presence and absence of P_2O_5 fertilizer. Accumulation of P in mycorrhizal fungi-inoculated maize was not different when treated with pyrolyzed biochar or P_2O_5 fertilizer. If root-mycorrhizal fungi interactions are simultaneously considered, biochar could be a valuable input for P_2O_5 availability and P uptake. Biochar influences plant access to soil P_2O_5 and requires careful management to improve P_2O_5 availability (Zwetsloot et al. 2016).

Biochar coupled with fertilizers can be applied at lower application rates to achieve benefits in plant growth and nutrition, as well as soil biological fertility. A study was investigated for the comparative influences of biochar, including mineral fertilizer with microbes and their combination on mycorrhizal colonization, growth and nutrition of wheat in a glasshouse experiment and sorghum in field conditions. Higher mycorrhizal colonization than mycorrhizae alone was observed showing that biochar has significant role in increasing mycorrhizal colonization. The results showed that biochar could increase mycorrhizal colonization, plant growth and nutrient uptake of wheat, particularly N, P, K, S and Zn. The field experiment confirmed that biochar application at a rate of 300 kg ha^{-1} could increase the yield of irrigated sorghum on a loam soil and provide better applied P use efficiency compared to a water-soluble fertilizer alone. These results indicated that biochar-based fertilizers might increase the resilience and sustainability of dry land cropping in environments such as in Western Australia and warrant further field evaluation (Blackwell et al. 2015).

13.7 Future Aspects and Concluding Remarks

The future challenge on the use of mycorrhizal fungi in the production of vegetables will be to optimize combinations of crop plant and mycorrhizal fungi inoculum, inoculation methods and soil or substrate properties for mycorrhiza establishment and use. Under field conditions, several experiments were performed to understand the potential contribution of mycorrhizae on horticultural plant growth and nutrient uptake. These facts show that mycorrhizal inoculation is necessary for healthy, effective and well-grown seedling production. In the near future, the effect of mycorrhizae on plant physiology is going to be the key mechanism for healthy food for all living organisms.

The response of mycorrhizal inoculation depends on soil, plant species, inoculums, the method of inoculation and other ecological factors. After several years, experiments revealed that under field conditions, selected mycorrhizal spores and also indigenous mycorrhiza successfully infected plant roots resulting in improvements. Under field conditions plants depend on mycorrhizal inoculation but also depend on P supply, and there are differences from year to year. With high P application, mycorrhizal dependence significantly declines. Both in greenhouse and field conditions, mycorrhiza-inoculated plants generally were mycorrhizal dependent.

In order to manage indigenous mycorrhizae under long-term field conditions, the effect of soil management and crop rotation will be important. Our early experiment showed that crop rotation had a positive effect on the number of mycorrhizal spores developed. Because of soilborne disease, soil solarization and compost technology also can be part of soil and crop management systems by using mycorrhizal inoculation for better plant nutrition and healthy plant growth. In general, horticultural plants such as melon, watermelon, green and bell peppers, eggplant, marrow, cucumber and citrus are more mycorrhizal-dependent plants. After several years of field experiments, it has been concluded that for field crops soil and plant management systems, but also for horticultural plants, mycorrhizal inoculation is more practical and advised to be used. For future, the research direction will be focusing on soil and crop management systems and using mycorrhizal-inoculate horticultural seedlings for large agricultural practice.

References

Abdullahi R, Lihan S, Edward R (2014) Isolation of indigenous arbuscular mycorrhizal fungi and selection of host plant for inoculum production. Int J Biosci 5:116–122

Aliasgharzad N, Bolandnazar SA, Neyshabouri MR, Chaparzadeh N (2009) Impact of soil sterilization and irrigation intervals on P and K acquisition by mycorrhizal onion (*Allium cepa*). Biologia 64:512–515. doi:10.2478/s11756-009-0072-0

Ansori A, Gholami A (2015) Improved nutrient uptake and growth of maize in response to inoculation with thiobacillus and mycorrhiza on an alkaline soil. Commun Soil Sci Plant Anal 46:2111–2126. doi:10.1080/00103624.2015.1048251

Aranda E, Scervino JM, Godoy P et al (2013) Role of arbuscular mycorrhizal fungus Rhizophagus Custos in the dissipation of PAHs under root-organ culture conditions. Environ Pollut 181:182–189. doi:10.1016/j.envpol.2013.06.034

Badda N, Yadav K, Aggarwal A et al (2015) Consortium effect of arbuscular mycorrhizal fungi and *Pseudomonas fluorescens* with various levels of superphosphate on growth improvement of cotton (*G. arboreum* L.) J Nat Fibers 12:12–25. doi:10.1080/15440478.2013.879085

Barea JM, Pozo MJ, Azcon R et al (2005) Microbial co-operation in the rhizosphere. J Exp Bot 56:1761–1778. doi:10.1093/jxb/eri197

Baum C, El-Tohamy W, Gruda N (2015) Increasing the productivity and product quality of vegetable crops using arbuscular mycorrhizal fungi: a review. Sci Hortic 187:131–141. doi:10.1016/j.scienta.2015.03.002

Bettoni MM, Mogor ÁF, Pauletti V et al (2014) Growth and metabolism of onion seedlings as affected by the application of humic substances, mycorrhizal inoculation and elevated CO_2. Sci Hortic 180:227–235. doi: http://dx.doi.org/10.1016/j.scienta.2014.10.037

Blackwell P, Joseph S, Munroe P et al (2015) Influences of biochar and biochar-mineral complex on mycorrhizal colonisation and nutrition of wheat and sorghum. Pedosphere 25:686–695

Bowles TM, Barrios-Masias FH, Carlisle EA et al (2016) Effects of arbuscular mycorrhizae on tomato yield, nutrient uptake, water relations, and soil carbon dynamics under deficit irrigation in field conditions. Sci Total Environ 566:1223–1234. doi:10.1016/j.scitotenv.2016.05.178

Cely MVT, de Oliveira AG, de Freitas VF et al (2016) Inoculant of arbuscular mycorrhizal fungi (Rhizophagus Clarus) increase yield of soybean and cotton under field conditions. Front Microbiol 7:9. doi:10.3389/fmicb.2016.00720

Chandrasekaran M, Kim K, Krishnamoorthy R et al (2016) Mycorrhizal symbiotic efficiency on C-3 and C-4 plants under salinity stress - a meta-analysis. Front Microbiol 7:1246. doi:10.3389/fmicb.2016.01246

Charron G, Furlan V, Bernier-Cardou M et al (2001) Response of onion plants to arbuscular mycorrhizae. Mycorrhiza 11:187–197. doi:10.1007/s005720100121

Duan JF, Tian H, Drijber RA et al (2015) Systemic and local regulation of phosphate and nitrogen transporter genes by arbuscular mycorrhizal fungi in roots of winter wheat (*Triticum aestivum* L.) Plant Physiol Biochem 96:199–208. doi:10.1016/j.plaphy.2015.08.006

Duponnois R, Plenchette C (2003) A mycorrhiza helper bacterium enhances ectomycorrhizal and endomycorrhizal symbiosis of Australian acacia species. Mycorrhiza 13:85–91

Frey Klett P, Garbaye JA, Tarkka M (2007) The mycorrhiza helper bacteria revisited. New Phytol 176:22–36

Gastol M, Domagala-Swiatkiewicz I, Bijak M (2016) The effect of mycorrhizal inoculation and phosphorus application on the growth and mineral nutrient status of apple seedlings. J Plant Nutr 39:288–299. doi:10.1080/01904167.2015.1109114

IJdo M, Cranenbrouck S, Declerck S (2011) Methods for large-scale production of AM fungi: past, present, and future. Mycorrhiza 21:1–16

Ingrid L, Sahraoui ALH, Frederic L et al (2016) Arbuscular mycorrhizal wheat inoculation promotes alkane and polycyclic aromatic hydrocarbon biodegradation: microcosm experiment on aged-contaminated soil. Environ Pollut 213:549–560. doi:10.1016/j.envpol.2016.02.056

Jansa J, Rezacova V, Smilauer P et al (2016) Root colonization of bait plants by indigenous arbuscular mycorrhizal fungal communities is not a suitable indicator of agricultural land-use legacy. Agric Ecosyst Environ 231:310–319. doi:10.1016/j.agee.2016.07.013

Karagiannidis N, Nikolaou N (1999) Arbuscular mycorrhizal root infection as an important factor of grapevine nutrition status. Multivariate analysis application for evaluation and characterization of the soil and leaf parameters. Agrochimica 43:151–165

Kayama M, Yamanaka T (2014) Growth characteristics of ectomycorrhizal seedlings of *Quercus glauca*, Quercus Salicina, and Castanopsis Cuspidata planted on acidic soil. Trees-Struct Funct 28:569–583. doi:10.1007/s00468-013-0973-y

Li XX, Zeng RS, Liao H (2016) Improving crop nutrient efficiency through root architecture modifications. J Integr Plant Biol 58:193–202. doi:10.1111/jipb.12434

Lira-Saldivar RH, Hernandez A, Valdez LA et al (2014) *Azospirillum brasilense* and *Glomus* intraradices co-inoculation stimulates growth and yield of cherry tomato under shadehouse conditions. Phyton-Int J Exp Bot 83:133–138

Liu SL, Guo XL, Feng G et al (2016) Indigenous arbuscular mycorrhizal fungi can alleviate salt stress and promote growth of cotton and maize in saline fields. Plant Soil 398:195–206. doi:10.1007/s11104-015-2656-5

Marschner P (2012) Marschner's mineral nutrition of higher plants. Academic, London

Maya MA, Ito M, Matsubara Y (2014) Tolerance to heat stress and anthracnose in mycorrhizal cyclamen. In: Chomchalow N, Supakamnerd N, Sukhvibul N (eds) International symposium on orchids and ornamental plants

Moran RE (2014) Growth and yield of 'Honeycrisp' apple trees with preplant inoculation with mycorrhizae and soil-incorporated compost. J Am Pomol Soc 68:2–13

Mostafavian SR, Pirdashti H, Ramzanpour MR et al (2008) Effect of mycorrhizae, thiobacillus and sulfur nutrition on the chemical composition of soybean [*Glycine max* (L.)] Merr. Seed. Pak J Biol Sci 11:826–835

Nadeem SM, Ahmad M, Zahir ZA et al (2014) The role of mycorrhizae and plant growth promoting rhizobacteria (PGPR) in improving crop productivity under stressful environments. Biotechnol Adv 32:429–448. doi:10.1016/j.biotechadv.2013.12.005

Nicolas E, Maestre-Valero JF, Alarcon JJ et al (2015) Effectiveness and persistence of arbuscular mycorrhizal fungi on the physiology, nutrient uptake and yield of Crimson seedless grapevine. J Agric Sci 153:1084–1096. doi:10.1017/s002185961400080x

Oliveira RS, Rocha I, Ma Y et al (2016) Seed coating with arbuscular mycorrhizal fungi as an ecotechnological approach for sustainable agricultural production of common wheat (*Triticum aestivum* L.) J Toxicol Environ Health Part A 79:329–337. doi:10.1080/15287394.2016.1153448

Olsson PA, Thingstrup I, Jakobsen I et al (1999) Estimation of the biomass of arbuscular mycorrhizal fungi in a linseed field. Soil Biol Biochem 31:1879–1887. doi:10.1016/S0038-0717(99)00119-4

Ortas I (2003) Effect of selected mycorrhizal inoculation on phosphorus sustainability in sterile and non-sterile soils in the Harran plain in South Anatolia. J Plant Nutr 26:1–17. doi:10.1081/pln-120016494

Ortas I (2008) Field trials on mycorrhizal inoculation in the eastern mediterranean horticultural region. In: Feldmann F, Kapulnık Y, Baar J (eds) Mycorrhiza works. Hannover, Germany, pp 56–77

Ortaş İ (2010) Effect of mycorrhiza application on plant growth and nutrient uptake in cucumber production under field conditions. Span J Agric Res 8(S1):116–122

Ortas I (2012a) Do maize and pepper plants depend on mycorrhizae in terms of phosphorus and zinc uptake? J Plant Nutr 35:1639–1656. doi:10.1080/01904167.2012.698346

Ortas I (2012b) The effect of mycorrhizal fungal inoculation on plant yield, nutrient uptake and inoculation effectiveness under long-term field conditions. Field Crop Res 125:35–48. doi:10.1016/j.fcr.2011.08.005

Ortas I (2015) Comparative analyses of Turkey agricultural soils: potential communities of indigenous and exotic mycorrhiza species' effect on maize (*Zea mays* L.) growth and nutrient uptakes. Eur J Soil Biol 69:79–87. doi:10.1016/j.ejsobi.2015.05.006

Ortas I, Akpinar C (2011) Response of maize genotypes to several mycorrhizal inoculums in terms of plant growth, nutrient uptake and spore production. J Plant Nutr 34:970–987. doi:10.1080/01904167.2011.555580

Ortas I, Coskan A (2016) Precipitation as the most affecting factor on soil-plant environment conditions affects the mycorrhizal spore numbers in three different ecological zones in Turkey. Acta Agric Scand Sect B Soil Plant Sci 66:369–378. doi:10.1080/09064710.2015.1132005

Ortas I, Ustuner O (2014a) Determination of different growth media and various mycorrhizae species on citrus growth and nutrient uptake. Sci Hortic 166:84–90. doi:10.1016/j.scienta.2013.12.014

Ortas I, Ustuner O (2014b) The effects of single species, dual species and indigenous mycorrhiza inoculation on citrus growth and nutrient uptake. Eur J Soil Biol 63:64–69. doi:10.1016/j.ejsobi.2014.05.007

Ortaş I, Varma A (2007) Field trials of bioinoculants. In: Varma A, Oelmuller R (eds) Advanced techniques in soil microbiology. Springer, Berlin, pp 397–413

Ortas I, Kaya Z, Çakmak I (2001) Influence of VA-mycorrhiza inoculation on growth of maize and green pepper plants in phosphorus and zinc deficient soils. In: Horst WJ, Schenk MK, Burkert A et al (eds.) Plant nutrition: Food security and sustainability of agro-ecosystems through basic and applied research. Kluwer Academic Publishers, Dordrecht, pp 632–633

Ortas I, Ortakci D, Kaya Z (2002a) Various mycorrhizal fungi propagated on different hosts have different effect on citrus growth and nutrient uptake. Commun Soil Sci Plant Anal 33:259–272. doi:10.1081/css-120002392

Ortas I, Ortakci D, Kaya Z et al (2002b) Mycorrhizal dependency of sour orange in relation to phosphorus and zinc nutrition. J Plant Nutr 25:1263–1279. doi:10.1081/pln-120004387

Ortas I, Sari N, Akpinar C (2003) Effect of mycorrhizal inoculation and soil fumigation on the yield and nutrient uptake of some Solanaceas crops (tomato, eggplant and pepper) under field conditions. Agricoltura Mediterranea 133:249–258

Ortas I, Sari N, Akpinar C et al (2011) Screening mycorrhizae species for increased growth and P and Zn uptake in eggplant (*Solanum melongena* L.) grown under greenhouse conditions. Eur J Hortic Sci 76:116–123

Ortas I, Sari N, Akpinar C et al (2013) Selection of arbuscular mycorrhizal fungi species for tomato seedling growth, mycorrhizal dependency and nutrient uptake. Eur J Hortic Sci 78:209–218

Ozdemir G, Akpinar C, Sabir A et al (2010) Effect of inoculation with mycorrhizal fungi on growth and nutrient uptake of grapevine genotypes (*Vitis* spp.) Eur J Hortic Sci 75:103–110

Pandey R, Singh B, Nair TVR (2005) Impact of arbuscular-mycorrhizal fungi on phosphorus efficiency of wheat, rye, and triticale. J Plant Nutr 28:1867–1876. doi:10.1080/01904160500251381

Parniske M (2008) Arbuscular mycorrhiza: the mother of plant root endosymbioses. Nat Rev Microbiol 6:763–775

Pellegrino E, Opik M, Bonari E et al (2015) Responses of wheat to arbuscular mycorrhizal fungi: a meta-analysis of field studies from 1975 to 2013. Soil Biol Biochem 84:210–217. doi:10.1016/j.soilbio.2015.02.020

Revillini D, Gehring CA, Johnson NC (2016) The role of locally adapted mycorrhizas and rhizobacteria in plant-soil feedback systems. Funct Ecol 30:1086–1098. doi:10.1111/1365-2435.12668

Rodriguez A, Sanders IR (2015) The role of community and population ecology in applying mycorrhizal fungi for improved food security. ISME J 9:1053–1061

Roy-Bolduc A, Hijri M (2012) The use of mycorrhizae to enhance phosphorus uptake: a way out the phosphorus crisis. J Biofert Biopest 2:104. doi:10.4172/2155-6202.1000104

Shiferaw B, Smale M, Braun H-J et al (2013) Crops that feed the world 10. Past successes and future challenges to the role played by wheat in global food security. Food Sec 5:291–317. doi:10.1007/s12571-013-0263-y

Şimşek D, Ortaş İ, Köse Ö et al (1998) The effect of mycorrhizal inoculation on growth and nutrient uptake of tomato, eggplant, pepper plants under field conditions. International symposium on arid region soils
Sylvia DM, Chellemi DO (2001) Interactions among root-inhabiting fungi and their implications for biological control of root pathogens. Adv Agron 73:1–33
Vosátka M, Látr A, Gianinazzi S et al (2012) Development of arbuscular mycorrhizal biotechnology and industry: current achievements and bottlenecks. Symbiosis 58:29–37
Zai XM, Hao ZP, Wang H et al (2014) Arbuscular mycorrhizal fungi (AMF) on growth and nutrient uptake of beach plum (*Prunus maritima*) under salt stress. Appl Mech Mater 618:268–272
Zai XM, Zai Y, Zhang HS et al (2015) Characterising the rhizospheric soil niches of beach plum (*Prunus maritima*) colonised by arbuscular mycorrhizal fungi and/or phosphate-solubilising fungi when grown under NaCl stress. J Hortic Sci Biotechnol 90:469–475
Zhu XC, Song FB, Liu TD et al (2010) Arbuscular mycorrhizae reducing water loss in maize plants under low temperature stress. Plant Signal Behav 5:591–593. doi:10.4161/psb.11498
Zou YN, Srivastava AK, Wu QS (2016) Glomalin: a potential soil conditioner for perennial fruits. Int J Agric Biol 18:293–297
Zwetsloot MJ, Lehmann J, Bauerle T et al (2016) Phosphorus availability from bone char in a P-fixing soil influenced by root-mycorrhizae-biochar interactions. Plant Soil 408:1–11

MIX
Papier aus verantwortungsvollen Quellen
Paper from responsible sources
FSC® C105338

If you have any concerns about our products,
you can contact us on
ProductSafety@springernature.com

In case Publisher is established outside the EU,
the EU authorized representative is:
Springer Nature Customer Service Center GmbH
Europaplatz 3, 69115 Heidelberg, Germany

Printed by Libri Plureos GmbH
in Hamburg, Germany